W9-BFM-180

TEXTBOOK OF BIOPHARMACEUTICS AND CLINICAL PHARMACOKINETICS

TEXTBOOK OF BIOPHARMACEUTICS AND CLINICAL PHARMACOKINETICS

Sarfaraz Niazi, Ph.D.

Associate Professor of Pharmacy
College of Pharmacy
University of Illinois at the Medical Center
Chicago, Illinois

Appleton-Century-Crofts
New York

79 80 81 82 83 / 10 9 8 7 6 5 4 3 2 1

Prentice-Hall International, Inc., London
Prentice-Hall of Australia, Pty. Ltd., Sydney
Prentice-Hall of India Private Limited, New Delhi
Prentice-Hall of Japan, Inc., Tokyo
Prentice-Hall of Southeast Asia (Pte.) Ltd., Singapore
Whitehall Books Ltd., Wellington, New Zealand

Library of Congress Cataloging in Publication Data

Niazi, Sarfaraz, 1949-
Textbook of biopharmaceutics and clinical pharmacokinetics.

Includes index.
1. Biopharmaceutics. 2. Pharmacokinetics.
I. Title. [DNLM: 1. Biopharmaceutics. 2. Kinetics.
3. Pharmacology. QV38.3 N577t]
RM301.4.N52 615'.7 79-10869
ISBN 0-8385-8868-9

Design: Meryl Sussman Levavi

PRINTED IN THE UNITED STATES OF AMERICA

To the memory of my father

Contents

Preface

The study of biopharmaceutics, pharmacokinetics, and clinical pharmacokinetics has become an essential part of various health education curricula. Briefly, these disciplines provide insight into the selection of the proper dosage form, route of administration, and dosage regimen for a successful drug therapy. Quite often the selection of a drug depends on the factors that affect its delivery to a site of action. Therefore, this book is designed to aid in the study of principles needed to initiate and successfully terminate drug therapy once a diagnosis has been made. The necessary depth and breadth of familiarity with these principles must be defined in terms of the given student population. This book specifically addresses undergraduate students of pharmacy and other health-related fields.

Chapters 2 through 4 concern the introduction of a given drug molecule into the body. They present a discussion of dosage formulation, evaluation, and route of administration, with special attention to underlying principles. Once the drug molecules reach the general circulation, the pharmacologic response is largely determined by their distribution to the various parts of the body. Knowledge of the principles of distribution is helpful in understanding the onset of action and the intensity of response. The termination of a drug action, which is almost as important as the initiation of an action, is brought about mainly by excretion and biotransformation, and to a lesser extent by tissue redistribution. A clear understanding of these principles is essential in designing a dosage regimen which will reduce the risk of toxic response.

Chapters 2 through 6 are based primarily on the physicochemical, physiologic, and pharmacologic principles. However, the behavior of drug molecules in the body is also governed by well-defined mathematical principles, a quantitative aspect which can be studied if the mathematical modeling of these processes is made; this is the topic of Chapter 7. Pharmacokinetic principles allow rather precise estimation of drug concentration versus time in the various parts of the body. However, even with the same dose, different individuals show different concentrations. The source of these variations is studied in Chapter 8 in terms of the absorption, distribution, biotransformation and excretion of drugs, as they relate to the various physiologic and pharmacologic principles. The last chapter gives the application of all the principles mentioned in this book: the selection of an appropriate dosage regimen. Conventional distribution- and disposition-dependent dosage adjustments are discussed along with the newer concept of direct plasma concentration monitoring as a guide to therapeutic management.

The main purpose of this book is to familiarize the student with the basic principles of biopharmaceutics and pharmacokinetics and their clinical applications. Whereas most of the principles discussed in this book should be thoroughly mastered, some of the material is included only to illustrate the complexity of various concepts. The desired level of learning can be ascertained from the questions listed at the end of each chapter.

This book is a direct result of many years of teaching the principles of biopharmaceutics and pharmacokinetics to undergraduate pharmacy students. It is this experience which has decided the essential depth and breadth of this book. Although some of my colleagues may not agree with the chosen scope of the book, as applied to undergraduate teaching, I hope that most of them will find it relevant and reasonably current. The rate with which the literature is expanding in this area of study is phenomenal; however, obsolence should not overtake this text for at least a few years, since the major thrust of the book is in teaching basic principles.

A great many scientists deserve my gratitude for their kind permission to reproduce their work in this book. I must also thank those authors whose ideas may have been adapted without specific recognition in this book. My colleagues and graduate students have been of enormous help to me in writing this text, and I must single out Dr. Paul A. Laskar and Mr. Krishna Vishnupad for their special efforts. I would also like to express especial gratitude for the warm encouragement of Dean August P. Lemberger. It is indeed a privilege to know an educator of his high caliber.

TEXTBOOK OF BIOPHARMACEUTICS AND CLINICAL PHARMACOKINETICS

THE FAMILY CIRCUS® **By Bil Keane**

"How will that stuff get from down there up to my sore throat?"

CHAPTER 1

Introduction to Biopharmaceutics and Clinical Pharmacokinetics

The use of drugs in the treatment of disease states is clearly a multifaceted process. First, a pharmacologically active molecule has to be synthesized, isolated, or extracted from various possible sources and its use rationalized in terms of its toxicity and potential benefits. Second, a dosage form of this drug has to be formulated that will contain and deliver a recommended dose, through an appropriate route of administration, to the site of action or a target tissue. Finally, a dosage regimen must be selected to provide an effective concentration in the body as determined by the various physiologic, pathologic, and clinical needs. The merit of each of these facets cannot be established independently since it is the successful integration of these facets which results in successful drug therapy. For example, an analgesic with a high therapeutic index can be of little use if it undergoes rapid decomposition in the gastrointestinal tract and is too irritating to be administered parenterally. Similarly, a well-designed intrauterine device for controlled delivery of drugs is useless unless it can carry a pharmacologically active drug.

Even if an active drug is successfully administered and made available to the target tissue, an optimum dosage regimen is essential to achievement of the desired therapeutic goals. Briefly, a large number of factors play an important role in determining the activity of a drug and some of these are listed in Scheme 1.1 to demonstrate some measure of the complexities of successful drug therapy.

A variety of disciplines are involved in understanding the complicated events that take place during the long process of an active chemical structure becoming an active drug. For example, the principles of physics, physical chemistry, chemical engineering, and mathematics are essential in the formulation of a dosage form; an in-depth knowledge of physiology, pharmacology, and anatomy is needed in selecting an appropriate route of administration; and the principles of kinetics, therapeutics, and analytical chemistry are essential to providing an effective concentration of the drug at the target tissue or the site of action. It is only when all of these principles are integrated that a successful therapeutic application results. The first such approach was made by Torsten Toerell, when he published his paper on the distribution

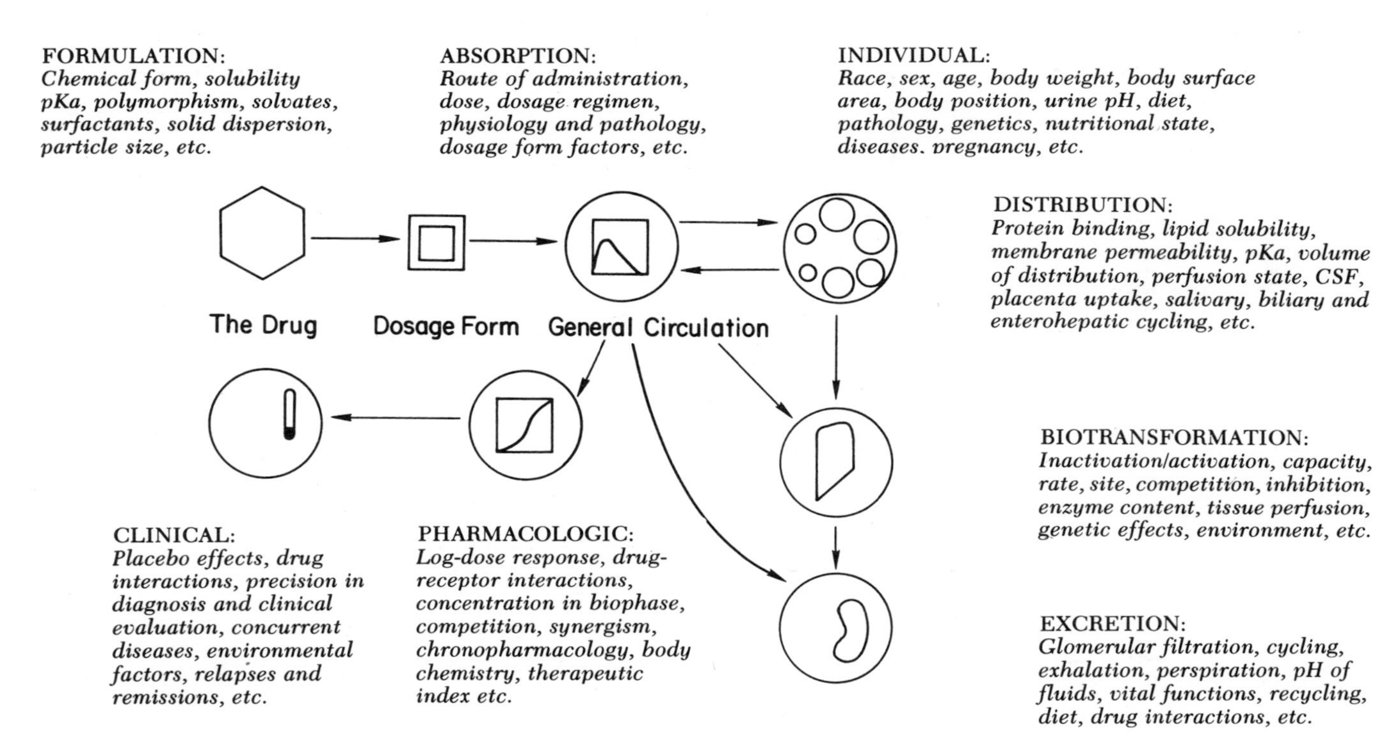

Scheme 1.1 The factors determining the activity of drugs.

kinetics of drugs in the late 1930s.[1] However, the major breakthrough in defining and developing new disciplines which combine the knowledge of the various areas mentioned above has come within the last 15 to 20 years. These new disciplines have reached a point where they can form the basis of a comprehensive textbook on the subject.

BIOPHARMACEUTICS is a pharmaceutical science encompassing the study of the relationship between the nature and intensity of biologic effects and the various formulation factors—such as the chemical nature of the drug, inert formulation additives, and the pharmaceutical processes used to manufacture the dosage forms (more appropriately termed the DRUG DELIVERY SYSTEMS).

The nature and intensity of biologic effects are generally proportional to the total amount of drug made available to the body. The rate of drug delivery, and therefore the effectiveness of a dosage form, is judged by the efficiency with which it can make the drug available to the body. For example, Figure 1.1 presents a situation in which one of two formulations is totally ineffective since it fails to provide the minimum effective concentration in the body. If the two formulations produced identical blood levels, they would be considered BIOEQUIVALENT, or equally bioavailable. The differences in BIOAVAILABILITY, or the extent of drug absorption from a dosage form, are the result of differences in formulation as well as of differences in the physiologic and pathologic states of the patients. It is indeed difficult and often impossible to assess the bioavailability of a drug in clinical use. Therefore, various correlates have been developed to estimate the bioavailability of drug products through in vitro or in vivo techniques. Generally, the goal of biopharmaceutical studies is to develop a dosage form that will provide consistent bioavailability at a desirable rate, which need not always be a rapid one. The importance of consistent bioavailability can be very well appreciated if a drug has a narrow therapeutic index whereby a slightly higher bioavailability

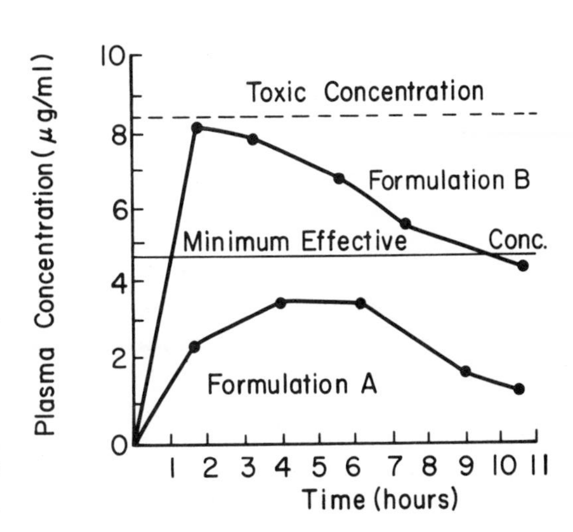

Figure 1.1. Plasma concentration profiles following administration of identical dose in two formulations.

than expected will result in a toxic response and a slightly lower bioavailability will result in a less than minimum effective concentration in the body.

The absorption of drugs is governed by well-defined physical, chemical, and biologic principles which also control the fate of the drug molecules once they enter the general circulation of the body. It is therefore possible to establish a mathematical model to describe the absorption and disposition (distribution and elimination) in the body. These mathematical aspects are discussed under the heading of PHARMACOKINETICS, which is defined as the science of the quantitative analysis of interaction between organism and drug. Basically, the purpose of pharmacokinetics is to study the time course of drug and metabolite concentrations in biologic fluid, tissues, urine, feces, sweat, breath, milk, etc., and to propose a suitable mathematical model to account for the absorption, distribution, biotransformation, and excretion processes. Recently, attempts to quantitate pharmacologic responses have also been made with some success and these studies also fall under the purview of pharmacokinetics.

There are several applications of pharmacokinetic studies, including bioavailability measurements, effects of body physiology and pathology on drug absorption and disposition, dosage adjustments in disease states, correlation of pharmacologic responses with dose, and evaluation of drug interactions, among many others. The most important application of pharmacokinetics comes in the clinical setting where the pharmacokinetic parameters (mathematical constants to describe a model) of a drug are used to develop a therapeutic model which can be used to individualize the drug regimen and thus provide the most effective drug therapy. This application is referred to as CLINICAL PHARMACOKINETICS, defined as the application of pharmacokinetics to the safe and effective therapeutic management of the individual patient. This involves initial design of a drug dosage regimen, including dose, dosing interval, route of administration, and dosage form as well as refinement and readjustment of the dosage regimen where necessary, based usually on serial monitoring of drug concentration in plasma or other fluids but sometimes based on direct assessment of clinical response. It also includes the diagnostic work-ups to determine the reasons for an unusual drug response, extent of patient complicance, bioavailability problems, medication errors, drug interactions, unusual distribution and elimination kinetics, or certain pharmacogenetic effects such as unusual receptor site sensitivity or abnormal levels of metabolic enzymes.[2] A general idea of the scope of this discipline can be obtained from some of the questions it answers: Can phenytoin be administered once a day rather than three times a day? What dosage adjustments of gentamicin sulfate are necessary for patients with impaired renal function? How many days of aspirin administration are required to elicit the maximum antiinflammatory effect produced by a given regimen? Should an asthmatic child who metabolizes theophylline rapidly receive a higher-than-average dose at the usual six-hour dosing interval, or should he receive the average dose at less than the six-hour interval? When relative to the time of dosing, should one obtain a blood sample to monitor digoxin concentrations in the plasma of a cardiac patient? What is the likelihood that a hypertensive patient will react with a lupus-like syndrome to hydralazine hydrochloride?[3]

The intent of this book is to summarize the basic principles of biopharmaceutics and pharmacokinetics and their applications in dosage form design and clinical situations. The recurring theme in this book is the demonstration of the applications of these principles rather than the philosophy of their existence. This book also contains current data on selected commonly used drugs and should thus serve as a continued source of reference.

References

1. Teorell T: Kinetics of distribution of substances administered to the body I: The extravascular modes of administration. Arch Int Pharmacodyn Ther 57:205, 1937
2. Levy G: An orientation to clinical pharmacokinetics. In Levy G (ed): Clinical Pharmacokinetics. Washington, Am Pharm Assoc, 1974, p 1
3. Gibaldi M, Levy G: Pharmacokinetics in clinical practice I: Concepts. JAMA 235:1864, 1976

CHAPTER 2

Delivery of Drugs: Introduction and Formulation Factors

The long journey from the discovery of an active drug molecule to the production of an effective drug product begins with the selection of its appropriate physical and chemical form, which is formulated in a dosage form suitable for a specific route of administration (Fig. 2.1). The ultimate goal is the delivery of active chemical species to a target tissue or the site of action. The choices of the physicochemical form of the drug, the dosage form, and the route of administration are interdependent and determined by the following factors:

1. *Onset of Action*. The administration of drugs directly into the general circulation will, in most instances, result in an immediate onset of action, which may be highly desirable in life-threatening conditions. However, the inconvenience of intravascular administration, especially in outpatient situations, and possible effects of instantaneous flushing of the drug into the vital organs may exclude these routes for many drugs. For example, sodium bicarbonate and epinephrine are administered intravenously for immediate effect in cardiac arrest; carbenicillin is administered intravenously (Geopen) for immediate response and orally (Geocillin) for less critical applications; propranolol (Inderal) is generally administered orally and its intravenous use is discouraged because of possible side effects.
2. *Duration of Action*. It is possible to provide a sustained delivery of drugs by using intramuscular depots or slowly dissolving dosage forms. Timed-release aspirin, testosterone oily intramuscular depots, or triamcinolone acetonide intrasynovial depots are some examples in which the route of administration is a crucial part of drug therapy because the rate of drug delivery is extremely important. Often a desirable rate of drug delivery can only be achieved through specialized dosage form designs, e.g., Occusert ophthalmic inserts or Progestasert contraceptive devices, which provide drug delivery at the site of action at a highly controlled rate.
3. *Nature of the Drug*. Not all drugs can be administered by all routes of administration. For example, intravenous administration requires that the drug be present in a solution system miscible with blood to avoid precip-

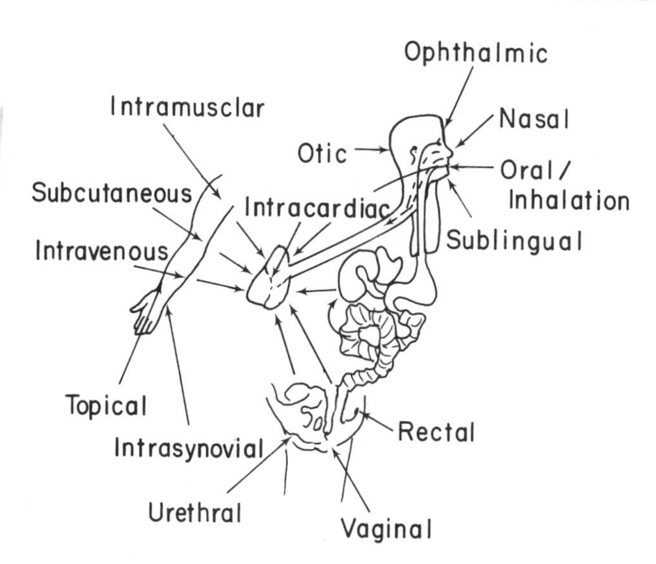

Figure 2.1. Routes of drug administration. (After Benet LZ: Biopharmaceutics, Basis for design of drug products. In Ariens EJ (ed), Drug Design. New York, Academic, 1973, vol. 5, p. 5)

itation in the veins. This will exclude several potent drugs if an appropriate solution of these drugs cannot be formulated. Similarly, highly irritant or chemically unstable drugs cannot be administered by the oral route, e.g., various penicillins and prostaglandins. More than 80 percent of drugs are administered in solid dosage forms since these are the most stable, easily handled and formulated, and provide a convenient mode of administration. However, such physico-chemical properties as low aqueous solubility of a drug may make this a less desirable dosage form when a rapid onset of action is needed.

4. *Bioavailability*. The physicochemical nature of certain drugs may exclude the oral route of administration because of poor bioavailability, as with spectinomycin sulfate, neomycin sulfate and sulfaguanidine. Biotransformations in the gastrointestinal tract often exclude this route for various drugs, e.g., salicylamide. Except for direct vascular administration, all routes of administration result in less than 100 percent delivery of the drugs to the general circulation. Although this is not the most important consideration for such drugs as aspirin or ascorbic acid, potent drugs with a low therapeutic index may show a great degree of variability in their pharmacologic response and toxic effects as a result of variations in bioavailability (Fig. 2.2). Doses of potent drugs are generally small, and therefore an excess of drug in a dosage form to circumvent low bioavailability can result in abnormally high absorption and resultant toxic response.
5. *Convenience of Administration*. It is neither convenient nor desirable to administer drugs intravenously for such purposes as occasional headache relief, appetite curtailment, or contraception. On the other hand, oral administration, though most convenient, may not be suitable for nauseated

patients or in pediatric practice. The route of administration should generally be determined by the length of therapy and the need to provide an exact dose with every administration, as with anticonvulsant and cardiotonic drugs.

6. *Route of Administration*. Each dosage form must conform to the specific requirements for administration through various routes, e.g., a solution for intravenous use, an aerosol for inhalation, an ointment for topical use, etc.

For those routes of administration whereby instantaneous drug delivery is achieved, few rate-limiting steps in the clinical response are possible. But for the majority of other routes of administration the rate-limiting steps are many, as shown in Schemes 2.1 and 1.1.

The major rate-limiting step in the absorption of drugs is their dissolution in the aqueous phase surrounding the site of administration. Formulation efforts have been directed towards both increasing and decreasing the dissolution rates, which are generally governed by the following well-known Noyes-Whitney equation (Fig. 2.3):[1]

$$dC/dt = (K \cdot D \cdot S/h)(C_s - C_t) \qquad \text{(Eq. 2.1)}$$

where dC/dt = dissolution rate
K = dissolution constant
D = diffusion coefficient
h = diffusion layer thickness
$(C_s - C_t)$ = difference between saturation solubility, C_s, and the concentration at time t. Concentration gradient.
S = surface area

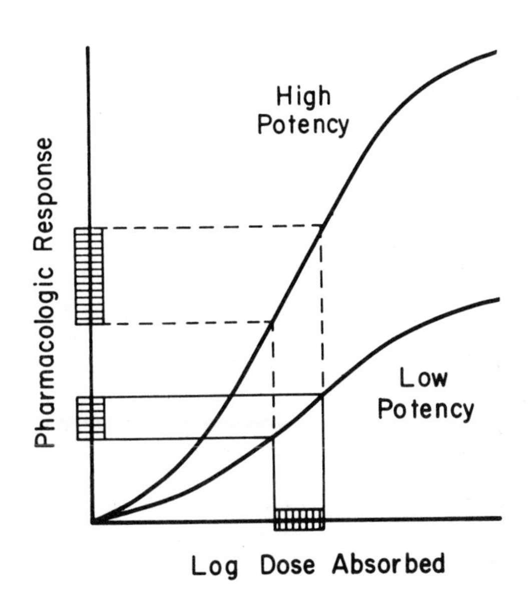

Figure 2.2. Log dose-response curves for high and low potency drugs. An identical variation in the bioavailability results in a significantly higher variation in the pharmacologic response for high potency drugs compared with low potency drugs.

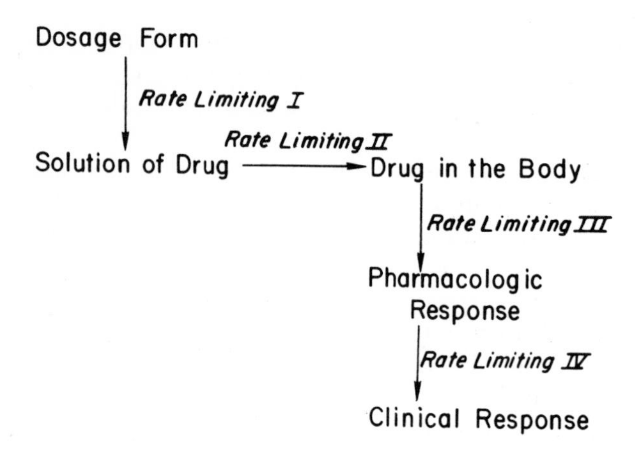

Scheme 2.1. The four categories of rate limiting steps in the clinical response as a result of dosage form administration.

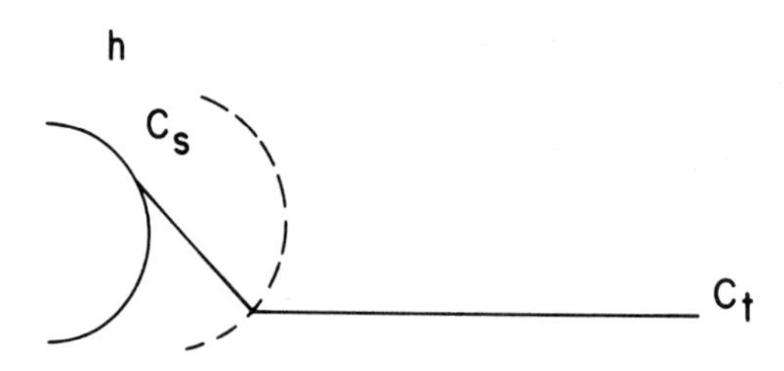

Figure 2.3. Diffusion layer dissolution model. The concentration in the diffusion layer around the particles is equal to the saturation solubility. The difference between the saturation solubility and bulk concentration (C_t) is the driving force for the dissolution of the particle.

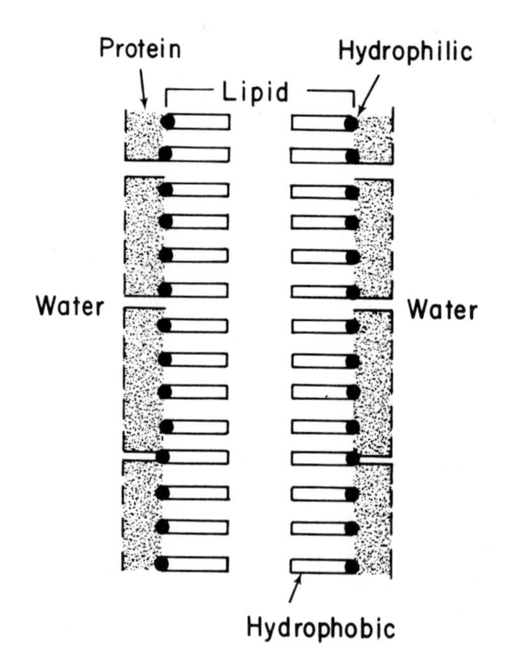

Figure 2.4. The Davson-Danielli diagram of the cell membrane, showing a double layer of lipid molecules covered by a protein coat. The hydrophobic end of each molecule is directed inward, while its hydrophilic end faces outward toward the protein coat. (From Davson and Danieli: The Permeability of Natural Membranes, 1952. Courtesy of Cambridge University Press)

It is therefore obvious that a variety of formulation parameters can be exploited to adjust the dissolution rates of drugs—these will be discussed later in this chapter. It should, however, be pointed out that dissolution alone is not sufficient to provide for the absorption of drugs. The drug molecules must have the characteristics required for crossing the various lipoid layers or membranes in order to reach the general circulation (Fig. 2.4).[2] Lack of sufficient aqueous solubility is usually the rate-limiting step in the dissolution process, and lack of sufficient lipophilic properties is the usual rate-limiting step in the penetration of the lipid barriers. Attempts to rectify dissolution problems can therefore lead to problems in membrane transport, and vice versa. A fine balance between the hydrophilic and lipophilic properties is needed to provide optimum delivery of drugs to the site of action.

The bioavailability of drugs is a prime consideration in formulation approaches, especially in attempting to increase bioavailability through physicochemical manipulations. The variations in bioavailability extend to almost all classes of pharmacologic categories and chemical structures (Table 2.1), making the chemical modifications required for optimum bioavailability difficult to summarize. Besides chemical modifications, formulation manipulations also significantly affect bioavailability. For example, almost 60-fold differences have been reported in the rates of absorption of formulations of spironolactone.[16–18] In its recent statement on bioavailability and bioequiva-

Table 2.1. BIOAVAILABILITY OF SOME COMMONLY USED DRUGS

DRUG	ROUTE OF ADMINISTRATION	PERCENT ABSORBED INTACT
Ampicillin[3]	po (anhydrous)	53
	po (hydrate)	49
	im (sodium salt)	85
Carbenicillin[4]	im	64
	po (indanyl sodium)	50
Cloxacillin[5]	po	77
Colistimethate[6]	po	0
Dicloxacillin[7]	im	75
Dicloxacillin[5]	po	80
Digoxin[8,9]	po	45–100
Gentamicin[10]	po	0.2
Kanamycin[11]	po	0.7
	im	40–80
Lincomycin[12]	po	30
Methacycline[13]	po	58
Nafcillin[5]	po	50
Neomycin[11]	po	0.6
Oxacillin[5]	po	67
Penicillin G[14]	po (sodium salt)	<30
Propranolol[15]	po	0.4
Tetracycline[13]	po	77

lency requirements the Food and Drug Administration has recognized several drugs with potential bioequivalency problems.[19] These drugs are listed in Table 2.2. Most of these drugs are highly potent, with steep log-dose-response curves (Fig. 2.2). Some exhibit an "all or none" effect, making the utmost importance of their bioavailability considerations.

Table 2.2 DRUGS WITH POTENTIAL BIOEQUIVALENCY PROBLEMS

Acetazolamide	Hydrochlorothiazide	Promethazine
Acetyldigitoxin	Hydroflumethiazide	Propylthiouracil
Alseroxylon	Imipramine	Pyrimethamine
Aminophyllin	Isoproterenol	Quinethiazide
Aminosalicylic acid	Liothyronine	Quinidine
Bendroflumethiazide	Menadione	Rauwolfia serpentina
Benzthiazide	Mephenytoin	Rescinnamine
Betamethasone	Methazolamide	Reserpine
Bishydroxycoumarin	Methyclothiazide	Salicylazosulfapyridine
Chlorambucil	Methylprednisolone	Sodium sulfoxone
Chlorodiazepoxide	Methyltestosterone	Spironolactone
Chlorothiazide	Nitrofurantoin	Sulfadiazine
Chloropromazine	Oxtriphylline	Sulfadimethoxine
Cortisone acetate	Para-aminosalicylic acid	Sulfamerazine
Deserpidine	Para-methadione	Sulfaphenazole
Dexamethasone	Perphenazine	Sulfasomidine
Dichlorphenamide	Phenacemide	Sulfasoxazole
Dienestrol	Phensuximide	Theophylline
Diethylstilbestrol	Phenylaminosalicylate	Thioridazine
Dyphylline	Phenytoin	Tolbutamide
Ethinyl estradiol	Phytonadione	Triamcinolone
Ethosuxmide	Polythiazide	Trichlormethiazide
Ethotoin	Prednisolone	Triethyl melamine
Ethoxzolamide	Primidone	Trifluoperazine
Fludrocortisone	Probenecid	Triflupromazine
Fluphenazine	Procainamide	Trimeprazine
Fluprednisolone	Prochlorperazine	Trimethadione
Hydralazine	Promazine	Uracil mustand
		Warfarin

Efforts to increase bioavailability are prompted by the following:

1. *Cost*. The unavailable component goes to waste. Although this may not be a very important factor for such drugs as aspirin or sulfonamides, the hormones, steroids, and antibiotics are considerably more expensive. For example, a recent formulation change which increased the bioavailability of griseofulvin resulted in an approximately 25 percent decrease in the over-the-counter prices of its bioavailable dose.
2. *Local Effects*. Drugs which are incompletely absorbed stay at the site of administration for a longer period of time and can cause local irritation

(Table 2.3) and modify the habitat—e.g., gastrointestinal flora.[20] Gastrointestinal irritation is an extremely important consideration since it can result in excessive blood loss, nausea, vomiting, and cramps. Possible sterilization of the gastrointestinal tract is also important since it can cause serious vitamin deficiencies. On the other hand, lack of absorption is desirable if the purpose is to produce local effect or to sterilize the gut, as with the use of neomycin sulfate.

Table 2.3. DRUGS WHICH CAN CAUSE GASTROINTESTINAL IRRITATION

Acetohexamide	Indomethacin
Acetylsalicylic acid	Isosorbide dinitrate
Allopurinol	Mefenamic acid
Aminophylline	Metronidazole
Amisometradine	Nalidixic acid
Azapetine	Nitrofurantoin
Bethanechol	Oxyphenbutazone
Biperiden hydrochloride	Phenformin
Bismuth sodium triglycollamate	Phenylbutazone
Chloral hydrate	Phenytoin
Chlorphenoxamine	Potassium chloride
Chlorpropamide	Procyclidine
Chlorthalidone	Rauwolfia serpentina
Cycrimine	Reserpine
Dithiazanine	Sulfamethoxypyridazine
Doxycycline	Sulfapyrazine
Ferrous gluconate	Thiabendazole
Ferrous sulfate	Tolazamide
Furazoladine	Tolbutamide
Griseofulvin	Triametrine

3. *Systemic Action*. Drug products with low bioavailability are more likely to show variations in pharmacologic response between individuals and within an individual. This is caused by the excess drug contained in the dosage form to compensate for the lower bioavailability.

A large number of physical, chemical, and technical modifications go into the formulation of a dosage form. Some of these are discussed here to provide an appreciation of the complexities involved.

CHEMICAL FORM

An appropriate selection of the chemical form of a drug is very important in achieving desired bioavailability. Chemical modification can involve changing the chemical structure of a drug to a form which is significantly different from the active drug entity.[21] This form can, however, provide a similar therapeutic response since within the body it breaks down into the active entity. Ideally, a drug molecule should have sufficient aqueous solubility for dissolution; an optimum oil/water partition coefficient to provide diffusion through several bilipid layers; and stable chemical groups which will interact with the receptor site. Such an ideal molecule does not usually exist in nature, however, and chemical modifications are generally directed toward that part of the molecule which is responsible for the hindrance in the overall absorption process. For example, it is desirable to restrict the absorption of a sulfonamide if it is to provide a local action in the gastrointestinal tract. This can be achieved by the synthesis of such chemical forms as succinylsulfathiazole, phthalylsulfathiazole, phthalylsulfacetamide, and salazosulfapyridine, with a free carbonic acid structure which can ionize in the gut. These chemical modifications, which lead to an ionized species, decrease the lipid/water partition coefficient sufficiently to restrict the absorption of the sulfonamides. The antibacterial activity is unfolded when the amide links are broken down by hydrolysis, thus releasing the free and active sulfonamide structures. On the other hand, a decrease in the ionization will result in better absorption, as demonstrated for such ganglionic blocking agents of the onium type as hexamethonium. By switching to tertiary amines, such as mecamylamine and pempidine, one obtains drugs that are more steadily and completely absorbed.[22]

Increasing lipid solubility through chemical modifications is further exemplified by doxycycline, a derivative of tetracycline. This compound is more efficiently absorbed from the intestine than tetracycline partly because of a better lipid solubility and partly because of a decreased tendency to form poorly soluble complexes with calcium and phosphate. This facilitated absorption decreases the risk of disturbances in the intestinal flora and of intestinal superinfection.

Table 2.4. INFLUENCE OF LIPID SOLUBILITY OF BARBITURATES ON THEIR ABSORPTION IN RAT COLON

BARBITURATE	ABSORPTION	PARTITION COEFFICIENT CHLOROFORM/WATER
Barbital	12	0.7
Aprobarbital	17	4.9
Phenobarbital	20	4.8
Butalbital	23	10.5
Butethal	24	11.7
Cyclobarbital	24	13.9
Pentobarbital	30	28.0
Secobarbital	40	50.7
Hexethal	44	>100

(After Schanker: J Med Pharm Chem 2:343, 1960)

Chemical changes related to lipid solubility and its effect on gastrointestinal absorption are best exemplified by barbiturates, in which an increase in lipid solubility is directly related to absorption from the colon (Table 2.4).[22] Similarly, the increased lipid solubility of several vitamins have also been related to better absorption. For example, the thiamine molecule is rearranged by opening the thiazole ring so that its quarternary structure is lost.[23] Other examples include such derivates as riboflavin-2¹,3¹,4¹,5¹-tetrabutyrate,[24] pyridoxine-triamino benzoate, 2-O-benzoyl-1-ascorbic acid,[25] and ascorbyl palmitate,[26] among several others.

The absorption of erythromycin is made more efficient by using its estolate form in the suspensions instead of the ethyl succinate ester.[27] The estolate form is also absorbed better than the base, salts, and other esters of erythromycin.

To enhance the absorption of 6-azauridine—an antimetabolite used in cancer chemotherapy and in the treatment of psoriasis—its triacetyl derivative has been successfully used.[28] Chemical modifications have also been introduced into lincomycin and griseofulvin to enhance their intestinal absorption. In the highly water-soluble antibiotic lincomycin, lipophilizing moieties such as propionate, stearate, and ethyl carbonate are used for this purpose,[29,30] while in the very poorly water-soluble griseofulvin, hydrophilizing moieties such as succinate and oxime are introduced to facilitate gastrointestinal uptake.[31]

For topical application of corticosteroid compounds, lipophilizing moieties such as valerate are used to enhance skin absorption.[32] Other examples include the esters of salicylic acid (Rheumacyl, Salenol, Tranvesin) used as external antirheumatic rubs, and esters of nicotinic acid (Nicotafuryl, Rubriment, Nicotherm) used as rubefacients.[33]

The aqueous solubility of drugs can be increased by such modifications as sulfacarboxychysoidine,[34,35] a sulfonamide designed on the basis of insight gained with prontosil and pontosil rubrum. The introduction of dicarbonic acid and sugars into the chemical structure increases the aqueous solubility of tuberculostatic, thiosemicarbazone, and isonicotinic acid hydrazide.[33] The hemisuccinates of estriol,[33] oxazepam,[36] griseofulvin,[31] erythromycin, and chloramphenicol also provide increased aqueous solubility. Additional examples are listed in Table 2.5.

Many important drugs are weak acids or bases. Salts of acidic or basic drugs have solubility characteristics different from those of acids and bases and thus tend to show different bioavailability. Sodium or potassium salts of weak acids dissolve much more rapidly than the corresponding free acid, regardless of the pH of the dissolution medium. The same is usually true of the hydrochloride salts or other strong acid salts of weak bases, such as tetracycline hydrochloride or atropine sulfate.

Salt formation, however, does not necessarily result in an enhancement of dissolution rate and bioavailability. For instance, the aluminum salt of aspirin dissolves so slowly in the gastrointestinal tract that the absorption of the drug is incomplete. This slow dissolution is attributed to the formation of a water

Table 2.5. EXAMPLES OF STRUCTURE MODIFICATIONS ENHANCING WATER SOLUBILITY OF DRUGS

COMPOUND	*DERIVATIVE WITH HIGHER WATER SOLUBILITY*
Tetracycline	Rolitetracycline, pipacycline, tetralysine
Theophylline	Diprophylline, soluphyllin
Theobromin	Isobromin
Prednisolone	Soludacortin, ultracorten, corticosol, magnadelt
Deoxycortone	Docaquosum, diethylstilbesterol, idroestril
Testosterone	Testosterone phosphate
Sulfanilamide	Rubiazole II, amebesid, septosil, glucosyl sulfanilamide
Menadiol	Menadiol diphosphate, menadiol disulfate hemodol T
Tocopherol	α-tocopheryl hemisuccinate
Chloramphenicol	Chloramphenicol hemisuccinate/hemiphthalate
Estriol	Estriol hemisuccinate
Phenetidine	Phenetidine hemisuccinate
Oxazepam	Oxazepam hemisuccinate
Hydroxydione	Hydroxydione dihemisuccinate

insoluble aluminum compound on the surface of the solid particles.[37] Phenobarbital tablets prepared from the free acid have a much higher dissolution rate than tablets made from the sodium salt in acidic media. This is attributed to the fact that the tablets containing the free acid disintegrate rapidly in the acidic media, whereas tablets containing the sodium salt do not disintegrate rapidly but only swell and dissolve slowly from the surface.[38] The free base of the experimental antihypertensive compound Su-17770 exhibits a greater dissolution rate than the monohydrochloride salt in 0.1 NHCl—this is attributed to the common ion effect.[39]

The use of acidic and basic salt forms to increase dissolution rates and bioavailibility depends on the mechanism of dissolution involved. For example, bases such as tetracyclines are freely soluble in alkaline media but their corresponding salts show a decrease in solubility above pH 3. This phenomenon is attributed to the common ion suppression of the solubility product equilibrium.[40] Therefore, higher dissolution rates are possible using chlortetracycline than with its hydrochloride salt in acidic media.

Sodium salts of barbiturates are readily soluble in water, which explains the higher rates of absorption of heptabarbital from capsules containing the sodium salt. In general, salts will show higher dissolution rates than the corresponding nonionic drugs at any pH, even though the final equilibrium solubility of the drug and its salt are the same. This is explained by the fact that if precipitation were to occur, the general result would be a suspension of fine particles with characteristics appropriate for rapid redissolution (Scheme 2.2). This applies to barbiturates because their sodium salts are precipitated in the acidic medium of the stomach. They are nevertheless absorbed at a much higher rate than the corresponding free acids. For example, heptabarbital given as a free acid in tablet form shows a peak plasma

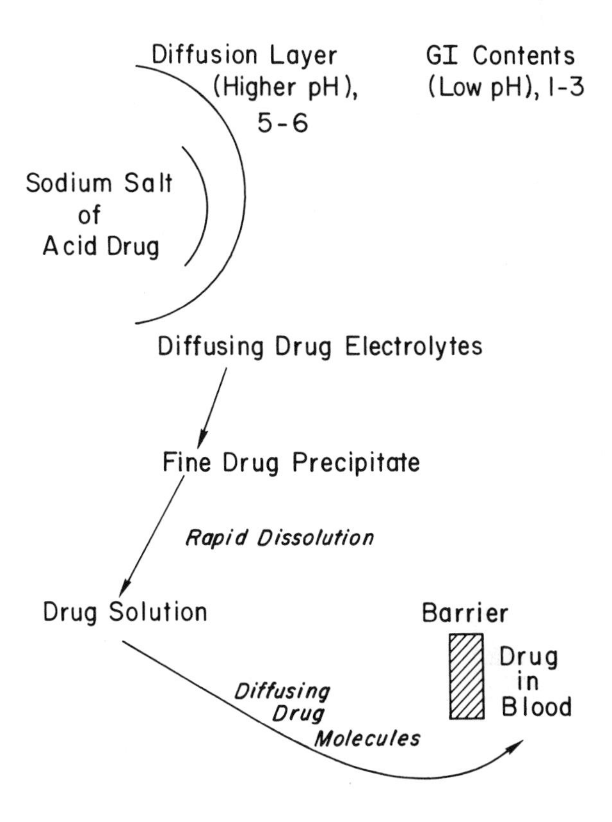

Scheme 2.2. Dissolution model of a buffered formulation. The buffering agent provides higher pH in the diffusion layer increasing the dissolution of weakly acidic drug, which precipitates and redissolves in the gastrointestinal tract. In the absence of buffering agent the pH of the diffusion layer will be acidic as a result of the dissolution of weak acid.

concentration from 1.5 to 5 hours after administration, whereas its sodium salt is absorbed more rapidly, showing a peak concentration from 0.4 to 2 hours after oral administration.[41] However, the bioavailability of heptabarbital is about 17 percent lower in the salt formulation compared to the base. This is attributed to the possibility of the precipitation of drug in relatively gross crystalline particles when using the salt form.

The formation of a salt may therefore not always be advantageous, as is

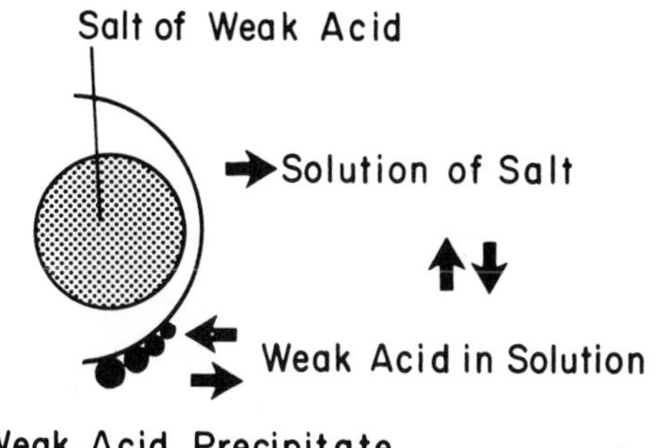

Scheme 2.3. Surface precipitation model for the dissolution of a weak acid in acidic media.

also demonstrated with benzphetamine pamoate[42] and sodium warfarin.[43] The general model for the release of an amine in acid solution from a pellet made from a weak acid salt of the amine, where the weak acid would precipitate on the surface of the pellet, is as shown in Scheme 2.3. In several other examples involving weak acids such as penicillin, phenobarbital, iopanoic acid, and para-aminosalicylic acid, dissolution and absorption rates have been increased as a result of salt formation.

Another approach to the use of salt formation involves additives which provide an alkaline pH around the dissolving particles of weakly acidic drugs. This is best exemplified by the buffered aspirin formulation (Scheme 2.2) in which the sodium bicarbonate content provides the alkaline pH. Similarly, sodium phosphate also provides an alkaline pH upon its hydrolysis in the gastrointestinal tract.

IONIZATION CONSTANT

The IONIZATION CONSTANT of weakly acidic and basic drugs is an important parameter since it determines their aqueous solubility, dissolution rates in the gastrointestinal tract, and the rate of transport across lipoidal layers.

The total solubility, S_t, of a drug is dependent on the relative contribution of the solubility of its un-ionized form, S_u, and the ionized form, S_i:

$$S_t = S_u + S_i \qquad \text{(Eq. 2.2)}$$

Whereas the solubility of the un-ionized form is a thermodynamic parameter which is constant under given temperature and pressure conditions, the fraction of the ionized form in the solution changes as a function of solution pH. Consider the example of a weak acid:

$$HA \overset{K_a}{\rightleftharpoons} A^- + H^+ \qquad \text{(Eq. 2.3)}$$

where

$$K_a = [A^-]\,[H^+]\,/\,[HA] \qquad \text{(Eq. 2.4)}$$

or

$$A^- = K_a\,[HA]/[H^+] \qquad \text{(Eq. 2.5)}$$

and therefore, for weak acids:

$$S_t = S_u\,(1 + K_a/H^+) \qquad \text{(Eq. 2.6)}$$

Using a similar approach for weak bases:

$$S_t = S_u\,(1 + H^+/K_a) \qquad \text{(Eq. 2.7)}$$

Thus the total solubility of weak acids will decrease and that of weak bases increase as H^+ increases in the solution or the pH (= $-\log H^+$) decreases. A simple way to remember this is that acids have higher solubility in alkaline pH and bases have higher solubility in acidic pH. The ionization constant is often expressed as pK_a, analogous to pH:

$$pK_a = -\log K_a \qquad \text{(Eq. 2.8)}$$

The effect of pH on the total solubility is dependent on the relative acidity or basicity of the compounds as expressed by the following conversion of Equations 2.6 and 2.7:
For monobasic acids:

$$S_t = S_u\,(1 + 10^{\mathrm{pH} - pK_a}) \qquad \text{(Eq. 2.9)}$$

For monoacid bases:

$$S_t = S_u\,(1 + 10^{pK_a - \mathrm{pH}}) \qquad \text{(Eq. 2.10)}$$

The following calculations illustrate the relationship of pH and pK_a with solubility:

pH − pK_a	S_t/S_u (acid)	S_t/S_u (base)
−2.0	1.01	101
−1.5	1.03	32.6
−1.0	1.10	11.0
−0.5	1.32	4.16
0.0	2.0	2.0
+0.5	4.16	1.32
+1.0	11.0	1.10
+1.5	32.6	1.03
+2.0	101	1.01

In applying the equations given above to in vivo situations in man, one should remember that the stomach contents are usually in the pH range of 1 to 3 and the contents of the upper small intestine, where most drug absorption occurs, have a pH range of 5.5 to 7.0, and are not "alkaline," only less acidic. Weakly basic compounds will therefore generally dissolve faster in the gastric fluids, and weakly acidic compounds will dissolve faster in the intestinal fluids. For example, salicylic acid with a pK_a of 3.0 shows an approximately 16-fold increase in its dissolution rate when the surrounding pH is changed from 1.5 to 6.8.[44] However, for weak acids a linear relationship between $1/H^+$ and its dissolution rate or for weak bases a linear relationship between the dissolution rate and H^+ may not always be possible, especially around neutral pH ranges. This is due to the mechanism of dissolution, which involves formation of a diffusion layer saturated with the dissolving compound, weak acid or base, which results in pH values which may not correspond to the pH values of the bulk medium. For example, dissolution of salicylic acid will result in a pH around the dissolving particles lower than the bulk pH of the intestinal fluids. It is this pH of the diffusion layer that determines the actual rate of dissolution. As discussed earlier, inclusion of such agents as sodium bicarbonate in aspirin formulations results in a higher pH in the diffusion layer, increasing the dissolution rates.

It is generally agreed that the un-ionized form of a drug is the most suitable for gastrointestinal absorption. Thus the efficiency of absorption of a weakly acidic or weakly basic compound will change as the dosage form passes through various pH conditions in the gastrointestinal tract. This theory is also referred to as the pH-partition theory and holds true for a variety of drugs. However, if one takes into account the large differences in the absorption surface areas of the stomach, the small intestine, and the colon, it seems logical to assume that most of the drugs will show sufficient absorption—especially from the upper part of the small intestine—if an equilibrium is always established between the un-ionized absorbable species and the ionized form of the drug. For example, in situ studies show that at pH 6.8, where salicylic acid is almost 100 percent ionized, the absorption is very fast from the rat intestine (50 percent absorbed in 7 minutes).

A knowledge of the ionization constant and the chemical nature of drugs is extremely important not only in establishing dissolution rates and rates of gastrointestinal absorption but also in predicting and avoiding various incompatibilities when drugs are mixed in such extemporaneous preparations as intravenous admixtures (Table 2.6).

Table 2.6. IONIZATION CONSTANTS OF DRUGS IN COMMON USE

DRUG	*NATURE*	pK_a
Acetazolamide	Acid	7.2
Allopurinol	Base	9.4
Amiloride	Base	8.7
Aminocaproic acid	Ampholyte	4.4, 10.7
Aminopyrine	Base	5.0
Aminosalicylic acid	Acid	3.2
Amitriptyline	Base	9.4
Amphetamine	Base	9.8
Amphotericin B	Ampholyte	5.5, 10.0
Ampicillin	Acid	2.5, 7.2
Amylobarbital	Acid	7.7
Antipyrine	Base	1.4
Apomorphine	Base	7.2, 8.9
Aspirin	Acid	3.5
Atropine	Base	9.8
Barbital	Acid	7.8
Benzylpenicillin	Acid	2.8
Betahistine	Base	3.5, 9.7
Bishydroxycoumarin	Acid	5.7
Bupivacaine	Base	8.1
Burimamide	Base	7.5
Butobarbital	Acid	7.9
Caffeine	Base	0.8
Carbenicillin	Acid	2.6, 2.7
Cephalexin	Acid	5.2, 7.3
Cephalothin	Acid	2.5

Table 2.6. (Continued)

DRUG	*NATURE*	pK_a
Chlorambucil	Base	8.0
Chlordiazepoxide	Base	4.6
Chlormethaziole	Base	3.2
Chloroquine	Base	8.4, 10.8
Chlorpheniramine	Base	9.2
Chlorpromazine	Base	9.3
Chlorpropamide	Acid	4.8
Chlortertacylcine	Base	3.3, 7.4, 9.3
Clindamycein	Base	6.9
Cloxacillin	Acid	2.7
Codeine	Base	6.0
Cyclizine	Base	8.2
Cyclobarbital	Acid	7.3
Cytarabine	Base	4.3
Dapsone	Acid	1.2-2.5
Demethylchlor-tetracycline	Base	7.2, 9.4
Desipramine	Base	9.5
Dextromoramide	Base	7.0
Dextropropoxyphene	Base	6.3
Diamorphine	Base	7.6
Diazepam	Base	3.3
Dicoumarol	Acid	5.7
Dihydrocodeine	Base	8.8
Diphenhydramine	Base	8.3
Diphenylhydantoin	Acid	8.3
Disodium Cromoglycate	Acid	2.0
Doxycycline	Ampholyte	3.4, 7.7, 9.7
Droperidol	Base	7.6
Emetine	Base	5.8, 6.6
Ephedrine	Base	9.6
Ergometrine	Base	7.3
Ergonovine	Base	7.3
Erythromycin	Base	8.8
Eserine	Base	8.5
Ethacrynic acid	Acid	3.5
Ethambutol	Base	6.5, 9.0
Ethosuxamide	Acid	9.3
Ethyl biscoumacetate	Acid	3.1
Fenfluramine	Base	9.9
Flucloxacillin	Acid	2.7
5-Fluorouracil	Base	8.1
Frusemide	Acid	3.7
Heroin	Base	7.6
Hexobarbital	Acid	8.2
Homatropine	Base	9.7

(Continued)

Table 2.6. (Continued)

DRUG	*NATURE*	pK_a
Hydrochlorothiazide	Acid	7.9, 9.2
Hyoscine	Base	8.1
Ibuprofen	Acid	4.4
Imipramine	Base	9.5
Indoramine	Base	7.8
Iprindole	Base	8.2
Isoprenaline	Base	8.6
Isoxsurpine	Base	8.0, 9.8
Levallorphan	Base	45
Levodopa	Amino acid	2.3, 8.7, 9.9
Lignocaine	Base	7.9
Lincomycin	Base	7.6
Lorazepam	Ampholyte	1.3, 11.5
Lysergide	Base	3.3, 7.8
Mecamylamine	Base	11.2
Mefenamic acid	Acid	4.2
Mepacrine	Base	7.7, 10.3
Meperidine	Base	8.7
Metaraminol	Base	8.6
Methadone	Base	8.6
Methicillin	Acid	2.8
Methohexital	Acid	7.9, 8.3
Methotrexate	Acid	4.8, 5.5
Methotrimeprazine	Base	9.2
Methoxamine	Base	4.8
Methylamphetamine	Base	10.0
Morphine	Base	7.9, 8.1, 9.9
Nalidixic acid	Acid	6.7
Nalorphine	Base	7.8
Nitrazepam	Base	3.2, 10.8
Nitrofurantoin	Acid	7.2
Novobiocin	Acid	4.3, 9.1
Orciprenaline	Base	8.9, 11.8
Orphenadrine	Base	8.4
Oxazepam	Ampholyte	1.7, 11.6
Oxyphenbutazone	Acid	4.7
Oxytetracycline	Base	3.3, 7.3, 9.1
Papaverine	Base	6.4
Penicillamine	Base	1.8, 7.9, 10.5
Pentobarbital	Acid	8.1
Perphenazine	Base	7.8
Pethidine	Base	8.7
Phenobarbital	Acid	7.2
Phenoxymethyl-penicillin	Acid	2.7
Phenylbutazone	Acid	4.5
Phenytoin	Acid	8.3

Table 2.6. (Continued)

DRUG	*NATURE*	pK_a
Physostigmine	Base	8.5
Piperazine	Base	5.7, 9.8
Practolol	Base	9.5
Prilocaine	Base	7.9
Probenecid	Acid	3.4
Procainamide	Base	9.2
Procaine	Base	8.8
Prochlorperazine	Base	8.1
Promazine	Base	9.4
Promethazine	Base	9.1
Propanolol	Base	9.45
Pyrazinamide	Base	0.5
Pyrimethamine	Base	7.2
Quinacrine	Base	7.7, 10.3
Quinalbarbital	Acid	7.9
Quinidine	Base	4.3, 8.4
Quinine	Base	4.3, 8.4
Reserpine	Base	6.1
Salbutamol	Base	9.3, 10.3
Salicylazosulfa-pyridine	Acid	0.6, 2.4, 9.7 11.8
Salicylic acid	Acid	3.0
Secobarbital	Acid	7.9
Succinylsulfa-thiazole	Acid	4.5
Sulfadiazine	Acid	6.3
Sulfadimethoxine	Acid	6.3
Sulfafurazole	Acid	4.9
Sulfamethiazole	Acid	5.4
Sulfamethoxy-pyridazine	Acid	6.7
Sulfamethoxazole	Acid	6.0
Sulfasalazine	Acid	0.6, 2.4 9.7, 11.8
Sulfathiazole	Acid	7.1
Sulfinpyrazone	Acid	7.1
Terbutaline	Base	10.1
Tetracycline	Base	3.3, 7.8, 9.7
Theophylline	Base	0.7
Thiopental	Acid	7.6
Tolazoline	Base	10.3
Tolbutamide	Acid	5.4
Tranexamic acid	Ampholyte	4.3, 10.6
Trifluoperazine	Base	8.1
Trimethorpim	Base	6.4
Vinblastine	Base	5.4, 7.4
Vincristine	Base	5.0, 7.4

POLYMORPHISM AND AMORPHISM

Variations in the arrangement of molecules or ions to form a crystal are referred to as **POLYMORPHISM**. This may not be apparent from the shape of the crystals. Each polymorph has a certain thermodynamic energy associated with it as a result of strains in the bonds of the lattice structure, and therefore one polymorph may be more stable than the others. At any given temperature and pressure only one crystal form of a drug will be stable, and other forms will convert to this form. When the conversion is relatively slow, the polymorph is said to be metastable and can be used in dosage forms. The various polymorphic forms are chemically indistinguishable. However, they differ significantly in physical properties, such as density, melting point, solubility, and dissolution rates. For example, riboflavin exists in several polymorphic forms with a 20-fold difference in their aqueous solubility. Amorphous forms in which no internal crystal structure exists exhibit the greatest difference, making these forms unstable towards conversion to a crystalline structure.

Based on the Noyes-Whitney equation,[1] the order of dissolution rates for the crystal forms can be arranged as:

$$\text{amorphous} > \text{metastable} > \text{stable}$$

It has been suggested that almost 40 percent of all organic compounds can exist in various polymorphic forms: sometimes in as many as five different forms, as in the case of cortisone acetate.

The use of a certain polymorphic form to increase bioavailability is based on the premise that the dissolution rate is the rate-limiting step in the absorption of the drug. This premise, however, may not be applicable to all drugs, especially those which are absorbed by an active process e.g., various vitamins.

In a recent study, the two polymorphs of tetracycline were shown to give different bioavailability upon intraduodenal administration to avoid the effects of food and gastric emptying in rabbits and rats. Similar studies on chlortetracycline and oxytetracycline failed to show such differences in the blood levels, which was attributed to rapid transformation of the polymorphic forms in the gastrointestinal tract. Differences in bioavailability in humans have also been demonstrated for tetracycline, with an approximately 33 percent difference in the total amount excreted in 10 hours.[45]

The transformation of a higher energy polymorph to low energy stable polymorphs has often been reported in aqueous media, as with sulfathiazole, but in a recent study it was shown that formulation additives such as polyethylene glycol 4000 can retard this conversion sufficiently to make differences in the dissolution rates and bioavailability (Fig. 2.5).[46] An example in which the amorphous form provides much better bioavailability than the crystalline form is provided by novobiocin when administered as a suspension.[47] The conversion of the amorphous form to the crystalline form has here been stabilized by the addition of methylcellulose to the suspension.

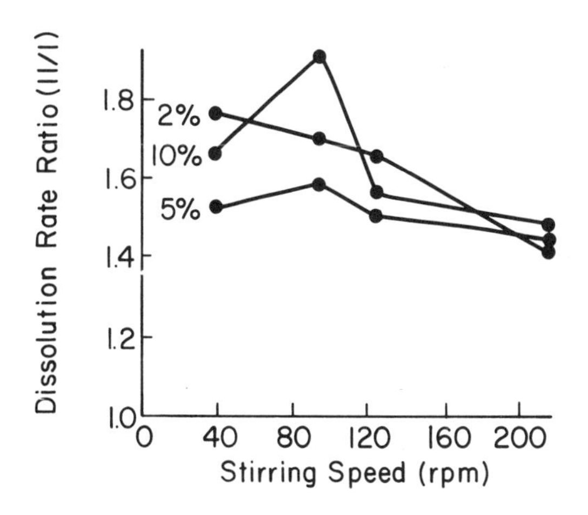

Figure 2.5. Relationship between the dissolution rate ratios of the two sulfathiazole polymorphs and the stirring speed at different percentage compositions of polyethylene glycol 4000. When no polyethylene glycol is added, this ratio is equal to one. (From Niazi: J Pharm Sci 65:302, 1976)

Recent studies have shown differences in the bioavailability of different polymorphic forms of various drugs, including aspirin, barbital, estrone, sulfonamides, chloramphenicol palmitate, chlorodiazepoxide hydrochloride, methylchlormethane, cholesteryl palmitate, adiphenine hydrochloride, benzocaine picrate, meprobamate, erythromycin, and others.

The choice of an appropriate polymorphic form is well demonstrated by insulin, where a mixture of amorphous and crystalline forms provides an initial fast absorption from the amorphous insulin and a sustained effect from the crystalline forms. Besides affecting bioavailability, polymorphic forms also affect other physicochemical properties. For example, theorbroma oil suppositories require appropriate selection of high melting forms to avoid liquefaction at room temperature; cortisone acetate suspensions can be stabilized with the correct choice of the polymorphic form,[48] and penicillin G shows a high degree of chemical instability in amorphous form.[49]

SOLVATION

During seeding, crystals may incorporate one or more of the molecules of the solvent into their structure and the resultant forms are referred to as SOLVATES. The solvates themselves may exist in various polymorphic forms and are referred to as PSEUDO-POLYMORPHS. Examples of pseudo-polymorphs include mercaptopurine, fluprednisolone, and succinylsulfathiazole.

The number of drug solvates is well over 100 and some of the most common examples include steroids, antibiotics, sulfonamides, barbiturates, xanthines, and cardiac glycosides.

The use of solvates in increasing bioavailability depends on the premise that some solvates dissolve faster than their corresponding asolvates. The anhydrate forms generally dissolve faster than their corresponding hydrates in aqueous media. However, the relationship becomes much more complex when alcoholates or other non-aqueous solvates are dissolved in water.

One interesting example of increased bioavailability through use of an asolvate is that of ampicillin, which exists in trihydrate and anhydrate forms. Earlier studies on the higher bioavailability of the anhydrate form were questioned on the basis of similar solubilities of the anhydrate and dihydrate forms in dilute hydrochloric acid. Later it was shown that both forms do indeed provide identical bioavailability.[50]

The monoethanol and hemiacetone solvates of tert-butylacetates of hydrocortisone and prednisolone show higher absorption rates than do the asolvate forms. These solvates also exhibit dimorphism and each form exhibits different absorption rates. The dimorphic solvates containing one mole of water, the tert-butylamine disolvate, and the three trimorphs of prednisolone show different dissolution rates which can be correlated to their pharmacologic responses from implanted pellets.[51]

Significant differences in the dissolution rates of anhydrous and hydrated forms of caffeine, theophylline, and glutethimide have been observed. For example, the anhydrous form of mercaptopurine was shown to dissolve twice as fast as the hydrate form in aqueous media (Fig. 2.6).[52]

An interesting application of solvation involves increasing the exposed surface area of powders upon solvation and desolvation cycling, as reported for griseofulvin and its chloroformate.[53] The first example of a drug-organic solvate exhibiting improved gastrointestinal absorption over the asolvate form is also that of griseofulvin chloroformate, which provides about 80 percent improvement in the bioavailability of griseofulvin.[54]

Hydrate forms are also used in the manufacture of effervescent tablets. The heat fusion method requires the blending of all the components with the inclusion of 15 to 25 percent of the acid ingredient as citric acid monohydrate. The batch is heated while it is mixed until the mole of water in the citric acid is released and becomes the granulating agent. The heat and the limited amount of water present guard against the propagation of the effervescent reaction.

Whereas the selection of an appropriate solvate or asolvate form can be advantageous in increasing the dissolution rates and bioavailibility in some

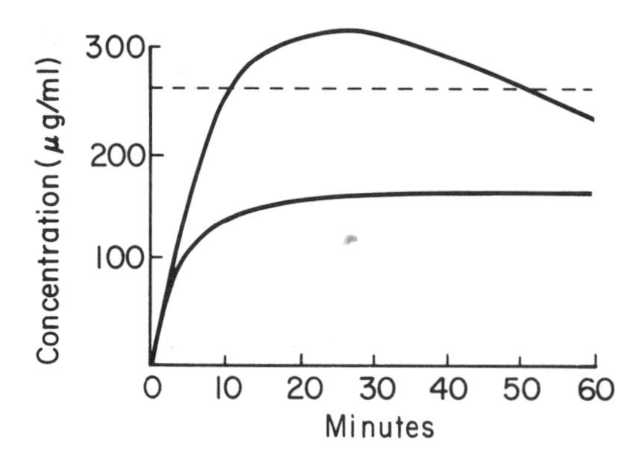

Figure 2.6. Dissolution rates of anhydrous (top) and monohydrate (bottom) forms of mercaptopurine in water. (From Huang and Niazi: J Pharm Sci 66:608, 1977)

instances, their use requires careful monitoring of manufacturing and storage conditions because of the possibility of inadvertent solvation or desolvation.

COMPLEXATION

A molecular complex consists of constituents held together by such weak forces as hydrogen bonds. The properties of drug complexes, such as solubility, molecular size, diffusivity, and lipid-water partition coefficient, can differ significantly from those of the drug itself, resulting in possible bioavailability variations. Complexation will generally increase the total solubility of a poorly water-soluble drug if the complex itself is soluble in aqueous media. If the complexation process is reversible, then:

$$\text{Drug} + \text{Complexing agent} \rightleftharpoons \text{Complex}$$

and the absorption rates and the extent of absorption will be increased for poorly soluble drugs. For drugs which are generally adequately absorbed, the complex formation may result in a slowing down of absorption, but the overall quantity of drug absorbed may not change.

The most frequently observed complex formation is between various drugs and macromolecules, as with gums, cellulose derivatives, high molecular weight polyols and nonionic surfactants. Mostly, however, these complexations are reversible with little effect on the bioavailability of drugs. But in those instances where the complex is insoluble in aqueous media, these interactions are clearly contraindicated, as with the complexation of amphetamine with sodium carboxymethylcellulose,[55] phenobarbital with polyethylene glycol 4000,[56] and of tetracycline with heavy metal ions.

The complex formation also occurs between drugs and complexing agents of comparable sizes, as with salicylic acid and caffeine; atropine and eosin B; diphenhydramine and methyl orange; and acid dyes and basic nitrogen compounds, resulting in increased lipid solubility of the dye. In all of these instances, however, no significant effect has been noted on bioavailability, due mainly to the dissociation of these complexes in the gastrointestinal tract.[57]

Several cases exist in which absorption rates have been increased as a result of complexation. For example, propylamide and certain other substituted propionamides form complexes with prednisone and prednisolone in lipid solvents enhancing their transfer across intestinal and synthetic lipoid barriers.[58–60] A fast-dissolving hydroquinone complex of digoxin was reported to double the rate of release, thereby increasing the overall bioavailability of digoxin.[61]

The complexation of caffeine with a large number of drugs is a well-known phenomenon which results in the increased aqueous solubility of drugs, as with ergotamine tartarate and benzocaine.[62–64] Other similar interactions occur between meta-aminobenzoic acid and tartaric acid, maleic acid and

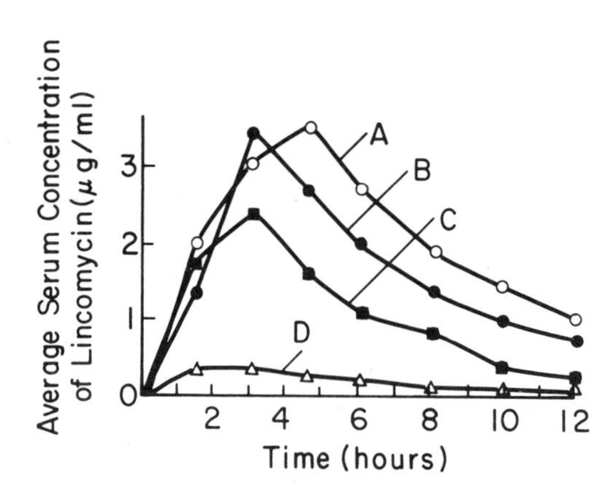

Figure 2.7. Average serum concentrations of lincomycin following a 0.5 g dose of lincomycin in capsule form taken orally. *A* = Lincocin alone: *B* = Kaopectate 2 hr *before* Lincocin: *C* = Kaopectate at *same time* as Lincocin: *D* = Kaopectate 2 hr *after* Lincocin. (From Wagner: Can J Pharm Sci 1:55, 1966)

creatinine,[65] and between various antihistamine compounds and beta-cyclodextrin.[66] The equimolar complexes between caffeine and citric acid result in an increased solubility of caffeine.

Besides the well-defined complexation processes, such other interactions as surface adsorption can also affect the bioavailability of drugs. Activated charcoal has long been used to decrease the bioavailability of drugs because of the large surface area it presents for adsorption by physical interaction through the network of π-bond carbon structure. Similarly, compounds such as attapulgite and pectin decrease the adsorption of drugs such as promazine and lincomycin (Fig. 2.7).[67–69] Generally, the absorption of drugs will be retarded by adsorption if they are administered with certain antacids and antidiarrheal preparations. The forces of adsorption are for the most part weak physical interactions, such as van der Waals forces, but often a chemical interaction also takes place. This is referred to as CHEMISORPTION.

Quite often the various diluents used in the pharmaceutical dosage form retard absorption by adsorbing the drugs and thereby slowing their release in an aqueous media, as is the case with riboflavin and thiamine.[70]

PARTICLE SIZE

Of all the possible manipulations of the physical properties of drugs to yield better absorption, the reduction of particle size is the most widely exploited. Increased absorption due to reduction of particle size is a result of increased dissolution, which is in turn the result of a larger surface area being exposed to the fluids in the gastrointestinal tract and other sites of administration. For example, the breakdown of a 3 mm cu particle into 1 mm cu particles results in a 300 percent increase in the exposed surface area.

Reduction in particle size can be achieved by several methods, including milling, grinding, precipitating the drug on an adsorbent, and dispersing the drug in an inert water-soluble carrier (referred to as a solid dispersion).

The solid dispersion formulation techniques have received great attention in the recent past and provide an innovative method of particle size reduction. If a hydrophobic drug is dispersed in a hydrophilic medium in a solid state, a faster release of the drug can be expected from this system since the rate-limiting steps in the dissolution of the drug will be fewer. The state of drug dispersion can vary from microcrystalline to molecular and thus a wide range of dissolution rates are possible. For example, dispersion of sulfathiazole in urea results in a monomolecular dispersion (solid solution) and the dissolution rates are increased by almost 700 times.[71] The dispersion of griseofulvin in polyethylene glycol 6000 results in an almost 100 percent increase in its bioavailability as compared to the micronized form of griseofulvin.[72] The solid dispersions are generally prepared by either fusing or dissolving the drug and the water soluble carrier and then solidifying the melt or solution by cooling or evaporation. The drugs also often coprecipitate, as with the sold dispersion of reserpine and deoxycholic acid.[73] Examples of drugs whose dissolution rates have been increased as a result of solid dispersion formulation include salicylic acid, reserpine, chloramphenicol, prednisone, salicylamide, pentaerythritol, and others.

Quite often solid dispersions can also be used to decrease the release of drugs, so as to provide sustained release as in the dispersion of chlorpheniramine in maleic anhydride copolymers.[74]

The conventional methods of particle size reduction have long been employed to improve the bioavailability of drugs, some of which are listed in Table 2.7. The drugs which have been most comprehensively studied are griseofulvin and digoxin, since for both of these drugs the dissolution rate is the rate-limiting step in the overall delivery to the general circulation. A recent study showed that decreasing the particle size of digoxin from 102 μ to between 7 and 13 μ resulted in a 100 percent increase in its bioavailability.[82]

The correlation between particle size and bioavailability is unique for each

Table 2.7. EXAMPLES OF DRUGS WHOSE BIOAVAILABILITY HAS BEEN INCREASED AS A RESULT OF PARTICLE SIZE REDUCTION

A, vitamin[75]	Medroxyprogesterone acetate[86]
4-Acetamidophenyl,2,2,2-tri-chlorethyl carbonate[76]	Nitrofurantoin[87,88]
Aspirin[77]	Phenobarbital[89]
Bishydroxycoumarin[78]	Phenacetin[90]
Chloramphenicol[79,80]	Procaine penicillin[91]
Cyheptamide[81]	Reserpine[92,93]
Digoxin[82,83]	Spironolactone[94]
Fluocinolone acetonide[84]	Sulfadiazine[95]
Griseofulvin[72]	Sulfasoxazole[96]
p-Hydroxypropiophenone[85]	Sulfur[97]
	Tolbutamide[98]

drug and a great deal of in-depth study is required to establish a rationale for particle size reduction.

The reduction in the particle size is, however, not always desirable. For example, nitrofurantoin, when administered in its fine particle size, causes more gastrointestinal irritation than when administered in its coarser size. This is due to the higher plasma and gastrointestinal concentrations resulting from use of a fine particle size. The use of the coarser size is therefore preferred even though this results in retarded absorption. When chemical instability is a problem the reduction of particle size is also contraindicated, as with penicillin G and erythromycin, which decompose in the gastrointestinal tract. Even in the solid state a small particle size means a greater surface area available for the adsorption of moisture, which can result in an increased rate of decomposition. The reduction of the particle size of hydrophobic drugs also leads to increased surface charges (static) resulting in the agglomeration of the particles, especially in an aqueous media because of thermodynamic repulsion. This results in a significant decrease in the effective or exposed surface area available for dissolution. This problem can usually be resolved by adding appropriate surfactants which will reduce the interfacial tension and allow penetration of water molecules through the pocket of hydrophobic air surrounding these particles.

SURFACE ACTIVITY

Surfactants have variable effects on the dissolution and absorption processes. The lowering of surface tension increases the dissolution rates by increasing the solubility of drugs if the concentration of the surfactant is above the critical micelle concentration, as shown in Scheme 2.4. The lowering of surface tension also increases the diffusion of free molecules in the medium, increasing the contact between free drug and the absorption surface. The surfactants can also increase the membrane permeability, allowing greater absorption of most chemical structures.

The overall effect of surfactants on the bioavailability of drugs is complex, since the molecules contained in the micelle are not available for absorption unless a quick equilibration between the free drug molecules and those inside the micelles can be established (Scheme 2.4). A number of drugs have surface active properties themselves (Table 2.8) and form their own micelles, thus facilitating the absorption process. Several cases exist where the use of surfactants in a formulation has resulted in increased absorption. Some of these cases are listed in Table 2.9.

The surface active properties of gastrointestinal contents primarily involve the various fluids, e.g., gastric juice and intestinal juice, and can be a major factor in the dissolution of several hydrophobic drugs. The most important component of gastrointestinal fluids is the high concentration of bile salts (100 to 200 mM) present, well above their critical micelle concentration (2 to 3 mM). The bile salt micelles solubilize polycyclic aromatic hydrocarbons,

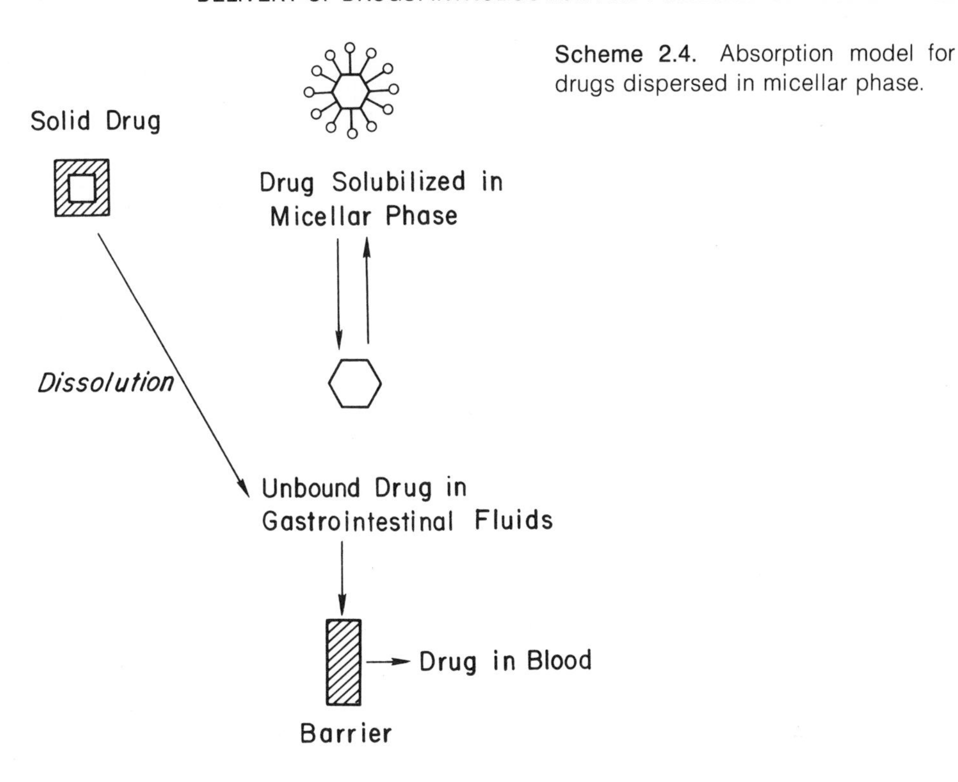

Scheme 2.4. Absorption model for drugs dispersed in micellar phase.

progesterone, cholesterol, esterone, griseofulvin, and reserpine, thus increasing their absorption. Enhancement of drug absorption after meals is often related to the increased flow of bile which can solubilize the drug molecules, as is demonstrated by griseofulvin.

The manipulations of formulations are too numerous and too diversified to be included under any single heading. In most instances, these involve an additive which changes the physical properties of the drug. For example, a

Table 2.8 EXAMPLES OF SURFACE-ACTIVE DRUGS WHICH FORM MICELLES

Potassium benzyl penicillin
Mixtures of penicillin and streptomycin salts
Amphetamine sulfate
Cyclopentamine hydrochloride
Ephedrine sulfate
Propoxyphene hydrochloride
Ionic derivatives of phenothiazines
Dyes
Quarternary ammonium salts of drugs
Liquoris

Table 2.9. EXAMPLES OF DRUGS WHOSE BIOAVAILABILITY HAS BEEN INCREASED DUE TO SURFACTANTS IN THE FORMULATION

DRUG	*SURFACTANT*
A, vitamin[99]	Sodium laurylsulfate
B-12, vitamin[100]	Polysorbate 80/85, G-1096
O-benzoylthiamine disulfide[101]	Sodium laurylsulfate
Cephaloridine[102]	Various
G-strophanthin[103]	Sodium laurylsulfate
Heparin[104,105]	Sodium laurylsulfate, dioctylsodium sulfosuccinate
Iodoform[106–108]	Polysorbate 80
Phenosulfonphthalein[109,110]	Dioctyl sodium sulfosuccinate
Riboflavin[111]	Sodium deoxycholate
Salicylamide[112]	Polysorbate 80
Salicylic acid[113]	Various
Spironolactone[114]	Polysorbate 80
Sulfasoxazole[115]	Polysorbate 80
Thiourea[116]	Alkylbenzone sulfonate, dodecyl trimethyl ammonium chloride

ground mixture of griseofulvin with cellulose has a higher bioavailability than the micronized form alone.[117] Dispersion methods such as simple blending, solvent deposition, ball milling, and mueller milling can have a marked influence on dissolution rate and bioavailability, as is demonstrated with digoxin. Although surfactants represent an important category of additives which increase the absorption of drugs, other additives are also very important. Magnesium oxide and hydroxide, for example, act by very specific mechanisms to increase the absorption of dicumarol by chelation.[118] One interesting case occurred when the replacement of calcium sulfate with lactose in the formulation of an anticonvulsant resulted in an outbreak of overdosing in patients maintained on the old formulation.[119] This was due to increased bioavailability as a result of the replacement of calcium sulfate.

The properties of the mediums in which the drug is formulated can significantly affect bioavailability profiles. For example, the rectal absorption of acetaminophen from an aqueous solution and from several polyethylene glycol base dosage forms was correlated with the dielectric constant of the medium.[120] The rate of absorption of phenolsulfonphthalein decreased when the drug was administered in high viscosity dosage forms.[121] Similar findings were reported for sodium phenobarbital and sodium salicylate, where increasing the viscosity of oral solution by adding methyl cellulose or sucrose resulted in sustained and retarded absorption.[122,123]

In summary, the various formulation factors, when appropriately utilized, can significantly affect the delivery of drugs to the body. An optimum delivery can only be achieved by carefully selecting and monitoring these factors.

References

1. Noyes AA, Whitney WR: The rate of solution of solid substances in their own solutions. J Am Chem Soc 19:930, 1897
2. Davson H, Danielli JF: The Permeability of Natural Membranes. Cambridge, England, Cambridge Univ Press, 1952, p 64
3. Loo JCK, Flotz EL, Wallick H, Kwan KC: Pharmacokinetics of pivampicillin and ampicillin in man. Clin Pharmacol Ther 16:35, 1974
4. Butler K, English AR, Knirsch AK, Korst J: Metabolism and laboratory studies with indanyl carbenicillin. Del Med J 43:366, 1971
5. Schumacher GE: Practical pharmacokinetic techniques for drug consultation and evaluation, I: use of dosage regimen calculations. Am J Hosp Pharm 29:474, 1972
6. Wright WW, Welch H: Chemical, biological, and clinical observations on colistin. Antibiot Ann 7:61, 1959–60
7. Doluisio JT, LaPiana JC, Dittert LW: Pharmacokinetics of ampicillin trihydrate, sodium ampicillin, and sodium dicloxacillin following intramuscular injection. J Pharm Sci 60:715, 1971
8. Hoffbrand BI: Digoxin bioavailability. Postgrad Med J 50 (6S):1, 1974
9. Johnson BF, Bye C, Jones G, Sabey GA: A completely absorbed oral preparation of digoxin. Clin Pharmacol Ther 19:746, 1976
10. Black J, Calensick B, Williams D, Weinstein MJ: Pharmacology of gentamicin, a new broad-spectrum antibiotic. Antimicrob Agents Chemother 3:138, 1963
11. Kunin CM: Absorption, distribution, excretion, and fate of kanamycin. Ann NY Acad Sci 132:811, 1966
12. Chow MSS, Ronfeld RA: Pharmacokinetic data and drug monitoring I: antibiotic and anti-arrhythmics. J Clin Pharmacol 15:405, 1975
13. Kunin CM: Comparative serum binding, and excretion of tetracycline and a new analogue, methacycline. Proc Soc Exp Biol Med 110:311, 1962
14. McDermott W, Bunn PA, Benoit M, Dubois R, Reynolds ME: The absorption, excretion, and destruction of orally administered penicillin. J Clin Invest 25:190, 1946
15. Evans GH, Nies AS, Shand DG: Disposition of propanolol V. Drug accumulation and steady-state concentration during chronic oral administration. Clin Pharmacol Ther 14:487, 1973
16. Gantt CL, Gochman N, Dyniewicz JM: Effect of a detergent on gastrointestinal absorption of a steroid. Lancet 1:486, 1961
17. Gantt CL, Gochman N, Dyniewicz JM: Gastrointestinal absorption of spironolactone. Lancet 1:1130, 1962
18. Bauer G, Rieckman P, Schaumann W: Influence of particle size and dissolution rate on gastrointestinal absorption of spironolactone. Arzneim Forsch 12:487, 1962
19. U.S. Government: Bioavailability and bioequivalency requirements. Fed Reg 42:1624, 1977
20. Physicians Desk Reference, 31st ed. Osadell NJ, Med Econ, 1977
21. Ariens EJ: Modulation of pharmacokinetics. In Ariens EJ (eds): Drug Design, Vol 2. New York, Academic, 1971, Chap 1
22. Schanker LS: On the mechanism of absorption from the gastrointestinal tract. J Med Pharm Chem 2:343, 1960
23. Kawasaki C: Modified thiamine compounds. Vitam Horm 21:69, 1963
24. Yagi K, Okuda J, Dmitrouskii AA, Honda R, Matsubara J: Studies on fatty acid esters of flavines I: Chemical synthesis of fatty acid esters of riboflavin. J Vitaminol 7:4, 1961
25. Nomura H, Sugimoto K: Synthesis of l-ascorbic acid acyl derivatives stabilized

against oxidation. Chem Pharm Bull 14:1039, 1966
26. Isler D: Developments in the field of vitamins. Experientia 26:225, 1970
27. Griffith R, Black HR: Comparison of blood levels following pediatric suspensions of erythromycin estolate and erythromycin succinate. Clin Med 76:16, 1969
28. Slavik M, Elis J, Raskova H, et al: Therapeutic effects of 6-azauridine-triacetate in psoriasis. Pharmacol Clin 2:120, 1970
29. Fletcher HP, Murray HM, Weddon TE: Absorption of lincomycin and lincomycin esters from rat jejunum, J Pharm Sci 57:2101, 1968
30. Magerlein BJ, Birkenmyer RD, Kagaw F: Lincomycin VI-4′-alkyl analogs of lincomycin. Relationship between structures and antibacterial activity. J Med Chem 10:355, 1967
31. Fischer LJ, Riegelman S: Absorption and activity of some derivatives of griseofulvin. J Pharm Sci 56:469, 1967
32. Katz M, Shaikh ZI: Percutaneous corticosteroid absorption correlated to partition coefficient. J Pharm Sci 54:591, 1965
33. Ippen H: Index Pharmacorum: Synonyms, Structure and Effects of Organic–Chemical Drugs. Stuttgart, Thieme, 1968
34. Durel P: Therapeutique Sulfamidee. Paris, Bailliere, 1940
35. Levaditi C: La chimiothérapie des infections microbiennes son mechanisme d'action. Pathol Microbiol 1:365, 1938
36. Walkenstein SS, Wiser R, Gudmunsen CH, Kimmel HB, Corradino RA: Absorption, metabolism, and excretion of oxazepam and its succinate half-ester. J Pharm Sci 53:1181, 1964
37. Levy G, Sabli BA: Comparison of the gastrointestinal absorption of aluminum acetylsalicylate and acetylsalicylic acid in man. J Pharm Sci 51:58, 1962
38. Solvang S, Finholt P: Effect of tablet processing and formulation factors on dissolution rate of the active ingredient in human gastric juice. J Pharm Sci 59:49, 1970
39. Lin SL, Lachman L, Swartz CJ, Heubner CF: Preformulation investigation I: Relation of salt forms and biological activity of an experimental antihypertensive. J Pharm Sci 61:1418, 1972
40. Mayazaki S, Nakano M, Arita T: A comparison of solubility characteristics of free bases and hydrochloride salts of tetracycline antibiotics in hydrochloric acid solutions. Chem Pharm Bull 23:1197, 1975
41. Breimer DD, deBior AG: Pharmacokinetics and relative bioavailability of heptobarbital and heptobarbital sodium after oral administration to man. J Clin Pharmacol 9:169, 1975
42. Morozowich W, Chulski T, Hamlin WE, et al: Relationship between in vitro dissolution rates, solubilities, and LD_{50}'s in mice of some salts of benzphetamine and etryptamine. J Pharm Sci 51:993, 1962
43. O'Reilly RA, Nelson E, Levy G: Physicochemical and physiologic factors affecting the absorption of warfarin in man. J Pharm Sci 55:435, 1966
44. Nelson E: Comparative dissolution rates of weak acids and their sodium salts. J Am Pharm Assoc 47:297, 1958
45. Miyazaki M, Arita M: Effect of crystal forms on the dissolution behavior and bioavailability of tetracycline, chlortetracycline, and oxytetracycline bases. Chem Pharm Bull 23:552, 1975
46. Niazi S: Effect of polyethylene glycol 4000 on dissolution properties of sulfathiazole polymorphs. J Pharm Sci 65:302, 1976
47. Mullin JD, Macek TJ: Some pharmaceutical properties of novobiocin. J Am Pharm Assoc 49:245, 1960
48. Macek TJ: US Patent 2,671,750, March 9, 1954
49. Macek TJ: The physical and chemical problems inherent in the formulation of dosage forms for new pharmaceuticals. Am J Pharm 137:217, 1965

50. Hill SA, Jones KH, Eager HS, Taskis CB: Dissolution and bioavailability of the anhydrate and trihydrate forms of ampicillin. J Pharm Pharmacol 27:594, 1975
51. Haleblian J, Koda R, Biles J: Comparison of dissolution rate of different crystalline phases of fluprednisolone by in vitro and in vivo methods. J Pharm Sci 60:1488, 1971
52. Huang ML, Niazi S: Polymorphic and dissolution properties of mercaptopurine. J Pharm Sci 66:608, 1977
53. Sekiguchi K, et al.: Studies on methods of particle size reduction of medicinal compounds, VI. Solvate formation of griseofulvin with benzene and dioxane. Chem Pharm Bull 24:1621, 1976
54. Bates TR, Fung H, Lee H, Tembo AV: Comparative bioavailability of anhydrous griseofulvin and its chloroform solvate in man. Res Commun Chem Pathol Pharmacol 11:233, 1975
55. Wagner JG: Biopharmaceutics: absorption aspects. J Pharm Sci 50:359, 1961
56. Singh P, Guillory JK, Sokoloski TD, Benet LZ, Bhatia VN: Effect of inert tablet ingredients on drug absorption I: Effect of PEG 4000 on intestinal absorption of four barbiturates. J Pharm Sci 55:63, 1966
57. Levy G, Matsuzawa T: Effect of complex formation on drug absorption I: Lipoid-soluble drug complexes. J Pharm Sci 54:1003, 1965
58. Hayton WL, Levy G: effect of complex formation on drug absorption XIII: Effect of constant concentrations of N,N-di-n-propylpropionamide on prednisolone absorption from the rat small intestine. J Pharm Sci 61:637, 1972
59. Hayton W, Guttman DE, Levy G: Effect of complex formation on drug absorption XI: complexation of prednisone and prednisolone with dialkylpropionamides and its effect on prednisone transfer through an artifical lipoid barrier. J Pharm Sci 61:356, 1972
60. Hayton WL, Levy G: Effect of complex formation of drug absorption XII: Enhancement of intestinal absorption of prednisone and prednisolone by dialkylpropionamides in rats. J Pharm Sci 61:362, 1972
61. Higuchi T, Ikeda M: Rapidly dissolving forms of digoxin:hydroquinone complex. J Pharm Sci 63:809, 1974
62. Zoglio MA, Maulding HV, Windheuser JJ: Complexes of ergot alkaloids and derivatives I: The interaction of caffeine with ergotamine tartarate in aqueous solution. J Pharm Sci 58:222, 1969
63. Connors KA, Mollica JA: Theoretical analysis of comparative studies of complex formation: solubility, spectral, and kinetic techniques. J Pharm Sci 55:772, 1966
64. Higuchi WI, Mir NA, Desai SJ: Dissolution rates of polyphase mixtures. J Pharm Sci 54:1405, 1965
65. Wurster DE, Kildsig DO: Effect of complex formation on dissolution kinetics of m-aminobenzoic acid. J Pharm Sci 54:1491, 1965
66. Monkhouse DC, Lach JC: Drug-excipient interactions. Can J Pharm Sci 7:29, 1972
67. Sorby DL: Effect of adsorbents on drug absorption I: Modification of promazine absorption by activated attapulgite and activated charcoal. J Pharm Sci 54:677, 1965
68. Sorby, DL, Liu G: Effect of adsorbents on drug absorption II: Effect of antidiarrheal mixtures on promazine absorption. J Pharm Sci 55:504, 1966
69. Wagner JG: Design and data analysis of biopharmaceutic studies in man. Paper presented at the American Pharmaceutical Association National Meeting, Dallas, Texas, 1966
70. Oser BL, Melnick D, Hochberg M: Physiologic availability of vitamins. Ind Eng Chem 17:405, 1945
71. Chiou WL, Niazi S: Phase diagram and dissolution rate studies of sulfathiazole-urea solid dispersions. J Pharm Sci 60:1333, 1971

72. Barret WE, Hannigan JJ: The bioavailability of griseofulvin-PEG ultramicrosize (GRIS-PEG) tablets in man. Curr Ther Res 18:491, 1975
73. Gibaldi M, Feldman S, Bates TR: Correlation of pharmacologic activity and dissolution rates of reserpine-deoxycholic acid dispersions. J Pharm Sci 57:708, 1968
74. Athanikar N: Design and evaluation of controlled release formulations prepared by the solid dispersion techniques. Ph. D. dissertation, University of Illinois, 1977
75. Morales S, Chung AW, Laus J, Menina A, Holt LE: Absorption of fat and vitamin A in premature infants. Pediatrics 6:644, 1950
76. Dittert LW, Adams HJ, Alexander F, et al: 4-acetamidophenyl 2,2,2-trichlorethyl carbonate particle size studies in animals and man. J Pharm Sci 57:1146, 1968
77. Dare JG: Particle size in relation to formulation. Aust J Pharm 45:S58, 1964
78. Lozinski E: Physiological availability of dicumarol. Can Med Assoc J 83:177, 1960
79. Kakemi K, Arita T, Ohashi S: The absorption and excretion of drugs XIII: The effects of metal salts and various particle sizes of chloramphenicol on rabbit blood level. Yakugaku Zasshi 82:1468, 1962
80. Sekiguchi K, Obi N, Ueda Y: Studies on absorption of eutectic mixtures II: Absorption of fused conglomerates of chloramphenicol and urea in rabbits. Chem Pharm Bull 12:134, 1962
81. Kraml M, Dvornik D, Cosyns L: Colorimetric determination of C-10 hydroxylated metabolites of cyheptamide: Application to absorption and enzyme studies in rat. J Pharm Sci 61:408, 1972
82. Journela AJ, Penikaenen PJ, Sothman A: Effect of particle size on the bioavailability of digoxin. J Clin Pharmacol 8:365, 1975
83. Shaw TRD, Carless JE: The effect of particle size on the absorption of digoxin. J Clin Pharmacol 7:269, 1974
84. Barett CW, Hadgreft JW, Caron CA: The effect of particle size and vehicle on the percutaneous absorption of fluocinolone acetonide. Brit J Dermatol 77:576, 1965
85. Foglia VG, Perhos JC, Montouri E: Relation of crystal size to estrogenic activity of parahydroxy-propiophenone. Endocrinol 57:559, 1955
86. Smith DL, Pulliam AL, Forest AA: Comparative absorption of micronized and nonmicronized medroxyprogesterone acetate in man. J Pharm Sci 55:398, 1966
87. Paul HE, Hayes KJ, Paul MF, Borgmann AR: Laboratory studies with nitrofurantoin. Relationship between crystal size, urinary excretion in the rat and man, and emesis in dogs. J Pharm Sci 56:882, 1967
88. Stoll RG, Bates TR, Swarbrick J: In vitro dissolution and in vivo absorption of nitrofurantoin from deoxycholic acid coprecipitates. J Pharm Sci 62:65, 1973
89. Miller LG, Findier JH: Influence of drug particle size after intramuscular dosage on phenobarbital to dogs. J Pharm Sci 60:1733, 1971
90. Prescott LF, Steel RF, Ferrier RL: The effect of particle size on the absorption of phenacetin in man. Clin Pharmacol Ther 11:496, 1970
91. Buckwalter FH, Dickinson LH: The effect of vehicle and particle size on the absorption, by the intramuscular route, of procaine penicillin G suspensions. J Am Pharm Assoc 47:661, 1958
92. Malone M, Hochman HI, Nieforth KA: Desoxycholic acid enhancement of orally administered reserpine. J Pharm Sci 55:972, 1966
93. Stupak EI, Bates TR: Enhanced absorption and dissolution of reserpine from reserpine-polyvinylpyrrolidone coprecipitates. J Pharm Sci 61:400, 1972
94. Bauer G, Rieckmann P, Schaumann W: Einfluss von teilchengrosse und losungsvermittlern aug die resorption von spironolacton aus dem magen-darmtrakt. Arzneim Forsch 12:487, 1962

95. Reihold JD, Philips FJ, Filipin HF: A comparison of the behavior of microcrystalline sulfadiazine with that of ordinary sulfadiazine in man. Am J Med Sci 210:141, 1945
96. Fincher JH, Adams AG, Beal HM: Effect of particle size on gastrointestinal absorption of sulfisoxazole in dogs. J Pharm Sci 54:704, 1965
97. Greengard H, Wooley JR: Studies on colloidal sulfur-polysulfide mixture absorption and oxidation after oral administration. J Biol Chem 132:83, 1940
98. Nelson E, Long S, Wagner JG: Correlation of amount of metabolite excreted and its excretion rate with available surface area of tolbutamide in dosage form. J Pharm Sci 53:1224, 1964
99. Fuchs B, Inglefinger FJ: The effect of detergent on intestinal digestion. Gastroenterol 27:802, 1954
100. Okuda K, Duran EV, Chow BF: Effect of physicochemical state of vitamin B_{12} preparation in digestive tract on its absorption. Proc Soc Exp Biol Med 103:588, 1960
101. Utsumi I, Kohno K, Takuechi Y: Surfactant effect on drug absorption III: Effects of sodium glycoholate and its mixtures with synthetic surfactants on absorption of thiamine disulfide compounds in rats. J Pharm Sci 63:676, 1974
102. Kreutler CJ, David WW: Normal and promoted GI absorption of water-soluble substances III: Absorption of antibiotics from stomach and intestine of the rat. J Pharm Sci 60:1835, 1971
103. Krause D: Forderung und sicherung der enteralen resorption von G-strophanthin durch natriumlaurysulfat. Arzneim Forsch 5:428, 1955
104. Engel RH, Riggi SJ: Effect of sulfated and sulfonated surfactants on the intestinal absorption of heparin. Proc Soc Exp Biol Med 130:879, 1969
105. Engel RH, Riggi SJ: Intestinal absorption of heparin facilitated by sulfated or sulfonated surfactants. J Pharm Sci 58:706, 1969
106. Riegelman S, Crowell WJ: The kinetics of rectal absorption III: The absorption of undissociated molecules. J Am Pharm Assoc 47:127, 1958
107. Riegelman S, Crowell WJ: The kinetics of rectal absorption I: Preliminary investigation into the absorption rate process. J Am Pharm Assoc 47:115, 1958
108. Riegelman S, Crowell WJ: The kinetics of rectal absorption II: The absorption of anions. J Am Pharm Assoc 47:123, 1958
109. Malik SN, Canaham DH, Gouda MW: Effects of suractants on absorption through membranes III: Effect of dioctyl sodium sulfosuccinate and poloxalene on absorption of a poorly absorbable drug, phenolsulfonphthalein, in rats. J Pharm Sci 64:987, 1975
110. Khalafallah N, Gouda MW, Khalil SA: Effect of surfactants on absorption through membranes IV: Effects of dioctyl dosium sulfosuccinate on absorption of a poorly absorbable drug, phenosulfonphthalein, in man. J Pharm Sci 64:991, 1975
111. Mayersoln M, Feldman S, Gibaldi M: Bile salt enhancement of riboflavin and flavin mononucleotide absorption in man. J Nutr 98:288, 1969
112. Yamada H, Yamamoto R: Biopharmaceutical studies on factors affecting rate of absorption of drugs I: Absorption of salicylamide in micellar solution. Chem Pharm Bull 13:1279, 1965
113. Shen W, Danti AG, Bruscato FN: Effect of nonionic surfactants on percutaneous absorption of salicylic acid and sodium salicylate in the presence of dimethyl sulfoxide. J Pharm Sci 65:1780, 1976
114. Gantt CL, Gochman N, Dyniewicz JM: Effect of a detergent on gastrointestinal absorption of a steroid. Lancet 1:486, 1961
115. Kakemi K, Arita A, Muranishi S: Absorption and excretion of drug XXXVII: Effect of nonionic surface-active agents on rectal absorption of sulfonamides. Chem Pharm Bull 13:976, 1965

116. Scala J, Mosher DE, Reller HH: The percutaneous absorption of ionic surfactants. J Invest Dermatol 50:371, 1968
117. Yamamoto K, Nakano M, Arita T, Nakai Y: Dissolution rate and bioavailability of griseofulvin from a ground mixture with microcrystalline cellulose. J Pharmacokin Biopharm 2:487, 1974
118. Akers MS, Lach JL, Fischer LJ: Alterations in absorption of dicumarol by various excipient materials. J Pharm Sci 62:391, 1973
119. Tryer JH, Eadie MJ, Sutherland JM, Hooper WD: Outbreak of anticonvulsant intoxication in an Australian city. Brit Med J 4:271, 1970
120. Pagay SN, Poust RI, Colaizzi JL: Influence of vehicle dielectric properties on acetaminophen bioavailability from polyethylene glycol suppositories. J Pharm Sci 63:44, 1974
121. Ashley JJ, Levy G: Effect of vehicle viscosity and an anticholinergic agent on bioavailability of a poorly absorbed drug (phenosulfonphthalein) in man. J Pharm Sci 62:688, 1973
122. Davison C, Guy JL, Levitt M, Smith PK: Distribution of certain non-narcotic analgetic agents in the CNS of several species. J Pharmacol Exp Ther 134:176, 1961
123. Malone MH, Gibson RD, Miya TS: A pharmacologic study of effects of various pharmaceutical vehicles on action of orally administered phenobarbital. J Pharm Sci 49:529, 1960

Questions

1. Comment on the statement, "Physicochemical properties of some drugs may exclude the oral route of administration."
2. Using log dose-response curves, show that bioavailability variations are more important with highly potent drugs.
3. Discuss the factors affecting the dissolution rates of drugs as summarized in the Noyes-Whitney equation.
4. Is the dissolution of drugs always the rate limiting step in the delivery of drugs to the general circulation?
5. What is the effect of hydrophobic:hydrophilic balance of chemical structure on the absorption of drugs?
6. What is the rationale behind attempts to increase the bioavailability of drugs?
7. Using chemical structures, show how a sulfonamide molecule can be restricted to the gastrointestinal tract and how the absorption of quarternary ammonium compounds can be increased.
8. List three examples with structures where introduction of a hydrophilic group to the molecule increases its absorption.
9. Using equations, show that the potassium and sodium salts of drugs dissolve more rapidly than the corresponding acid, regardless of the pH of the medium.
10. List and discuss three examples where the use of the salt form may not be preferable for increasing dissolution rates and bioavailability.
11. What is the surface coating model for the dissolution of the salts of weakly acidic drugs? Why has the use of aluminum aspirin been abandoned?

12. How does the ionization of a drug affect its absorption?
13. Derive Equation 2.7 for weakly basic drugs.
14. Suggest a possible mechanism for producing slower dissolution of a weakly acidic drug in an alkaline medium than would be expected from solubility considerations. What is the advantage of the use of buffered aspirin formulations?
15. Define the terms *polymorphism* and *amorphism*. Why does an amorphous form dissolve faster than a stable polymorphic form? Why does ferrous sulfate anhydrate dissolve slower in water than the corresponding hydrate form?
16. List five drugs whose dissolution rates have been increased as a result of polymorphic modifications.
17. Discuss the role of polymorphism in insulin therapy.
18. Define the term *solvation* and cite at least five examples where an asolvate shows higher dissolution than the corresponding solvate form.
19. How are hydrate forms utilized in the formulation of effervescent tablets?
20. What are the physical and chemical forces responsible for the complexation process? What are the chemical groups that most commonly undergo complexation? Why is caffeine an excellent complexing agent?
21. Under what conditions will complexation result in increased dissolution and bioavailability of drugs?
22. What makes activated charcoal a good adsorbent? What happens if lincomycin is administered with Kaopectate?
23. Using equations for finding the areas of various particle shapes, show that reduction of particle size results in an increased surface area.
24. What are the advantages of using solid dispersion formulations in decreasing the particle size as opposed to using conventional micronization methods?
25. What are the major disadvantages in decreasing the particle size?
26. What is the mechanism of surfactant-increased absorption of drugs?
27. How would the micellization of surfactant drugs affect their absorption? What is the role of bile salts in drug absorption?

CHAPTER 3

Delivery of Drugs: Dosage Forms and Their Evaluation

The purpose of formulation efforts is to design a dosage form with a suitable combination of the following attributes:

1. It contains the labelled amount of drug in an active form.
2. It is free from extraneous materials.
3. It consistently delivers the drug to the general circulation at an optimum rate and to an optimum extent.
4. It is suitable for administration through an appropriate route.
5. It is acceptable to patients.

The dosage form characteristics, such as particle size, salt form, solvent type, and dissolution rate, as well as the additives discussed in the previous chapter, all contribute to the dosage form design. The additives may be pharmacologically inert, as with tablet binders and lubricants, or they may have the function of modifying the absorption, the biotransformation, or the excretion of the primary therapeutic agents, as with enteric coating materials, probenecid with penicillin, etc.

In many instances, different dosage forms are available for a given drug, and an appropriate selection must be made based on the attributes listed above. The following discussion attempts to characterize various dosage forms and provides a rational basis for their selection.

DOSAGE FORMS

Solutions

Solutions are thermodynamically stable monomolecular dispersions of drug molecules in a liquid or solid phase. Absorption from aqueous solutions is generally very fast and complete from all sites of administration, provided that penetration through the absorption barrier (such as the gastrointestinal barrier) is not a rate limiting factor. Such rate limiting steps as disintegration and dissolution, due to dosage form factors, are minimal in the use of solutions. For example, potassium penicillin V gives higher blood levels than benzathine penicillin V when both are administered orally in tablet form, but solutions of the two drugs yield essentially equal blood levels of penicillin, suggesting the importance of dosage form factors in these formulations.[1]

Besides providing an adequate bioavailability, solutions are also convenient for administration to pediatric and geriatric patients. In some instances the use of solutions is a crucial part of the drug delivery. For example, calcium must be administered as a solution in its citrate form to achlorhydric patients, since the solid carbonate form will not dissolve sufficiently in the gastrointestinal tract without the presence of hydrochloric acid.[2] An analogous problem exists in the administration of sodium salts of weakly acidic drugs, which precipitate in the stomach in crystalline form. These crystals are usually very fine and redissolve quickly, but there is always a possibility of retarded absorption due either to precipitation as large particles or to coating of the particles with hydrophobic acid, as is demonstrated with such poorly water soluble drugs as warfarin and phenytoin. These drugs can therefore be absorbed better from a suspension dosage form than from a solution of their sodium salts.[3]

Quite often, solutions of poorly water-soluble drugs are affected by adding cosolvents, such as alcohol or propylene glycol, and by adding complexing agents, which form a water-soluble complex with the drug or with surfactants that solubilize the drugs. In all of these instances the drugs will precipitate because of the dilution effect and are subject to possible difficulty in redissolution.

Sometimes nonaqueous solutions provide better absorption than aqueous solutions, as is demonstrated by indoxole.[4] A solution of indoxole in oil, administered as an oil-in-water emulsion, shows three times better absorption than the aqueous solution.

The use of solid solutions is a novel application of dispersion techniques, whereby the drug is dispersed in a solid water-soluble vehicle, such as urea, succinic acid, or polyethylene glycols, which dissolves rapidly in water, releasing the macrocrystalline or monomolecular form of the drug. Although there is a large volume of data about the applications of the principles of solid dispersions and solutions, only one product is currently available which utilizes this concept, i.e., Gris-PEG, a dispersion of griseofulvin in polyethylene glycol.

Suspensions

Suspensions require the dissolution of particles before they can be absorbed. This dissolution process can be rate-limiting, depending on the aqueous solubility of the drug and the formulation additives involved. Thus there are many more factors that can affect drug absorption in the use of suspensions than are possible in the use of solutions. Generally, however, suspensions will provide better absorption than such other dosage forms as capsules and tablets, as demonstrated by trimethoprim and sulfamethoxazole combinations[5] and sulfadimethoxine.[6] Suspensions are also used when a slow release of the drug is desired, as with intramuscular administration of triaminocolone acetonide or with tetracycline ophthalmic suspensions. Since suspensions provide a large surface area, various antacid products are most effectively administered as suspensions, since the mode of action involves both the chemical neutralization of hydrochloric acid and its physical adsorption onto the suspended particles. It is interesting to note that a majority of official oral suspensions in current use involve antiinfective agents, e.g., pyranatel pamoate, pyruvinium pamoate, thiabenzadole, chloramphenicol palmitate, democlocycline, methacycline, oxytetracycline, penicillin, tetracycline, methenamine mandelate, nitrofurantoin, sulfonamides, trisulfapyrimidines, nystatin, etc. Most of the antiinfective agents are chemically unstable, can cause gastrointestinal irritation, and are often erratically absorbed from such solid dosage forms as tablets and capsules. The use of suspensions for these drugs provides an ideal mechanism for solving formulation problems related to these attributes. Consider a drug with a decomposition constant of 0.2 hr^{-1} and aqueous solubility of 4 g/liter. A 500 mg dose in aqueous solution will have a shelf life (10 percent decomposition) of 0.53 hours, whereas 500 mg suspended in 10 ml of saturated solution will have a shelf life of 6.25 hours, indicating a more than 1000 percent increase in the drug's stability.

The use of suspensions is also advantageous in pediatric or geriatric practice, where they can be accurately and conveniently administered using droppers, or oral syringes. Suspension dosage forms are utilized for all routes of administration except intravascular. Some examples are listed in Table 3.1.

Powders

The formulation and bioavailability problems associated with suspensions are also characteristic of powders, whereby the active ingredient is mixed with inert diluents and administered either directly or in a capsulated form. An additional problem therefore arises due to possible adsorption of drugs onto diluents, from which the drug may not be released quickly enough for adequate absorption. For example, only 40 percent of thiamine and 79 percent of riboflavin are available for absorption from capsules containing Fuller's earth, which adsorbs these drugs.[7] Similarly, calcium phosphate used as a diluent in tetracycline capsules reduces absorption by the formation of insoluble complexes.[8]

The particle size of powders is significant in their dissolution and bioavailability, as is demonstrated by such drugs as spironolactone and griseofulvin, the micronization of which leads to significantly higher absorption in humans. However, smaller particle powders have a greater tendency to adsorb moisture from the atmosphere, which results in possibly unstable preparations. Smaller particle size also means increased electrostatic charges on the particle surface, especially with hydrophobic drugs. This might result in

Table 3.1. SELECTED LIST OF DRUGS, THEIR COMMERCIAL DOSAGE FORM, AND ROUTES OF ADMINISTRATION AS SUSPENSIONS

DRUG	*BRAND NAME*	*ROUTE*
Tetracycline	Achromycin	Ophthalmic
Ampicillin	Alpen	Oral
Aluminum hydroxide	Amphojel	Oral
Penicillin G benzathine	Bicillin	Intramuscular
Betamethasone	Celestone/Soluspan	Intraarticular, intralesional, intramuscular, periarticular
Colistine sulfate	Coly-Mycin	Otic, oral
Cortisone acetate	Cortisone acetate	Intramuscular
Chlorothiazide	Diuril	Oral
Theophylline	Elixicon	Oral
Ferrous fumarate	Feostat	Oral
Acetyl sulfisoxazole	Gantrisin	Oral
Prednisolone acetate	Meticortelone	Intraarticular
Phensuximide	Milontin	Oral
Nystatin	Mycostatin	Oral
Diazoxide	Proglycem	Oral
Erythromycin estolate	Ilosone	Oral
Nalidixic acid	NegGram	Oral
Triamcinolone diacetate	Aristocort	Intralesional
Trisulfapyrimidines	Terfonyl	Oral

aggregation and the consequent loss of an effective or exposed surface area for dissolution.

An example in which smaller particle size is not always desirable, even though it does increase bioavailability, is in the use of nitrofurantoin. The use of a larger particle size is recommended to avoid the gastrointestinal

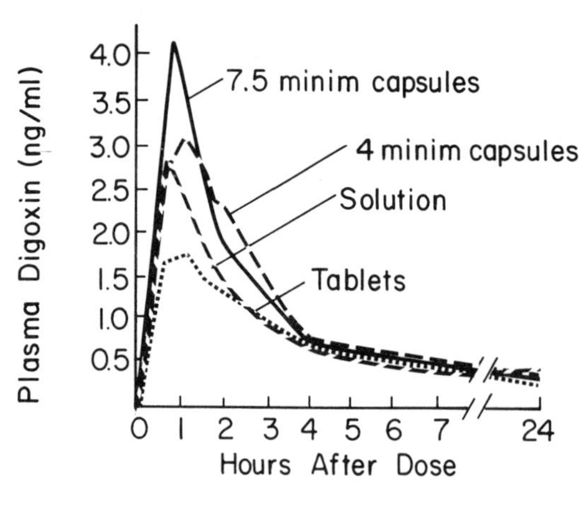

Figure 3.1. Comparison of mean plasma concentration curves after 500 mg of digoxin administered as tablets, or after 400 mg administered as a solution or as capsules. (From Johnson: Clin Pharmacol Exp Ther 19:746, 1976)

irritation and accompanying nausea which occurs very frequently with the use of fine particles in oral dosage forms.

When powders are administered in a gelatin capsule, the capsule shell itself may affect the absorption process. Hard gelatin capsules dissolve more readily in the gastrointestinal fluids than soft gelatin capsules. For example, the slow absorption of vitamin B_{12} from soft capsules may be attributed to the slow dissolution of the capsules themselves.[9] However, a recent study showed that soft capsules might produce an unexpected increase in the absorption of digoxin, with which absorption rates even higher than the solution dosage forms were obtained (Fig. 3.1).[10] This finding was attributed to the possible interaction of digoxin with the soft elastic capsule walls and also to the protection of digoxin against possible chemical decomposition from gastrointestinal fluids.

A large number of drugs are administered in powder form, such as iodochlorhydroxyquin and methylbenzethonium chloride, or contained in a capsule, as are phenytoin, chloramphenicol, erythromycin, tetracycline, lithium carbonate, quinine sulfate, chlordiazepoxide hydrochloride, cephalexin, and propoxyphene.

Tablets

Whereas solutions represent a state of maximum dispersion, compressed tablets have the closest proximity to particles. Complexities in dissolution and bioavailability are generally inversely proportional to the degree of dispersion—compressed tablets are thus most prone to bioavailability problems. This is primarily due to the smaller surface area exposed for dissolution until the tablets break down into smaller particles (Scheme 3.1). Factors responsible for the primary breakdown of tablets into granules and their subsequent breakdown into finer particles include such parameters as the concentrations of binder, disintegrant, and lubricant; the hydrophobicity of the drug and the adjuvants; the compression force applied; storage conditions; and so on. It can, therefore, be expected that a significant difference is always possible in the dissolution and bioavailability of various tablets.

The problem of tablet disintegration is well demonstrated by such drugs as dipyridamole,[11] thioridazine,[12] and digoxin,[13] which, when administered in tablet formulations, exhibited higher blood levels when the tablets were crushed before administration.

The disintegration test for tablets has long been used to detect ineffective products, as determined by a lack of disintegration into large particles within a given period of time. This test allows monitoring of batch-to-batch variations in the manufacturing process. However, adequate disintegration alone does not assure ultimate dissolution, which may be retarded by the adsorption of drug on hydrophobic lubricants in the formulation, the recrystallization of

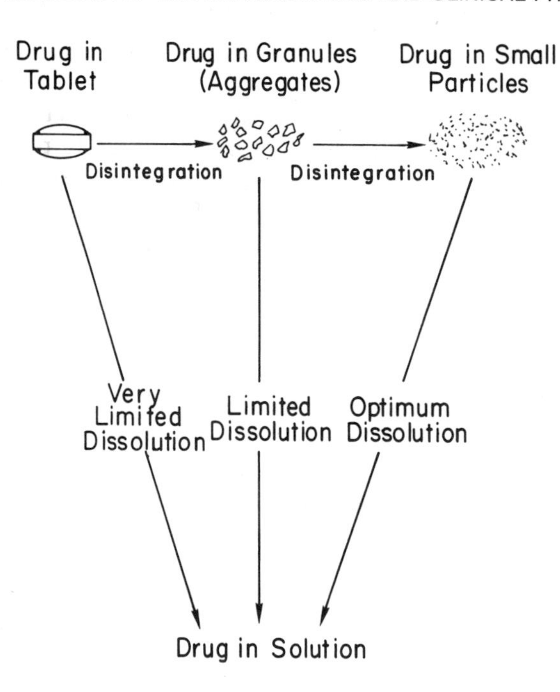

Scheme 3.1. Disintegration as rate limiting factor in the dissolution of drugs. (From Cadwallader: Biopharmaceutics and Drug Interactions, 1974. Courtesy of Roche Laboratories)

drugs, the presence of large primary granules, and the failure of these granules to break down further into finer particles. The importance of using smaller particles in tablet formulations is well demonstrated in the use of griseofulvin, with which the reduction of particle size has been consistently related to bioavailability. Recently, a solid dispersion of griseofulvin was formulated which contained ultramicrosize particles of the drug, resulting in an almost 100 percent improvement in its bioavailability compared to the micronized forms (Fig. 3.2).[14]

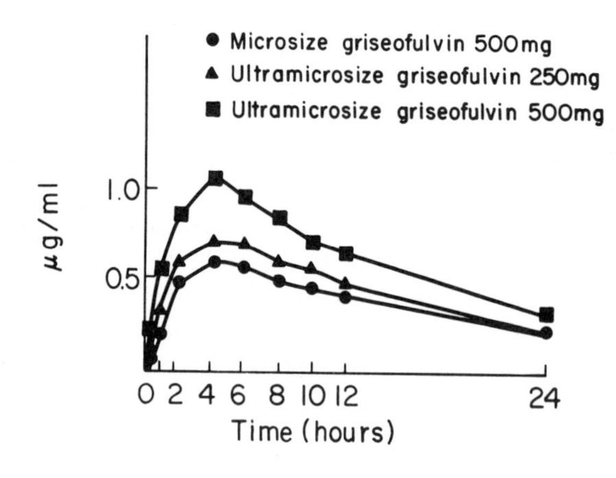

Figure 3.2. Plasma concentrations of griseofulvin (μg/ml) after a single oral dose in humans, comparing the absorption of microsize griseofulvin with ultramicrosize griseofulvin. (From Barrett and Bianchine: Curr Ther Res 18:501, 1975)

The coatings of tablets, which are applied for a variety of reasons, add another rate-limiting factor, since a coating must dissolve or disrupt before the tablet can disintegrate and the dissolution process begins. The sugar coating used to mask unpleasant taste, appearance, and odor, or to protect a tablet ingredient from decomposition during storage, consists of an application of poorly soluble polymers which can interfere with the disintegration of tablets. Film coatings are generally less problematic, but enteric coatings used to protect both the gastric mucosa from the drugs and the drugs from the gastric fluids give the most variable bioavailability, since their disintegration is often dependent on gastrointestinal pH and other highly variable physiologic and physicochemical factors.

Controlled-Release Dosage Forms

Unless specific formulation efforts are made to control the release of drugs, the rates of drug absorption are generally proportional to the amount of drug at the site of absorption. In many instances it is necessary to prolong the action of drugs by sustaining their absorption over a longer period of time.

The design of oral prolonged-action dosage forms includes such modifications as:

1. Barrier coating, whereby the drug diffuses out through a membrane within which it may be dissolved by the penetrating gastrointestinal fluids.
2. Fat embedment, which involves suspending the drug in a fatty medium in a solid dosage form from which the drug is released by erosion, hydrolysis of fat, and direct dissolution.
3. Plastic matrices, which allow leaching and diffusion of drugs from a solid plastic matrix which is left intact after the drug has been released.
4. Repeat action tablets, utilizing a double coating which releases an initial dose followed by another dose released either instantaneously or by slow diffusion.
5. Ion-exchange resins, which provide prolonged dissolution by the formation of drug salts with resins, which then react with either hydrochloric acid in the stomach or sodium chloride in the intestine to exchange the drug.
6. Hydrophilic matrices, utilizing hydrophilic gums for compression into tablets which undergo gelation and release the drug by diffusion.
7. Polymer resin beads, in which the drug is first dissolved or suspended in plastic monomers and then polymerized. The beads are then either filled into a capsule or compressed into tablets. Drug release is controlled by the dissolution and swelling of the resin and the diffusion of drug from the beads.
8. Soft gelatin depot capsules involve the dissolution or suspension of drugs in sponge-forming solutions and consequent filling into capsules which leave a solid skeleton upon diffusion of the drugs.

9. Drug complexes utilizing macromolecules provide prolonged release upon the hydrolysis of the complex.

The release of drugs administered parenterally can also be controlled by the following methods:

1. Pharmacologic methods: Intramuscular or subcutaneous administration instead of intravenous. Simultaneous administration of vasoconstrictors (adrenalin in local anesthetics, ephedrine in heparin solutions), blocking elimination of drugs through the kidney by simultaneous administration of a blocking agent such as probenecid with penicillin or p-aminosalicylic acid.
2. Chemical methods: Use of salts, esters, ethers, and complexes of the active ingredient with low solubility.
3. Physical methods: Selection of a proper vehicle giving prolonged release, as with the use of oleaginous solutions instead of aqueous solutions; the addition of macromolecules which increase the viscosity, such as carboxymethylcellulose, tragacanth, etc.; the use of swelling material to increase the viscosity of oleaginous solutions, as with aluminum monostearate; the addition of adsorbents; the use of a solution from which the drug is precipitated upon contact with body fluids; the use of aqueous and oleaginous suspensions; and the use of implants.

A newer category of dosage forms, termed Therapeutic Systems, has recently been marketed in this country.[16] A Therapeutic System is a dosage form that provides preprogrammed, unattended delivery of drugs at a rate, and for a given time period, designed to meet a specific therapeutic need. These systems have been developed for introducing drug substances both via the systemic circulation and directly to specific target organs. Many new drug delivery techniques have been developed, including:

1. Diffusion of drugs through rate controlling membranes.
2. Osmotic pumping.
3. Biodegradable polymer matrices.
4. Polymer-bound active species.

The Therapeutic Systems are composed of an active drug in a delivery module, which consists of a drug reservoir, which may be a single or multicompartment element; a rate controller; and an energy sour e to effect the release of the drug molecules through a delivery portal. The drug delivery module is housed in a "platform" which is compatible with the tissues and couples the system to the body site in which it is deployed. The platform may be either fixed or mobile within a defined area. Some examples include the ocular platform, which is designed so that it can float comfortably and inconspicuously in the tear film on the eye beneath the eyelid for controlled delivery of pilocarpine (Occusert); and the T-shaped progesterone impregnated polymer unit for intrauterine deployment for fertility control (Progestasert).

The osmotic drug delivery system resembles an ordinary tablet in appearance and is comprised of a solid core of drug surrounded by a semipermeable membrane with a single minute orifice. The membrane allows steady entry of water at a predetermined rate to dissolve the drug. Drug solution is then continuously pumped through the orifice, providing a constant rate of release.

Other novel ideas include a transdermal Therapeutic System consisting of a disc 0.2 mm thick and 2 cm in diameter, which is worn behind the ear like a tiny adhesive bandage and releases scopolamine for its antiemetic properties. The use of biodegradable polymers has also been suggested for implant systems for the controlled release of drugs.

The foregoing innovations are cited here to make the reader aware of the possibilities of bioavailability variation as a result of a large number of physicochemical and technologic implementations in the design of dosage forms. The complexities in the design of dosage forms necessitate the development of an elaborate system to evaluate dosage forms and systems on the basis of the attributes listed at the beginning of this chapter.

EVALUATION OF DOSAGE FORMS

It is not possible to predict that every dose administered will result in a consistent desirable therapeutic response. However, several tests can be conducted to assure some measure of reliability in dosage form functions. These include the following.

Chemical Content

It is essential that dosage forms contain the labelled amount of the active drug. Chemicals which are biologically active are also highly chemically reactive and can therefore undergo chemical decomposition reactions which result in a loss of content. For example, aspirin decomposes to salicylic acid and acetic acid. Salicylic acid is undesirable because it causes more gastrointestinal irritation than aspirin and also because it may not possess a therapeutic activity equivalent to aspirin (some recent data suggest otherwise). Para-aminosalicylic acid decomposes via decarboxylation to meta-amino phenol, resulting in discoloration and enhanced toxicity. Tetracycline converts to epianhydrotetracycline, which is very toxic to the kidneys.

Although not all chemical decomposition reactions result in a toxic product, a change in the color or the consistency of a preparation will quite often make it unacceptable to the patient. It is therefore necessary to provide a shelf life or expiration date on the products. A five year expiration date is rather common for a relatively stable product, and sometimes a shelf life of only a few days or weeks is assigned to highly reactive drugs or radio-labelled compounds.

In order to account for the loss of drug during shelf life, overage additions are often made at the time of manufacture. For example, urea peroxide products used for removing ear wax are labelled as containing 6.5 percent carbamide peroxide in anhydrous glycerol, but in a fresh product the carbamide peroxide content may be significantly higher than the labelled amount. This overage addition is necessary to compensate for the high reactivity of the active components. Such large overage additions cannot, however, be allowed for drugs used internally or for those for which a narrow plasma concentration fluctuation has to be maintained, as with cardiac glycosides and anticoagulants, which have narrow therapeutic indices.

Chemical decomposition is not the only way in which active ingredients are lost from dosage forms. For some compounds with a low boiling point the active principal can be lost by evaporation or volatilization. For instance, nitroglycerin tablets, if dispensed in a plastic container, have been reported to lose up to 80 percent of their active components in two years. Nitroglycerin tablets are therefore required by law to be dispensed in the original glass container and even then there can be a significant drug loss within the unopened container.[17]

The container also plays an important role in determining dosage form effectiveness. For example, the increasing use of plastic large-volume parenteral containers has created an unanticipated problem of loss of drugs due to adsorption onto the plastic and absorption through it, resulting in a significant loss of drugs such as vitamin A.[18]

Chemical content evaluations are therefore fundamental in determining dosage form effectiveness. Hundreds of drugs have been recalled by the FDA due to subpotent or in some instances superpotent products, making potency one of the primary criteria in the evaluation of dosage forms.

Content Uniformity

The chemical equivalence testing described above is generally performed on a large number of dosage form units (e.g., 20 tablets) at one time. This testing determines the average amount of active ingredient(s). It will not, however, reveal variations in drug content among the units. For example, the oral contraceptive Ortho-Novum 1/50 contains 1 mg of norethindrone and 0.05 mg of mestranol per tablet. What if one tablet contains 0.1 mg of mestranol while another tablet contains none? Although the two tablets combined will pass the chemical equivalence test, a course of therapy with tablets of this quality might result in an unanticipated pregnancy. The problem of content uniformity, therefore, exists for all products containing minute amounts of active ingredients, as is shown in Table 3.2.

The problems of content uniformity arise mainly from the mixing of small amounts of drugs into large batches where a uniform distribution must be assured. Again, the FDA has recalled many products in the last few years due to noncompliance with the USP content uniformity requirements of 5 percent.

Table 3.2. SELECTED EXAMPLES OF TABLETS CONTAINING SMALL AMOUNTS OF ACTIVE DRUG COMPONENT

DRUG	*AVAILABLE TABLET STRENGTH (mg)*
Atropine sulfate	0.3
Colchicine	0.5
Dexamethasone	0.25
Diethylstilbesterol	0.1
Ethinyl estradiol	0.05
Digitoxin	0.05
Digoxin	0.125
Reserpine	0.1

Presence of Contaminants

Contaminants are defined as any undesirable substances contained in a formulation.[19] Contamination of the drug product may occur during processing from impurities in raw materials, heavy metal ions from manufacturing equipment, microorganisms, or chemical decomposition products, which may be toxic as noted above or inactive, as the product of reaction between isoproterenol and bisulfite preservatives. Another source of contamination is dust spreading during the manufacturing process, when several products are handled simultaneously in a manufacturing facility. Although the presence of contaminants may not always be deleterious, it is always desirable to have as few as possible to prevent changes in the physical or esthetic appearance of a product as well as unanticipated adverse reactions.

Disintegration Test

The disintegration test ascertains the time required for a compressed tablet to break up into granules. The first official disintegration test was included in *Pharmacopeia Helvetica* 5 in 1934. Since then most official pharmacopeias have included these tests to formulate a basis for prediction of the availability of drugs from dosage forms. Up until the 1950s, disintegration was the key word and any dosage form that disintegrated within a prescribed time was assumed to provide adequate bioavailability.

A large number of formulation factors can affect the rate of tablet or capsule disintegration, including:

Diluents or fillers
Manufacturing methods, such as dry or wet granulation, etc.

Compression pressure or pressure in capsulation
Hardness
Concentration of disintegrant and the method of its addition
Types and concentrations of lubricants, surfactants, and binders
Drug properties such as particle size, surface characteristics, solubility, and crystallinity
Composition and properties of capsule shell
Type and composition of coating
Age of finished product and storage conditions

The United States Pharmacopoeia (USP) disintegration method involves a basket-rack assembly which is moved up and down 30 times a minute. At specific times, the number of tablets or capsules disintegrated is determined. The disintegration time allowed varies from five minutes to one hour. For example, aspirin tablets have a time limit of five minutes.

The present USP and National Formulary (NF) disintegration tests measure only the physical break-up of the tablet or capsule, which may not necessarily correlate with drug bioavailability. In order for a drug to be absorbed, it must be present in a solution form. It is possible that the particles from disintegrated tablets might not further disintegrate or dissolve and thus no bioavailability assurance can be obtained from formulations meeting only the official disintegration tests.

Dissolution Test

A dissolution test is much more discriminating than the disintegration test. It is a better estimate of bioavailability, though it is still not foolproof. Dissolution rate tests can be used to predict bioavailability if these two conditions are met:

1. The dissolved drug remains free and intact in the gastrointestinal tract. If the dissolved drug complexes with a component of the gastrointestinal tract, and if drug decomposition occurs in the gastrointestinal tract, then the dissolution test cannot be a very good index of bioavailability.
2. Absorption is not the rate limiting step. If the solution formed is quickly absorbed, then the amount absorbed can be correlated with the in vitro dissolution rate. However, when absorption is slow or limited, bioavailability may not be proportional to the dissolution rate.

The formulation factors listed as affecting the disintegration rates also affect the dissolution rates. A large volume of data has been reported which correlates various formulation factors and the dissolution rates. For example, the particle size of a drug is most clearly related to the drug's dissolution rate. Addition of surfactants quite often substantially increases the dissolution rates of hydrophobic drugs by the removal of air pockets around the particles,

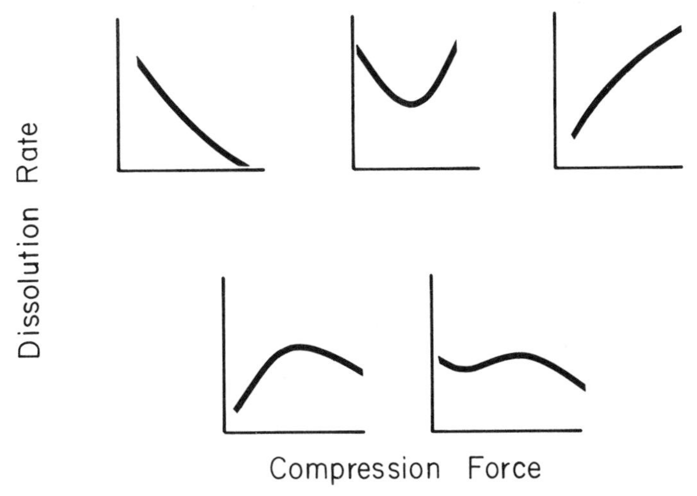

Figure 3.3. Reported relationships between compression force and dissolution rates.

thus facilitating the contact of the dissolution medium with the drug. An important source of surface activity is the gastric fluid, where the surface tension varies between 38 and 52 dynes/cm. This lower surface tension allows better wetting of particles and promotes dissolution. The primary cause of surface activity in the gastric fluids is the reflux of intestinal contents into the stomach. The intestinal fluids have significant surface activity, as may be expected because of their lecithins, bile salts, etc.

The fillers and diluents used in a formulation have a significant effect on its dissolution. If the drug is hydrophobic, a hydrophilic filler will tend to enhance dissolution, especially if this filler is at the same time a disintegrant. Starch has hydrophilic properties and is an effective disintegrant and thus proves to be an excellent filler.

The lubricants used may have varying effects. If the granule particles are hydrophilic and disintegrate quickly, a surface active lubricant will have little effect. If the granule particles are less hydrophilic and do not disintegrate as quickly, a surface active lubricant may enhance dissolution. The use of such hydrophobic lubricants as stearates decreases dissolution rates, but this effect is minimal if their concentration is less than 1 percent.

The effect of compression pressure on dissolution rates is the most difficult to predict. For example, the relationships shown in Figure 3.3 have all been reported in the literature. Dissolution rates will generally decrease with increasing compression pressure due to a closer binding of the granules to each other. At higher pressure a crushing of the granules and perhaps even of the drug crystals would occur, resulting in an increased surface area and an increased dissolution rate. A further increase in pressure may make the

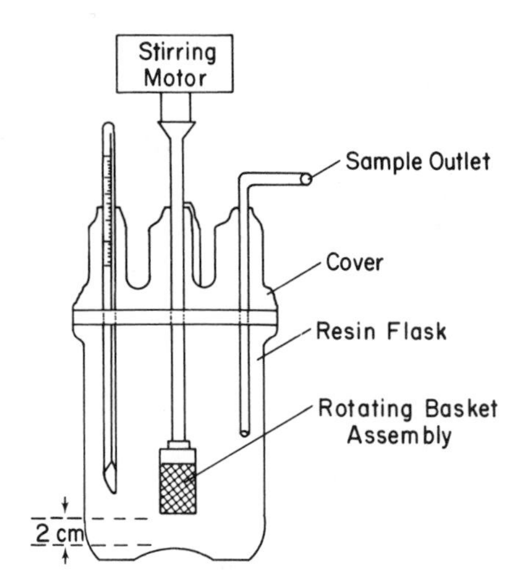

Figure 3.4. Apparatus for dissolution testing. (From Anon.: The National Formulary, 14th ed., 1975. Courtesy of the American Pharmaceutical Association)

bonding more important than the crushing, resulting in a decrease in the dissolution rates. Where the bonding is not significant, a direct increase in the dissolution rates can be expected with increasing compression pressure.

The effect of tablet storage on the dissolution rate can also be important and reports have been made suggesting both increasing and decreasing dissolution rates.

Briefly, the large number of formulation factors that can alter dissolution rates make dissolution rate studies necessary to detect batch-to-batch and within-batch variations during the manufacturing processes. Several methods are available to study dissolution rates, including the rotation basket method, as suggested in the USP and NF. The monographs describe the specific temperature, the dissolution medium (distilled water, simulated gastric fluid, or simulated intestinal fluid), the rotation speed of the basket (60 to 150 rpm), and the percentage of drug to be dissolved as an endpoint. These conditions are determined by the intrinsic properties of the drug and its dissolution behavior. Products which meet USP requirements must dissolve within the time prescribed. For example, prednisolone tablets must dissolve to the extent of 60 percent in 10 minutes at 100 rpm; and tolbutamide to the extent of 60 percent in 30 minutes at 105 rpm. Table 3.3 lists the drugs for which a dissolution test is required according to the USP or NF. This list will continue to grow with each revision of these compendia, as more data become available relating bioavailability and dissolution rates.

Other dissolution methods include Levy's beaker method, basket-rack assembly, the Pernarowski method, and the membrane filtration method, among a variety of methods currently in use (Figs. 3.4 and 3.5). An immediate source of concern, therefore, is the reproducibility of a dissolution test in various laboratories and with the various methods of dissolution in current use. In order to reduce the large degree of variation which exists it may be

Table 3.3. DRUGS FOR WHICH OFFICIAL COMPENDIA REQUIRE A DISSOLUTION TEST

Acetohexamide	Nitrofurantoin
Digoxin	Phenylbutazone
Ergotamine tartarate and caffeine tablets	Prednisolone
Hydrochlorothiazide	Prednisone
Lithium carbonate	Sulfamethoxazole
Meprobamate	Sulfisoxazole
Methaqualone	Theophyllin, ephedrine hydrochloride, and phenobarbital tablets
Methylprednisolone	Tolbutamide

necessary to formulate "dissolution standards"—tablets or capsules standardized and distributed by the official agencies. These "dissolution standards" should, therefore, be used to compare the dissolution rates reported from different laboratories and from different dissolution procedures. Although no such standards are available to date, these will most surely be available within a few years.

Dissolution rates are generally described in terms of the Noyes-Whitney equation (Equation 2.1), where the dissolution rates are proportional to the difference between the saturation concentration and the concentration in the dissolution medium, the so-called concentration gradient. The concentration

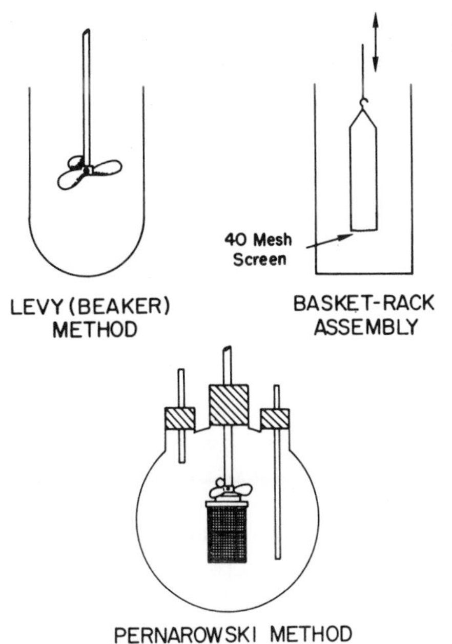

Figure 3.5. In vitro methods of determining dissolution rate.

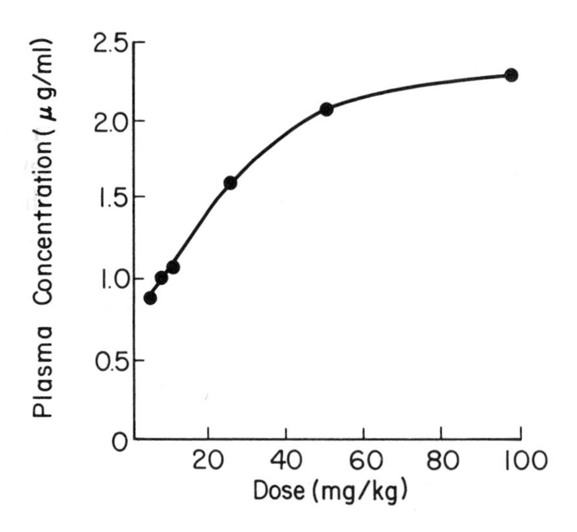

Figure 3.6. Mean blood levels of griseofulvin five hours after a single oral dose in humans. (After Grin and Denic: Acta Med Iugosl 19:53, 1965)

in the dissolution medium is often so low that it can be overlooked, making dissolution rates directly proportional to the saturation solubility, among such other factors as surface area, diffusion coefficient, and diffusion layer thickness. The condition whereby $C_t \leq 0.1C_s$ is referred to as the sink condition, which exists quite frequently in the absorption of drugs. This is due mainly to the fast absorption of drugs when absorption is not the rate-limiting step in bioavailability. However, in those instances where a large dose of hydrophobic drug is administered there may not even be sufficient fluid to dissolve the administered dose, and even if it is dissolved, the absorption may not be fast enough to create a sink condition retarding the overall availability of the drug. In these instances, the same amount of drug given in divided doses will provide a better bioavailability than a single large dose.

An interesting report on blood levels of griseofulvin as a function of dose suggests a nonlinear relationship between the dose and plasma concentration (Fig. 3.6).[20] A proportionally smaller increase in the plasma levels with increasing dose can be attributed to the possible lack of sufficient aqueous fluids in the gastrointestinal tract to dissolve the drug. The importance of the specific absorption sites for griseofulvin is also suggested.

Patient Acceptance

It is needless to say that a dosage form cannot be effective if it is not acceptable to the patient. The purpose is not necessarily to have an attractive dosage form but rather to have one which is free from objectionable odors or taste. The use of various flavoring agents in formulations can overcome such problems. Smell, appearance, and taste can also be masked by sugar coating, film coating, etc. Such considerations are especially important in pediatric use and are exemplified by children's chewable vitamin products, which have various shapes (animals, objects) and several colors and flavors.

The placebo effect associated with the administration of drugs cannot be ignored. It is a well studied fact that a patient's confidence in the quality of medication significantly contributes to drug therapy. For example, red-colored tablets are generally regarded as having greater potency than blue or yellow tablets. Patients maintained on one brand of a drug would often refuse to accept substitute brands, especially in psychiatric care. One of the major current problems in drug therapy is patient noncompliance with a prescribed regimen. An unappealing dosage form can only add to these complications.

Bioavailability

The in vitro methods of evaluating dosage forms provide only indirect evidence of the therapeutic utility of the drug in a given dosage form. It would be more desirable to employ a direct means of ascertaining the utility of a particular dosage form. In order to do this in vivo methods are needed which will reflect the true safety and efficacy of the dosage forms.

The estimation of the bioavailability of a drug in a given dosage form is direct evidence of the efficiency with which a dosage form performs its intended function. The FDA has recently suggested bioequivalency requirements which must be met for Federal approval of the marketing of drugs.[21] The following definitions are necessary to an understanding of these requirements:

Drug product: A finished dosage form that contains the active drug ingredient—generally, but not necessarily, in association with inactive ingredients.

Pharmaceutical equivalents: Drug products that contain identical amounts of the same active drug ingredient—i.e., the same salt or the ester of the same therapeutic moiety—in identical dosage form, but not necessarily containing the same inactive ingredients; and that meet the identical compendial or other applicable standard of identity, strength, quality, and purity, including potency and, where applicable, content uniformity, disintegration time, and/or dissolution rates.

Pharmaceutical alternatives: Drug products that contain an identical therapeutic moiety, or its precursor, but not necessarily in the same amount or dosage form or as the same salt or ester. Each drug product individually meets either an identical or its own respective compendial, or other applicable standards as mentioned above.

Bioequivalent drug products: Pharmaceutical equivalents whose rate and extent of absorption may not show a significant difference when administered at the same molar dose of the therapeutic moiety under similar experimental conditions, either in single dose or in multiple doses. Some pharmaceutical equivalents or pharmaceutical alternatives may be equivalent in the extent but not the rate of their absorption and yet may be considered bioequivalent because such differences in the rate of absorption are intentional and are reflected in the labeling, are not essential to the attainment of effective body drug concentration in chronic use, or are considered medically insignificant for the particular drug product studied.

Bioequivalency requirements: Requirements imposed by the FDA for in vitro and/or in vivo testing of specified drug products which must be satisfied as a condition of marketing.

The criteria and evidence needed to establish a bioequivalency requirement are based on information from clinical reports and well designed bioavailability studies, and on evidence that the drug product exhibits a narrow therapeutic ratio, e.g., there is less than a two-fold difference in median lethal dose and median effective dose values (or less than a two-fold difference in the minimum effective concentration and minimum toxic concentration in the blood).

The physicochemical evidence needed to establish a bioequivalency requirement includes low water solubility, e.g., less than 5 mg/ml, or, if dissolution in the stomach is critical to absorption, the volume of gastric fluids required to dissolve the recommended dose (gastric fluid content is assumed to be 100 ml for adults and is prorated for infants and children). The dissolution rates are also taken into consideration if less than 50 percent of the drug dissolves in 30 minutes using official methods. Also included under physicochemical evidence are particle size and surface area of the active drug ingredient, if surface area is critical in determining bioavailability. Certain physical structural characteristics of the active drug ingredient, e.g., polymorphism, solvation, etc., are also considered. Drug products which have a high ratio of excipients to active ingredient (e.g., greater than 5: 1) may also be subjected to bioequivalency requirements. Other evidence includes specific absorption sites or where the available dose is less than 50 percent of an administered dose. Drugs which are rapidly biotransformed in the intestinal wall or liver during absorption, and drugs which are unstable in specific portions of the gastrointestinal tract—requiring special coating or formulations—are also subjected to bioequivalency requirements, as are drugs which show dose-dependent absorption, distribution, biotransformation, or elimination. Bioequivalency requirements may include an in vivo test in humans or animals or an in vitro test that has been correlated with human in vivo bioavailability data.

For some dosage forms bioequivalency requirements can be waived, as with topical products, oral dosage forms not intended for absorption, inhalations, and solutions if there is sufficient evidence that the inactive ingredients do not affect absorption rates. Table 2.2 shows the drugs with potential bioavailability problems and for which bioequivalency must be demonstrated.

The distinction between bioavailability and comparative bioavailability is an important one. Any study that estimates the relative amount of an administered drug to reach the general circulation, and the rate at which this occurs, is considered a bioavailability study. Hence this category includes those studies in which a drug is administered orally and the extent of absorption

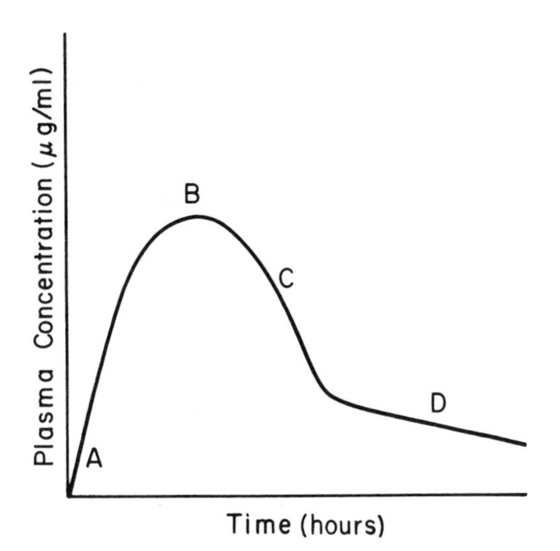

Figure 3.7. Distinct phases of plasma concentration profiles.

is compared with the results of intravenous administration (which provides 100 percent bioavailability). Comparative bioavailability studies involve determination of the relative bioavailability of an active drug in two or more formulations, without regard for the actual amount absorbed from each formulation.

The bioavailability of a drug is controlled by three principal factors, namely: (1) the rate and extent of release of the drug from the dosage form, (2) its subsequent absorption from the solution state, and (3) the biotransformation during the process of absorption.

In all quantitative determinations of bioavailability, concentration is measured in such various body fluids as the blood, plasma, and urine. Plasma concentrations following the oral administration of a drug assume four sequential phases (Fig. 3.7):

1. absorption $>$ elimination
2. absorption $\approx$ elimination
3. absorption $<$ elimination
4. absorption $= 0$; elimination > 0

The exact shape of the plasma concentration profile will depend on the relative rates of absorption and elimination. For example, the plasma concentration profiles may be quite different with different routes of administration (Fig. 3.8). Intravenous and sometimes intramuscular routes yield an early peak due to fast or almost instantaneous absorption, whereas oral, subcutaneous, rectal, and other routes may show delayed peaks due to slower rates of absorption. It should be noted that the rate of elimination is considered constant since it depends primarily on the specific nature of the active drug

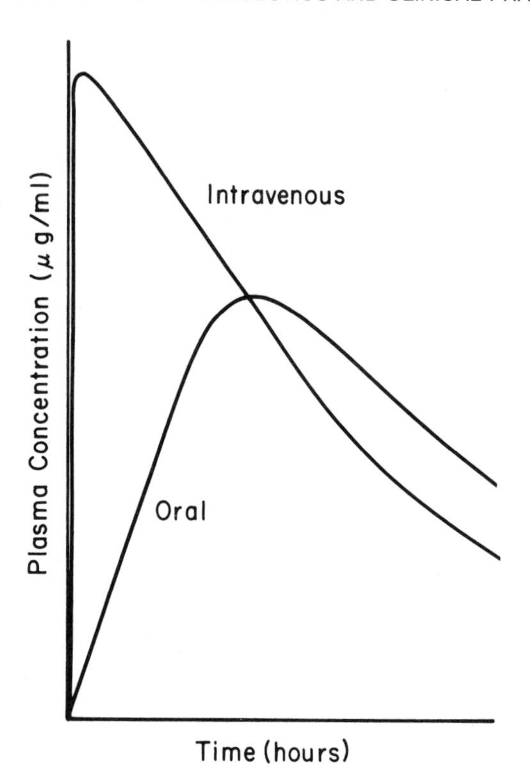

Figure 3.8. Effect of routes of administration on the plasma concentration profiles.

ingredient. Thus "phase 2," or the peaks, are determined only by the rates of absorption.

The estimation of bioavailability from plasma concentration profiles requires a thorough understanding of the nature of these profiles. For example, a higher or earlier peak does not necessarily mean greater overall absorption than from a product giving a smaller or delayed peak. The total absorption of drugs is, therefore, proportional not only to the plasma concentrations achieved but also to the length of time these concentrations persist in the blood or plasma. One parameter that characterizes this aspect is the area under the plasma concentration vs. time profile. The major contribution to the area under the curve for a fast absorbed formulation is due to the high peak concentration whereas for a slowly absorbed formulation, the area is mainly due to substained or prolonged plasma concentration. It should be noted that the area under the plasma concentration vs. time profile (AUC) is only proportional to the total amount of drug absorbed and cannot be used to determine the actual amount of drug administered unless it is compared with a known standard, whereby the extent of absorption is either measured by other methods or assumed to be 100 percent, as in the case of intravenous administration.

Many methods are available for the measuring of AUCs. Three of these are described as follows.

THE TRAPEZOIDAL RULE

This is the simplest of all the methods and involves the breaking up of the plasma concentration vs. time profile into several trapezoids (Fig. 3.9), calculating the areas of individual trapezoids, and then adding up these areas to arrive at a cumulative AUC:

$$\mathrm{AUC}_t = \frac{(C_0 + C_1)}{2}(t_1 - t_0) + \frac{(C_1 + C_2)}{2}(t_2 - t_1) + \cdots + \frac{(C_{n-1} + C_n)}{2}(t_n - t_{n-1}) \quad \text{(Eq. 3.1)}$$

$$= \sum_{i=1}^{n} \frac{(C_{i-1} + C_i)}{2}(t_i - t_{i-1}) \quad \text{(Eq. 3.2)}$$

The units for the AUC are: concentration × time, e.g., μg · hr/ml or mg · min/liters. Consider the following example for the calculation of the AUC.

Time, hrs.	*C, μg/ml*	*Trapezoid area*	*Cumulative area*
0	0	—	—
1	2	1	1
3	15	17	18
5	5	20	38
7	1	6	44 μg · hr/ml

As a general rule, the larger the number of segments or trapezoids formed, the greater the accuracy. In other words, the closer the interval between each plasma concentration reading taken, the more accurate the result. Exact results will be obtained only when the number of trapezoids is infinite. If the plasma concentration values are quite far apart, a smooth curve may be drawn which then can be broken up into a large number of trapezoids.

The AUC is proportional to the dose absorbed only when the calculations are extended to the point where the plasma concentration approaches zero. This may not be possible in some instances, in which case the comparisons can either be made up to a given time or the plasma concentrations can be extended to follow the shape of the curve (Fig. 3.10), both of these approaches adding to the error in the bioavailability estimations.

INTEGRATION METHOD

The rate of change of plasma concentration (C) is described as:

$$dC/dt = \text{rate of absorption} - \text{rate of elimination} \quad \text{(Eq. 3.3)}$$

$$= k_a X_a - KX \quad \text{(Eq. 3.4)}$$

where k_a and K are absorption and elimination rate constants and X_a and X are the amounts of drug in the gastrointestinal tract and the body, respectively. An integration of this equation between limits of time for which the drug remains in the body, as reflected by the plasma concentration, gives:

$$C = A(e^{-Kt} - e^{-k_a t}) \qquad \text{(Eq. 3.5)}$$

and the total area under the curve (AUC), for which the total integral between time zero and infinity is given by:

$$\text{AUC} = A\left(\frac{1}{K} - \frac{1}{k_a}\right) \qquad \text{(Eq. 3.6)}$$

Thus, if C can be fitted to an equation which will allow calculation of the absorption and elimination rate constants, exact calculation of AUC can be made very easily. This approach is identical to the trapezoidal rule method described earlier, except that it allows the use of trapezoids whose time differential is approaching zero.

PHYSICAL METHODS

A variety of methods which utilize physical properties can be used to calculate AUCs. For example, if plasma concentration profiles are plotted on smooth texture paper, these can be cut out and weighed on an electronic balance. The weight of these cut-out plots will be proportional to the area under the curve. Again, in order to convert the weight into absolute bioavailability, a reference formulation is needed.

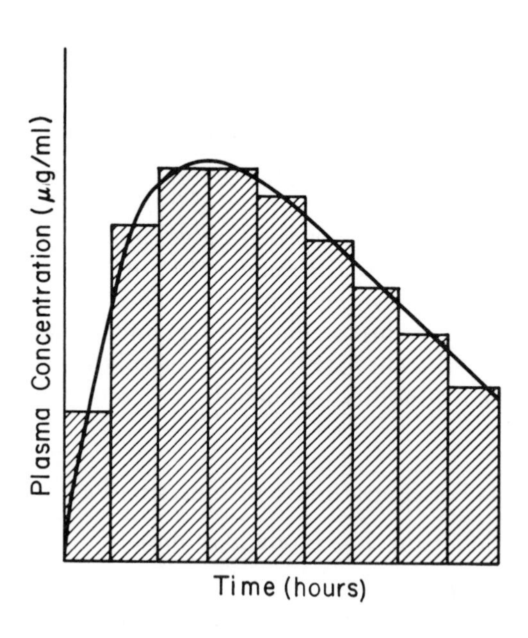

Figure 3.9. Application of trapezoidal rule in the calculation of area under the plasma concentration:time profile. Shaded area represents the calculated area.

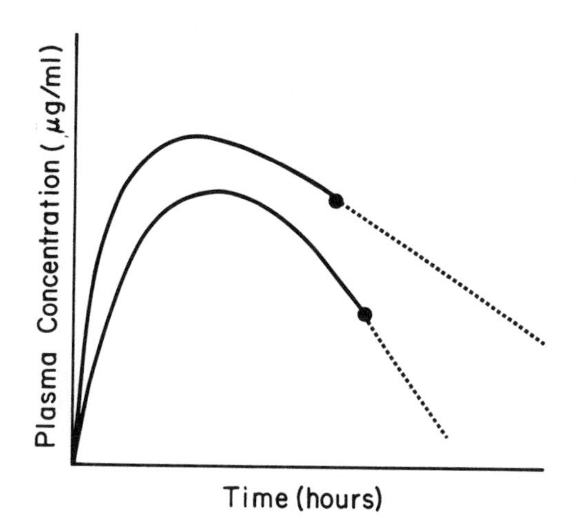

Figure 3.10. Extrapolations of plasma concentration profiles according to the shape of the curves.

Planimeters are also useful in calculating the areas under the curve. A planimeter is a precision instrument which allows the calculation of areas by tracing their outlines.

The degree of error is always higher when these methods are used since additional instrumental and human errors are introduced. However, in some instances these may be the only methods that can be used.

Once the AUCs are determined, the bioavailability estimations become simple.

ABSOLUTE BIOAVAILABILITY FROM BLOOD LEVEL DATA

Drugs which are not significantly biotransformed in the gastrointestinal tract or during their first passage through the liver allow direct estimation of bioavailability by comparing the blood levels following oral administration with those following intravenous administration (where 100 percent bioavailability is assumed):

$$F = \frac{\text{Dose}_{iv}\ \text{AUC}_{po}}{\text{Dose}_{po}\ \text{AUC}_{iv}} \qquad \text{(Eq. 3.7)}$$

The dose here refers to the actual dose administered as determined by assay and not by the labeled amount. This is necessary to account for possible variations in chemical equivalence. F indicates the absorption efficiency or the fraction of the orally administered dose which is absorbed. In those instances where identical doses are administered orally and intravenously, F is simply the ratio of AUC_{po} and AUC_{iv}.

ABSOLUTE BIOAVAILABILITY FROM URINARY EXCRETION DATA

Since all of the drug that enters the body is eventually eliminated in some way, mainly through the kidneys, the total amount of drug eventually eliminated in the urine, $X_u\infty$, can be correlated with the amount absorbed:

$$F = \frac{\text{Dose}_{iv}\,(X_u{}^{\infty})_{po}}{\text{Dose}_{po}\,(X_u{}^{\infty})_{iv}} \qquad \text{(Eq. 3.8)}$$

The assumption involved here is that a constant fraction of the available dose is eliminated through the kidneys whether the drug is given orally or intravenously. Another assumption is that $X_u{}^{\infty}$ represents the maximum amount that is eliminated. In order to assure this, the study must be conducted for a time long enough to allow complete elimination (generally, 5 to 7 disposition half-lives).

If the unchanged drug and all of its biotransformation products can be assayed, then the total amount of drug excreted may be used in bioavailability calculations. In some instances, a drug may be extensively biotransformed in the body, providing negligible concentrations of the unchanged drug in the urine. In such cases, the total amount of biotransformed product excreted in the urine may be substituted for the total amount excreted or the amount of unchanged drug excreted in the urine. However, a condition which must be met for this exercise is that the fraction of available drug which undergoes that specific biotransformation does not change. A common example is the use of salicylic acid and salicyl glucuronides, the biotransformation products of aspirin, to calculate aspirin's bioavailability. Some recent data, however, suggest dose-dependent biotransformation of aspirin at higher doses, making this approach less reliable.

ABSOLUTE BIOAVAILABILITY WITHOUT REFERENCE TO INTRAVENOUS DATA

For all of the absolute bioavailability assessments described above, a reference point is necessary to convert the areas under the curves into amounts. However, in some instances this reference point, which is generally intravenous administration, may not be needed. For example, if the total recovery of the drug from the various fluids analyzed is close to the dose administered, comparison with intravenous administration is only an exercise in futility.

An interesting approach has been proposed for cases when the elimination of drugs in the urine is disturbed. This involves increasing or decreasing urinary excretion by acidifying or alkalinizing the urine.[22] The change in renal clearance will provide a different AUC from which the bioavailability can be calculated as follows:

$$F = \frac{\Delta\,\text{Cl}_R\,(\text{AUC}_1)\,(\text{AUC}_2)}{\text{Dose}_{po}\,(\text{AUC}_1 - \text{AUC}_2)} \qquad \text{(Eq. 3.9)}$$

where ΔCl_R is the change in renal clearance, and AUC_1 and AUC_2 are the areas under the plasma concentration profiles for the two conditions. The term **RENAL CLEARANCE** will become more understandable after the discussion of pharmacokinetic parameters, but for the present purpose it is defined as the ratio between the urinary excretion rate and the plasma concentration and it has units of volume/time, e.g., ml/min or liters/hour.

The use of this method is limited, though, by the conditions which must be met before it can be applied. One of these conditions is that changes in

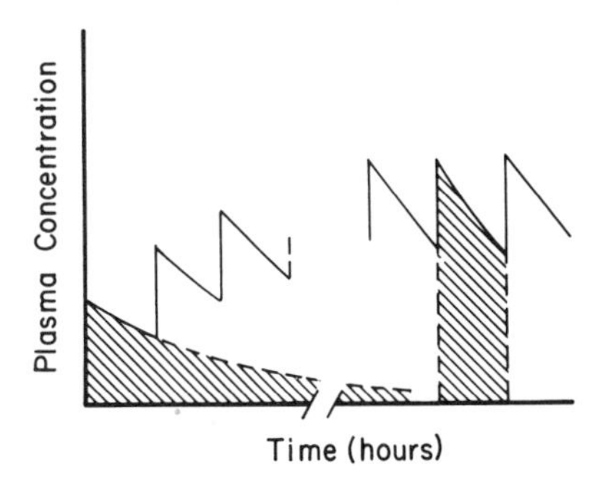

Figure 3.11. Bioavailability estimations using steady state levels of drugs.

renal clearance do not affect the distribution of the drug in the body or its elimination by other routes.

ABSOLUTE BIOAVAILABILITY FROM REPEATED DOSES

If a drug is administered repeatedly for a long period of time, the plasma concentrations reach a plateau, or a steady state (Fig. 3.11). This steady state is reached when the rate of elimination becomes equal to the rate of administration, e.g., 100 mg/6 hours. Thus at the steady state, during the dosing interval, the amount of drug eliminated from the body is equal to the available dose, and therefore the area under the curve during the dosing interval is equal to the area which will be obtained if a single dose is given and a plasma concentration profile is followed up to infinite time. The shaded areas in Figure 3.11 are therefore equal to each other.

The advantage of this method is that it eliminates the need to extrapolate the plasma concentration profiles to obtain the total AUC and also that it simulates the clinical application of the drug more closely than the single dose administration. However, it requires more time to complete the study as well as greater exposure of the study population to the effect of drug, which may not be desirable.

The steady state approach may be extended to include urinary excretion of the unchanged drug and/or biotransformation products, which may make the plasma concentration profiles unnecessary for the same reasons presented above for single dose studies.

COMPARATIVE BIOAVAILABILITY STUDIES

The main goal of comparative bioavailability studies is to estimate the relative efficiency of a dosage form in delivering the drugs to the body. A reference formulation may be chosen arbitrarily or it may be the formulation which is expected to provide highest bioavailability, such as a solution dosage form. A reliable and proven-effective formulation is generally used for comparison purposes, e.g., Lanoxin for comparison with various digoxin formulations.

If the subject population is carefully chosen and a sound experimental design is utilized, comparative bioavailability studies provide highly significant data in terms of product selection and cost-efficiency ratios.

Although it is desirable to conduct the studies for durations of time that

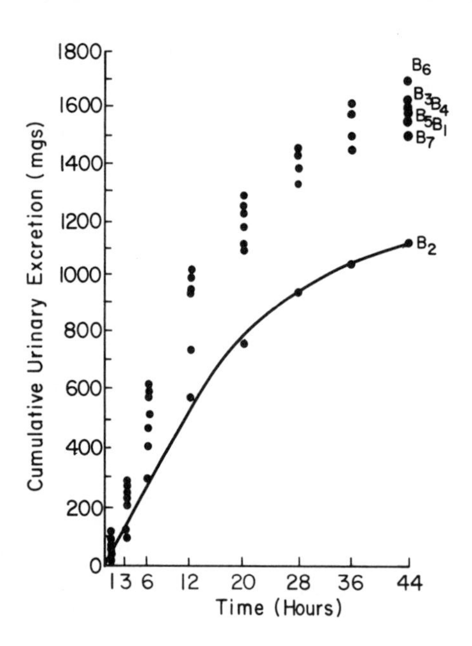

Figure 3.12. Cumulative urinary excretion of total sulfonamides from seven commercial suspensions. (From Niazi and Khan: Unpublished data, 1978)

allow complete elimination of the drug from the body, this is less crucial for comparative studies than it is for absolute bioavailability studies. For example, a recent study reported the bioavailability of trisulfapyrimidines from oral suspensions in humans.[23] The purpose of this study was to establish the relative efficiency of the bioavailabilities of three sulfonamides from suspension forms. Total urinary recovery, including biotransformation products, were estimated for up to 48 hours (Fig. 3.12). Although complete excretion was not studied, the necessary bioavailability comparisons were successfully made.

PHARMACOLOGIC EVALUATION OF BIOAVAILABILITY

The estimations of bioavailability discussed above are based on plasma and/or urine levels of the drug and/or its biotransformation products. It is understood in these calculations that these concentrations relate in some manner to the pharmacologic or clinical response of the drug. It is, therefore, desirable to measure bioavailability as a function of pharmacologic or clinical effect. In order to do so, a specific and discriminating test is needed. Some quantitative endpoint must be available which measures efficacy or quantitates the drug effect. For example, lowering of blood sugar by an antidiabetic agent, lowering of blood pressure by a hypotensive agent, weight loss produced by an anorexic agent, etc., would be appropriate measures. Less reliable measures, such as psychologic rating score and a physician's opinion of efficacy, cannot be of great value in these studies.

Before any comparative bioavailability testing is performed using pharmacologic or clinical response, a satisfactory dose-response curve should be obtained on one of the formulations to be included in the study (Figure 3.13). An assumption is made here that the amount absorbed is proportional to the dose administered.

The success of the application of this dose-response curve depends on two factors. First, the curve should be steep, indicating that significant changes in pharmacologic response occur with a small change in the dose, and secondly, the dose contained in the formulations should be such that the response lies between 20 percent and 80 percent to assure linear measurements. Responses falling beyond these ranges are more difficult to quantitate.

Few studies have reported the use of dose-response curves in bioavailability measurements, but the idea is certainly attractive and relevant to drug therapy.

Logic of Bioavailability

Erroneous conclusions can easily be made if the logic behind bioavailability studies is not clearly understood. For example, where concentrations are monitored in the biologic fluids, the specificity of the assay methods is of utmost importance. This is especially applicable to single dose studies in which small concentrations should be monitored in order to allow study of the complete elimination of the drug from the body.

It is generally assumed that the absorption rates of drugs are higher than the rates of elimination, but there can be exceptions, in which case the terminal plasma concentration profiles would represent both the absorption and elimination processes and the AUCs become more of a function of the elimination than the absorption.

The extrapolation of plasma or urinary concentrations always introduces

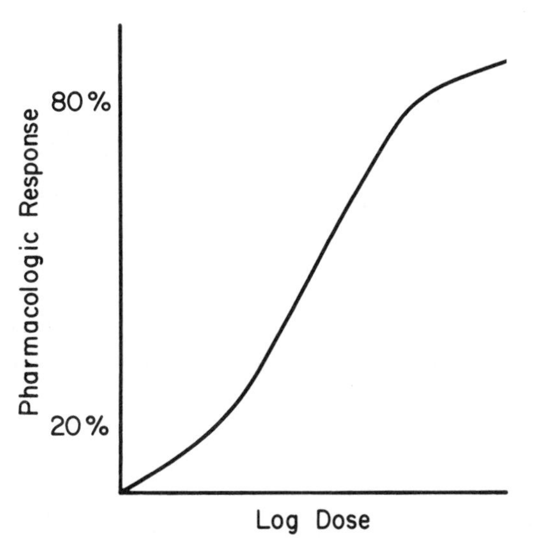

Figure 3.13. Typical log dose-response curve.

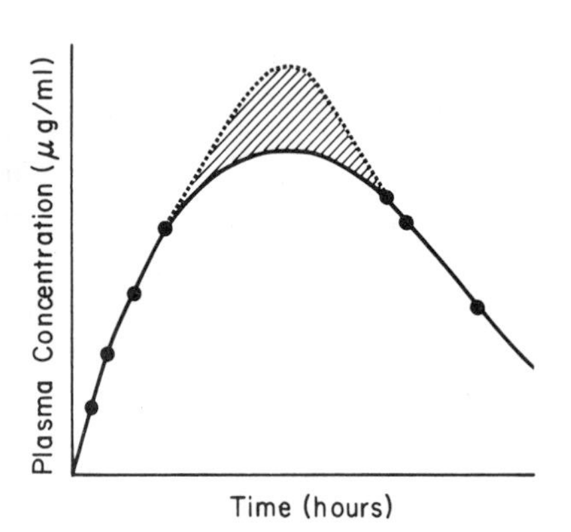

Figure 3.14. Error in AUC measurements due to insufficient data points.

some error in the calculations; it is desirable to extend the study to at least three elimination half-lives when plasma concentration is monitored, and for at least seven half-lives when monitoring urinary excretion.

One of the major sources of error in bioavailability studies is the lack of sufficient data points to characterize the plasma concentration profiles. For example, Figure 3.14 shows an instance where significant area can be lost if sufficient points are not collected during the peaking of the concentration. In general, there should be at least three data points before the peak occurs and at least four or five values after the peak, if possible.

A significant source of error which is most difficult to rectify is the variation among individuals of the elimination rates of a drug. The proportionality between AUC and bioavailability is based on the assumption that the elimination rates are invariant and thus any deviation from the norm will result in significant error (Fig. 3.15). Correction of this error can be made if the elimination rate constants are calculated for each subject and the AUC is corrected as follows:

$$AUC_{corrected} = AUC_{apparent}\,(K) \qquad \text{(Eq. 3.10)}$$

If a drug is eliminated fast, K will be large, accounting for possible underestimation of the $AUC_{apparent}$.

Since bioavailability estimations have economic and political motives, these data have been grossly misinterpreted in the past. One of the most common mistakes is to compare the data of different studies which may not be well matched in terms of the characteristics of the subject population, study conditions, or routes of administration. It is ironic that such cross-study comparisons are both very common and very misleading.

When identical concentrations are obtained in the plasma or serum following administration of equimolar doses from different formulations, these may be considered bioequivalent—the principle is referred to as the superimposition principle. In using this principle, one must choose a number of subjects in accordance with statistical criteria which will reduce the likelihood of both noted differences, which may not be clinically significant, and superimposition within the range of variation, which may be problematic. Thus, one must demonstrate an at least 20 percent difference in the means of values in order to make them clinically significant. This criterion can be applied to the concentration at each sampling time, to the peak concentration, and to the time of the peak concentrations and the AUCs.

It should be noted that just because a drug product meets compendial standards of purity and other criteria, its bioavailability is not assured. In fact, compendial requirements fall far short of assuring the efficiency of dosage forms in releasing drugs. The latest edition of USP and NF requires demonstration of sufficient dissolution for some drugs where evidence of dissolution affecting bioavailability has been suggested. A large number of drugs remain to be included in this list and it is hoped that eventually demonstration of bioavailability will become a compendial requirement. The costs of performing bioavailability studies make such requirements impractical for some drugs. However, without such requirements it is difficult to justify the rejection of a product on the grounds that its chemical equivalence varies by more than 10 percent, when its biologic equivalent is allowed to vary to any degree and both serve the same purpose.

It is true that awareness of the problems associated with the bioavailability of drugs is much greater today than it was ten years ago. The examples set by thiazide-potassium chloride products, which caused hundreds of stenosing ulcers of the small bowel, and the use of digoxin tablets without reference to bioavailability for several years make one wonder whether we really know enough about the logic of bioavailability.

Clinical Efficacy Trials

The ultimate evaluation of dosage forms lies in their clinical effectiveness. The FDA maintains a file on adverse drug reactions and unexpected bioa-

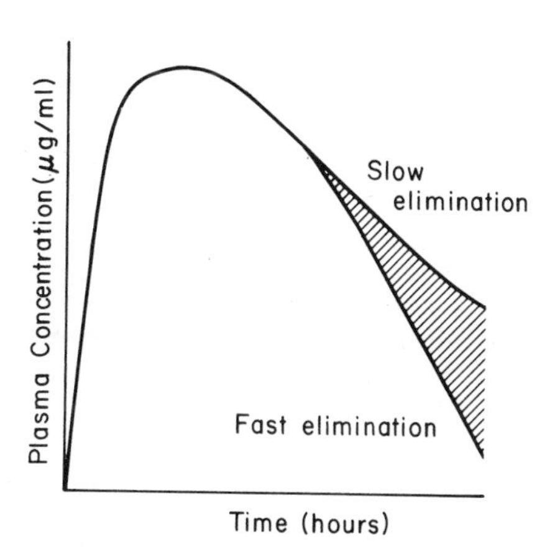

Figure 3.15. Error in AUC estimation as a result of variation in rates of drug elimination.

vailability observations. Regardless of the efficiency with which a drug product is designed and proven effective in a laboratory setting, it is only when clinical efficacy trials prove it to be meritorious that it becomes a standard product. Several commercial products enjoy this status.

References

1. Juncher H, Raaschou F: The solubility of oral preparations of penicillin V. Antibiot Med 4:497, 1957
2. Niepmann W: Experimentelle Untersuchungen zur intestinalen Calcium-Resortion in Abhangigkeit von der Magenaciditat beim Menschen. Klin Wochenschr 39:1064, 1961
3. Gibaldi M: Biopharmaceutics and Clinical Pharmacokinetics. Philadelphia, Lea and Febiger, 1977, p 44
4. Wagner JG, Gerard ES, Kaiser DG: Effect of dosage form on serum levels of indoxole. Clin Pharmacol Ther 7:610, 1966
5. Langlois Y, Gagnon MA, Te'treault L: A bioavailability study on three oral preparations of the combination trimethoprim-sulfamethoxazole. J Clin Pharmacol 12:196, 1972
6. Sakuma T, Daeschner CW, Yow EM: Studies on the absorption, distribution, excretion, and use of a new long lasting sulfonamide (sulfadimethoxine). Am J Med Sci 239:142, 1960
7. Oser BL, Melnick D, Hochbey M: Physiological availability of the vitamins. Ind Eng Chem 17:405, 1945
8. Bogen WP, Gavier JJ: Evaluation of tetracycline preparations. N Engl J Med 261:829, 1959
9. Chow BF, Hsu J, Okuda K, Grasbeck R, Horonick A: Factors affecting the absorption of vitamin B_{12}. Am J Clin Nutr 6:386, 1958
10. Johnson BF, Bye C, Jones G, Sabey GA: A completely absorbed oral preparation of digoxin. Clin Pharmacol Exp Ther 19:746, 1976
11. Mellinger JJ, Bohorfoush JG: Blood levels of dipyridamole (Persantine) in humans. Arch Int Pharmacodyn Ther 163:471, 1966
12. Mellinger JJ: Serum concentration of thioridazine after different oral medication forms. Am J Psychol 121:1119, 1965
13. Shaw TRD, Howard MR, Hamer J: Variations in the biological availability of digoxin. Lancet 2:303, 1972
14. Barrett WE, Bianchine JR: The bioavailability of ultramicrosize griseofulvin (Gris-PEG) tablets in man. Curr Ther Res 18:501, 1975
15. Ritschel WA: Peroral solid dosage froms with prolonged action. In Ariens J (ed): Drug Design, Vol 5. New York, Academic, 1973, p 38
16. O'Neill WP, Wolman AJ: Markets for therapeutic systems. Drug Cosmet Ind 120:28, 1977
17. Page DP, Carson NA, Buhr CA, et al: Stability study of nitroglycerin sublingual tablets. J Pharm Sci 64:140, 1975
18. Chiou WL, Moorhatch P: Interaction between vitamin A and plastic intravenous bags. JAMA 223:328, 1973
19. Flaum I: Contamination of pharmaceutical products. J Pharm Sci 67:1, 1978
20. Grin EI, Denic M: Investigations of human blood levels and their curative effect on tinea capitis. Acta Med Iugosl 19:53, 1965
21. U.S. Government: Bioequivalency requirements. Fed Reg 45(5):1634, 1977
22. Lalka D, Feldman H: Absolute drug bioavailability. Approximation without comparison to parenteral dose for compounds exhibiting perturbable renal clearance. J Pharm Sci 63:1812, 1974
23. Niazi S, Khan G: In vitro-in vivo evaluation of commercial trisulfapyrimidine suspensions. Unpublished data. 1978

Questions

1. What is an ideal dosage form?
2. How does probenecid affect the plasma concentration profile of penicillin?
3. Explain why warfarin and phenytoin may be better absorbed from a suspension dosage form than a solution.
4. Under what conditions might an elixir show retarded absorption as compared with the aqueous solution?
5. Cite three examples of the use of suspensions for prolonged release of drugs.
6. Why are a large number of antiinfective agents administered in suspension form?
7. Given a decomposition rate constant of 0.2 hr^{-1}, show that the shelf life of a solution containing 500 mg of this drug in 125 ml of solution is 0.53 hours.
8. Why does the addition of Fuller's earth in thiamine and riboflavin formulations decrease their bioavailability?
9. What are the differences between the release profiles of drugs in soft and hard gelatin capsules?
10. Sometimes there is no correlation between the rate of disintegration and rate of dissolution of tablets. Why?
11. Why does an enteric coating provide a highly erratic absorption?
12. Why is it necessary to prolong the absorption of drugs?
13. List the methods used to prolong the absorption of orally and parenterally administered drugs.
14. What is a Therapeutic System? What are its components?
15. How does an osmotic drug delivery system work?
16. What are the various causes of the loss of active ingredient from a dosage form?
17. Why must nitroglycerin tablets be dispensed in the original glass containers?
18. What are the disadvantages associated with the use of plastic large-volume parenteral containers?
19. Cite five examples where content uniformity is a major concern in the production of a dosage form.
20. What are the sources of inadvertent product contamination?
21. Under what conditions can a dissolution test be used to correlate with the bioavailability?
22. What is the source of the lower surface tension of gastric fluids?
23. Under what conditions will a hydrophilic lubricant increase the dissolution rates of granules?
24. Suggest possible reasons for the relationship between compression force and dissolution rates (Fig. 3.3).
25. Why are "dissolution standards" needed?

26. What is a sink condition and how does it affect the bioavailability of drugs? Explain the observations presented in Fig. 3.6.
27. Define the terms "drug product," "pharmaceutical equivalents," "pharmaceutical alternatives," and "bioequivalent drug products."
28. What are the physicochemical criteria used by the FDA to establish bioequivalency requirements?
29. Why is there a specific need for bioequivalency requirements for drugs which decompose or are biotransformed in the gastrointestinal tract or during absorption?
30. For which dosage forms can bioequivalency requirements be waived?
31. Distinguish between absolute and comparative bioavailability studies.
32. What are the factors determining the time and intensity of peak plasma concentration following oral administration of drugs?
33. How would you convert total AUC to the total amount of drug absorbed?
34. Compare the relative bioavailability of the following dosage forms containing identical amounts of an antipyretic agent from plasma concentration profiles:

Hours	*Product A*	*Product B*
0	0	0
1	1 μg/ml	0
2	3 μg/ml	1 μg/ml
3	6 μg/ml	2 μg/ml
5	6 μg/ml	3 μg/ml
8	4 μg/ml	2 μg/ml
12	2 μg/ml	1 μg/ml
24	0.1 μg/ml	0.1 μg/ml

35. A suppository of 300 mg of salicylic acid was administered to a subject and the following plasma concentration data were obtained:

Hours	*Plasma concentration*
1	5 μg/ml
2	10 μg/ml
3	15 μg/ml
5	15 μg/ml
8	9 μg/ml
24	0

Calculate the AUC and estimate the absolute bioavailability when the AUC following intravenous administration of an identical dose is 230 μg.hr/ml.

36. Plot the data in exercise 34 on a piece of paper and calculate the relative bioavailability by cutting and weighing the paper.

37. What are the assumptions involved in using cumulative urinary excretion as a measure of bioavailability? How long must a urinary excretion study be run to obtain bioavailability data?
38. Under what conditions should biotransformation products not be used to estimate bioavailability?
39. When during multiple dosing does the AUC during the dosing interval become equal to the total AUC following administration of a single dose?
40. What are the advantages and disadvantages of using multiple dose study to ascertain bioavailability?
41. What is the rationale behind using pharmacologic response to estimate bioavailability? Is this approach applicable to all drugs? Why?
42. What are the major sources of error in bioavailability estimations?
43. What is the superimposition principle?
44. Do official compendia assure a desirable therapeutic response from a dosage form? Would it ever be possible to guarantee that every dose administered would provide consistent response?

CHAPTER 4

Delivery of Drugs: Routes of Administration

The intensity, onset, and often the nature of pharmacologic response is determined by the route of administration. The dose of a drug is often dependent on the route of administration (Table 4.1) as well as on the more specific site of administration. For example, administration of a 20 μg dose of arecoline via the cephalic vein in dogs produces a prompt and intensive hypotensive effect; a dose of 2 mg is required for the same effect when arecoline is administered via the splenic vein.[1] In rats, a 50 μg/kg dose of nitroglycerin administered via the jugular vein produces a hypotensive effect which cannot be obtained even with a dose of 5000 μg/kg when nitroglycerin is administered via the portal vein.[2] Similarly, the rates of drug absorption vary significantly upon intramuscular administration, depending on the site; decreasing in the order: deltoid, vastus lateralis, gluteal.[3]

Drug action variation due to the route of administration is a result of, among other factors, the passage of drug molecules through a variety of organs before reaching the general circulation (See Fig. 2.1). The drugs which are absorbed from the intestine pass through the liver where a significant biotransformation can take place, often decreasing the apparent bioavailability. Similarly, all drug molecules entering the general circulation pass through the lungs, where they can be eliminated from the body if they are sufficiently volatile. These processes are referred to as the **FIRST-PASS EFFECTS**, representing the first passage of drug molecules through an organ system.

The routes of drug administration can be broadly classified as enteral or nonenteral depending on whether or not the drug is administered through the gastrointestinal tract.

GASTROINTESTINAL ADMINISTRATION

When a drug is introduced into the gastrointestinal tract and is present in a form which can be absorbed, the process of absorption may be categorized as either **PASSIVE DIFFUSION** or **ACTIVE TRANSPORT**.

Table 4.1. EXAMPLES OF ROUTE-DEPENDENT DOSING

DRUG	DOSE ($ROUTE_1$)	DOSE ($ROUTE_2$)
Morphine	75 mg (PO)	10 mg (IM)
Isoproterenol	3–20 mg (PO/SL)	0.02–0.15 mg (IV)
Bethanechol	5–30 mg (PO)	2.5–5 mg (SC)
Propranolol	80 mg (PO)	1–3 mg (IV)

Passive Diffusion

This process describes the movement of drug molecules from a region of high relative concentration to a region of lower relative concentration. It also includes the movement of ions from a region of high ionic charge of one type to a region of lower charge of the same type or of opposite charge:

$$\frac{\Delta X_a}{\Delta t} = -DA\,(C_{gut} - C) \qquad \text{(Eq. 4.1)}$$

where X_a = amount of drug at the absorption site
D = diffusion coefficient
A = area of absorption surface
C_{gut} = concentration of drug in the gastrointestinal tract
C = concentration of drug in the plasma

The driving force for passive diffusion is the concentration or the electrical gradient across the membrane which separates the gastrointestinal lumen from the circulating blood. The concentration gradient is, however, more appropriately viewed as the chemical potential, represented by the number of molecules or ions which are free to move across a membrane and not by the total concentration in the lumen or plasma. In many instances, the plasma concentration is much lower than the concentration in the gastrointestinal tract due to the rapid removal of the absorbed drug by the circulating blood, making the rate of transport across the membrane proportional to the chemical potential only in the gastrointestinal tract.

Apart from the concentration gradient, diffusion rates depend also on the permeability characteristics of the membrane. The gastrointestinal membrane acts like a lipid barrier (Fig. 2.4) which permits the passage of lipid-soluble drugs, but across which lipid-insoluble molecules diffuse only with difficulty or not at all. Smaller, lipid-insoluble but water-soluble molecules may pass across the membrane through numerous pores which are too small to be seen even with the aid of an electron microscope, but for which strong evidence exists.

Active Transport

ACTIVE TRANSPORT is a specialized process which requires the expenditure of energy. The various active transport processes found in the gastrointestinal tract are relatively structure-specific and serve primarily in the absorption of natural substances, such as monosaccharides, l-amino acids, pyrimidines, bile salts, and certain vitamins. However, there is evidence that certain drugs may also be absorbed by one of these active processes, if their chemical structures are sufficiently similar to that of the natural substrate. The anticancer drug 5-fluorouracil is an example of an actively transported drug. It is similar in structure to the natural substance, uracil, which is absorbed by means of the pyrimidine transport system.

Active transport is specific not only in terms of chemical structure but also with respect to direction (Fig. 4.1), transporting molecules mainly from the mucosal side to the serosal side of the gastrointestinal tract. The transport can also take place against the concentration gradient, i.e., from a region of lower concentration or activity to the region of higher activity. Since active transport involves enzymes, these can be saturated at higher concentrations of the drug, and that particular drug can compete with, and thus depress, the transport of another drug if both are transported by the same process. Since active transport processes consume energy, they can be inhibited by various metabolic poisons, such as fluoride or dinitrophenol, as well as by lack of oxygen.

Quite often an active transport of drug molecules occurs concomittantly with passive diffusion. Faster absorption rates can generally be expected at lower concentration in the gastrointestinal tract if active transport is taking place. At higher concentrations, passive diffusion becomes more important due to possible saturation of the active transport process (Fig. 4.2).

There are some variants of the two major types of transport processes described thus far. Water flux, in the same direction as drug movement, can

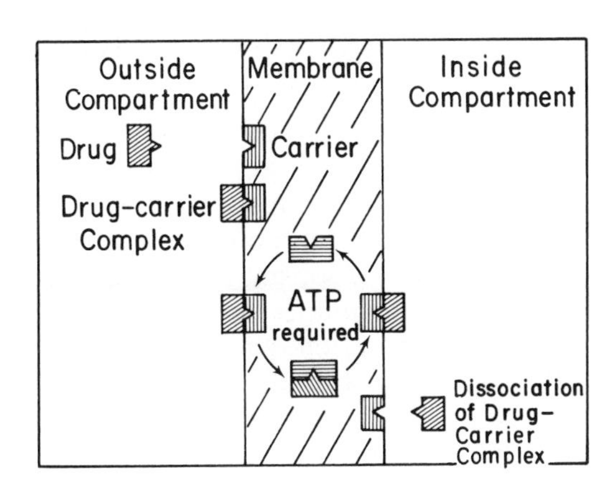

Figure 4.1. Mechanism of carrier-mediated transport of drugs. (From Ritschell: Handbook of Basic Pharmacokinetics, 1976. Courtesy of Drug Intelligence Publications)

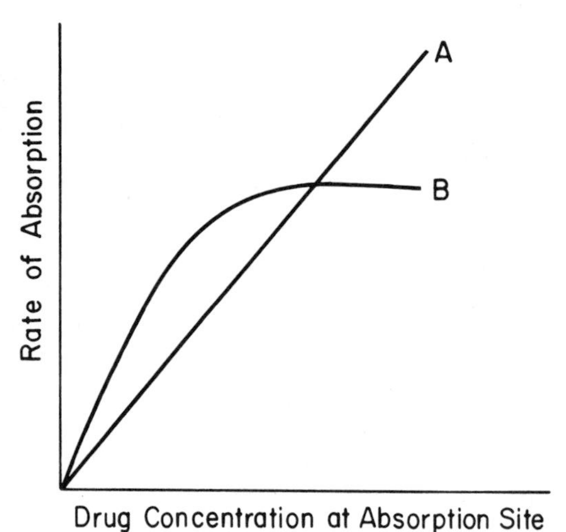

Figure 4.2. Relationship between drug concentration and absorption rate for a passive process (Curve A) and for a carrier-mediated process (Curve B). (From Gibaldi: Biopharmaceutics and Clinical Pharmacokinetics, 1977. Courtesy of Lea and Febiger)

increase the diffusion rate of a substance across the gastrointestinal membrane. This is known as **SOLVENT DRAG.**

Some substances are transported by a process which does not take place against a concentration gradient, but which involves a carrier which is subject to competition by other substances of similar structure and is affected by the metabolic inhibitors. This absorption process appears to be an active one and is referred to as **FACILITATED TRANSPORT**. The classical example of facilitated transport is the absorption of vitamin B_{12}. The vitamin B_{12} forms a complex with the intrinsic factor produced by the stomach wall and is transported in the form of this complex (Fig. 4.3).

The absorption of highly ionized compounds at gastrointestinal pH cannot be explained by passive diffusion or other mechanisms. A hypothesis has been suggested whereby highly ionized compounds (such as quarternary structures and sulfonic acids) form neutral complexes with other ions in the gastrointestinal tract (such as mucin) and these ion-pair complexes are then absorbed by passive diffusion, since the complex has both the required lipid and aqueous solubility (Fig. 4.3). This mechanism is referred to as **ION-PAIR TRANSPORT.**

Another mechanism of absorption is that of **PINOCYTOSIS**, a process of physical absorption whereby an invagination of the cell membrane engulfs the particulate or droplet material. It is the only transport mechanism whereby a drug does not have to be in aqueous solution in order to be absorbed. Only a few compounds are absorbed by this mechanism, including vitamins A, D, E, and K. Pinocytosis is of significant importance in the uptake of nutrients.

The gastrointestinal tract is composed of heterogeneous anatomic regions. As drug molecules descend through the gastrointestinal tract, they encounter different environments which vary in pH, enzymes, and fluidity of contents, as well as in the area available for absorption.

As discussed earlier, lipid-soluble forms of drug molecules, i.e., un-ionized forms, are absorbed from the gastrointestinal tract. However, attainment of a rapid equilibrium between ionized and un-ionized forms allows absorption of drugs with a wide range of ionization constants (Table 2.6). The ionization of drugs also determines the total solubility of drugs in the gastrointestinal fluids (Eq. 2.2), making it an important factor in the overall release and absorption of drugs from dosage forms.

While the differences between the pH of the gastric and the intestinal fluids can account to some extent for the different rates of absorption of certain drugs from these two areas, the main reason is the difference in the absorption surface areas. Anatomically, the small intestine is much better designed for absorption than the stomach. The intestinal mucosa is covered by numerous villi and microvilli, providing a large surface area of approximately 120 m^2 (the intestine without the villi and microvilli would have a surface area of only 4 m^2). The large intestine has no villi and little drug absorption takes place from this region.

As the drug passes through the small intestine, the consistency of gastrointestinal contents changes from fluid to paste due to the absorption of water. Thus the drug particles which have not been dissolved in the stomach or upper small intestine will encounter difficulty in their dissolution. Even if the drug is dissolved, it may not be absorbed quickly from the lower part of the intestine due to the retarded diffusion of molecules through pasty contents. Thus in addition to the differences in absorption rates in different regions of the gastrointestinal tract, due to pH differences and absorption surface areas, the consistency of the contents is also an important factor. In general, therefore, the upper part of the small intestine is the most important area for the absorption of drugs, whether acids or bases.

Except for the colon, all other regions of the gastrointestinal tract have

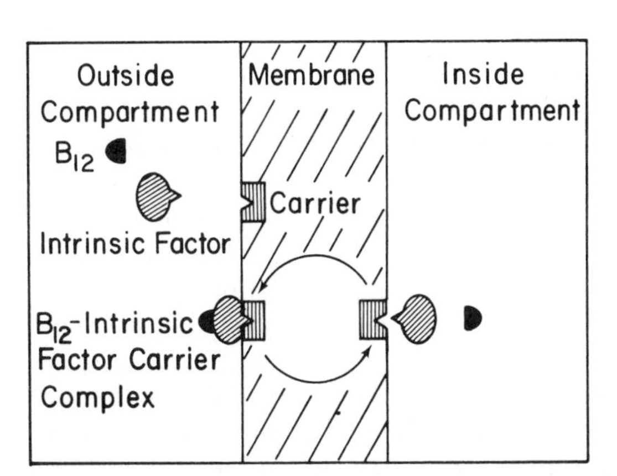

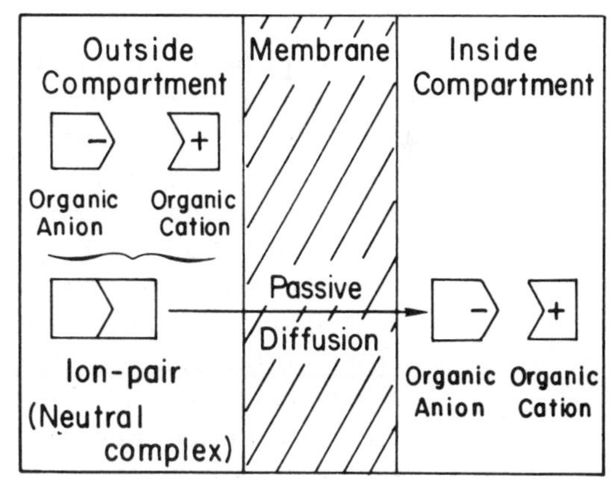

Figure 4.3. Mechanisms of facilitated and ion-pair transport of drugs. (From Ritschell: Handbook of Basic Pharmacokinetics, 1976. Courtesy of Drug Intelligence Publications)

areas for the specific transport of compounds. For example, iron absorption occurs mainly in the proximal part of the small intestine and decreases progressively in the more distal intestinal segments; the absorption of bile salts is limited to the distal ileal segments; riboflavin is absorbed only from the upper region of the intestine; thiamine absorption occurs mainly in the proximal regions; and vitamin B_{12} is absorbed from the ileum. Therefore, if a drug is absorbed primarily through a specific gastrointestinal area, it should not be administered by the rectal route.

For most drugs in general, and especially for those which are absorbed from a specific part of the gastrointestinal tract, the extent and rates of absorption are dependent on the rate of passage of contents through the gastrointestinal tract. Depending on the rate of passage, there may be only a limited time available for the dissolution of a solid particle and for the modification of its molecules into absorbable forms. This is especially critical if the optimum absorption site is the proximal section of the small intestine. The rate of passage of intestinal contents through the upper small intestine is higher than it is through the lower part. Thus, if a drug is not present in an absorbable form within indicated time limits, it may be propelled past its absorption site and excreted totally or in part in the feces.

The gastric emptying rate affects the absorption rate primarily through the pH differences between the stomach and the intestine. For example, weakly basic drugs such as amphetamine and codeine will be absorbed primarily from the small intestine rather than from the stomach, and any delay in gastric emptying will tend to delay the absorption and thus the therapeutic response. Slow gastric emptying can also affect the bioavailability of drugs that are unstable in gastric fluids, e.g., l-dopa, since the extent of degradation is proportional to the time for which the drug is exposed to the low pH and the enzymes of the stomach.

Some of the factors which affect the gastric emptying rate are as follows.

TYPE OF FOOD

The type of food will affect the stomach emptying rate significantly. For example, fats decrease the rate; proteins effect a lesser decrease; and carbohydrates retard gastric emptying the least. A fatty meal can therefore retard absorption rates of drugs and delay the onset of action. However, with such water-insoluble drugs as griseofulvin the absorption can be increased as a result of retarded gastric emptying. The reason is that griseofulvin passes slowly to the small intestine, and therefore the longer duration of contact of griseofulvin with the intestine results in a greater chance for it to be dissolved and absorbed through a specific region.

The effects of food are generally less on the total bioavailability than on the rate of absorption. For example, acetaminophen taken on an empty stomach shows an earlier peak plasma concentration than when taken with food. The total bioavailability, however, does not change.

A faster gastric emptying rate is desirable for drugs which are decomposed in the stomach. These should be taken either on an empty stomach or an hour before or two hours after meals (Table 4.2).

Table 4.2. EXAMPLES OF RELATIONSHIPS BETWEEN FOOD INTAKE AND DRUG REGIMENS

DRUGS TO BE TAKEN ON AN EMPTY STOMACH	*DRUGS TO BE TAKEN 1 HOUR BEFORE OR 2 HOURS AFTER MEALS*	*DRUGS TO BE TAKEN 1 HOUR BEFORE MEALS*
Piperazine citrate	Tetracycline	Anticholinergic agents
Bephenium hydroxynapthoate	Ampicillin	Methantheline bromide
Castor Oil	Cefazolin sodium	Mepenzolate bromide
Pentaerythritol tetranitrate	Sulfisoxazole	Pancrelipase
Lincomycin	Trimethoprim	Sitosterols
Isosorbide dinitrate	Demeclocycline hydrochloride	Chlordiazepoxide hydrochloride
	Dicloxacillin sodium	Fenfluramine hydrochloride
	Erythromycin	Phenmetrazine hydrochloride
	Penicillin	Propantheline bromide
	Cholestyramine	Glycopyrrolate
	Rifampin	Mazindol
	Methacycline	Diethylpropion hydrochloride
	Troleandomycin	Hexocyclium methylsulfate
	Nafcillin	Anisotropine methylbromide
	Oxytetracycline	
	Hetacillin	

VOLUME OF FLUID OR FOOD

The volume of fluids or food has a definite influence on the gastric emptying rate. The rate with which gastric contents leave the stomach is proportional to their volume. With small volumes, there is an initial lag time before gastric emptying begins, while with higher volumes, there is a initial phase of more rapid emptying. The fluid intake also affects the dissolution rate and forms the integral part of certain drug actions, e.g., the use of bulk laxatives.

OSMOTIC PRESSURE

The emptying rates are also dependent on the osmotic pressure of the liquids. For example, water leaves the stomach with a half life of about five minutes (a glass of water leaves the stomach in about 5 to 20 minutes). Hypertonic or hypotonic solutions generally leave the stomach at a slower rate than isotonic solutions (e.g., 0.9 percent sodium chloride).

ACIDITY

Gastric emptying is also retarded by increased acidity of gastric fluids. The use of antacid compounds therefore increases gastric emptying rates. An interesting application of this property is the administration of l-dopa with sodium bicarbonate. Since l-dopa decomposes in the stomach, its administration with sodium bicarbonate increases its bioavailability due to decreased decomposition in the stomach—the result of both decreased acidity and increased emptying of the contents into the intestine.

DIFFERENCES IN FOOD TEMPERATURES

Hot or cold foods or fluids prolong the gastric emptying. For example, water taken at 25 C leaves the stomach at one third the rate of water taken at 37 C.

Liquids of low VISCOSITY are emptied faster than liquids of higher viscosity. Solutions or suspensions of fine particles leave the stomach at a higher rate than lumpy substances.

PSYCHOLOGIC STATE

The psychologic state of an individual also affects gastric emptying. Depression, injury, and trauma lead to prolonged emptying. Agitation and excitement increase the peristaltic movement, thus increasing the rate of gastric emptying.

BODY POSTURE

Body posture can also significantly affect gastric emptying. Lying on the right side and standing may facilitate emptying, whereas the supine position may retard emptying.

DRUGS

A number of drugs are capable of affecting the gastric emptying rate, usually through some central mechanism, e.g., such anticholinergic drugs as atropine, antihistamines, tranquilizers, aspirin, and morphine derivatives. In some instances, change in emptying rates is brought about by a local effect, as is described for antacids. On the other hand, cholinergic drugs enhance gastric emptying.

In addition to the factors described above, the bioavailability of orally administered drugs can be affected by biotransformation brought about by the enzymes in the gastrointestinal tract, the microflora, the mucosa, and the various organs (e.g., the liver) through which the drug molecules pass before reaching the general circulation. For example, the chromotrophic activity of isoproterenol is about 1000 times greater when administered intravenously than through oral administration, largely due to the biotransformation of isoproterenol into an inactive sulfate during the transfer across the gut wall and passage through the liver.[4] Similarly, some of the steroids are also extensively biotransformed during absorption. Since biotransformation reactions require the presence of enzymes, the saturation of these enzymes at higher drug concentrations results in an apparent dose-dependent bioavailability. Such dose-dependent effects have been noted for l-dopa,[5] which undergoes biotransformation through both enzymatic and nonenzymatic processes in the gut; para-amino-benzoic acid and isoniazid, which are partly acetylated, sulfated, and glucuronidated; and para-amino hippuric acid,[6] which under-

Table 4.3. DRUGS WHICH CAN UNDERGO BIOTRANSFORMATION IN THE LUMEN OR DURING ABSORPTION IN THE MUCOSA

Acetylsalicylic acid	Meperidine
Aldosterone	Methadone
Aminobenzoic acid	α-Methyl dopa
Aminohippuric acid	Nitrates, organic
Chlorpromazine	Pentazocine
Cortisone	Progesterone
Dexamethasone	Propoxyphene
l-Dopa	Salicylamide
Estrogens	Stilbesterol
Hippuric acid	Sulfonamides
Hydrocortisone	Testosterone
Isoproterenol	Terbutaline

goes amide hydrolysis in the gut. Table 4.3 lists the drugs for which significant luminal or gut wall biotransformation is suspected.

The intestinal microflora also play an important role, causing biotransformation of such drugs as methotrexate, succinylsulfathiazole, and certain coumarin derivatives.[7]

A distinction can often be made between the biotransformation in the intestine and that in the liver during the first pass by administering the drugs either intraperitoneally or directly into the portal vein. Some mathematical approaches have also been used and will be discussed later. Table 4.4 lists the drugs which are suspected of first pass hepatic biotransformation.

Table 4.4. DRUGS FOR WHICH FIRST PASS HEPATIC BIOTRANSFORMATION IS SUSPECTED, POSSIBLY IN ADDITION TO GASTROINTESTINAL BIOTRANSFORMATION

Alprenolol	Pheniprazine
Desmethylimipramine	Propranolol
Dopamine	Reserpine
Lidocaine	Serotonin
Nortriptyline	Tryptophan
Oxyphenbutazone	

SUBLINGUAL/BUCCAL ADMINISTRATION

Some dosage forms are administered by placing them beneath the tongue or in the cheek pouch. A rapid absorption of drugs is thereby generally expected due to the high vascularity of this region. The pH of saliva is about 6, and drugs are absorbed by passive diffusion with a slightly higher requirement for lipid solubility than is needed for intestinal absorption.

A significant advantage of this route is that gastrointestinal degradation and biotransformation are bypassed along with hepatic first pass biotransformation. A variety of drugs can be administered by this route, including nitrates, such hormones as methyltestosterone, testosterone propionate, and oxytocin. Few studies have reported on the effective use of this route of administration. One such study reports significantly higher blood levels of methyltestosterone from sublingual tablets than are obtained from other routes.[8] Absorption properties of sympathomimetic amines, methadone, meperidine, lidocaine, chlorpheniramine, imipramine, desimipramine, and barbiturates have also been studied.[9–11]

RECTAL ADMINISTRATION

Some drugs are administered rectally either in suppository or solution form, e.g., retention enema. The solutions yield better absorption provided that they are retained for a sufficient length of time in the rectum. The suppositories are the most commonly used dosage forms for both local and systemic effect. Examples of drugs administered rectally for systemic action include aspirin, acetaminophen, indomethacin, theophylline, perchlorperazine, cyclizine, promethazine, and barbiturates.

The absorption mechanism mainly involves passive diffusion with no sites for active transport. The absorption rate and bioavailability are more erratic than is observed with oral administration, due to such added factors as the presence of feces retarding absorption or irritant suppository bases such as Carbowaxes causing early evacuation. The use of an enema before drug administration generally increases the absorption significantly.

The rectal route of administration is not suitable for irritant drugs such as tetracycline or penicillin. A large number of studies have attempted to develop an "ideal" base for suppositories or formulations for a microenema, but little has been reported regarding their comparative bioavailability in humans. Thus a conclusion cannot be drawn regarding the relative merits of this route of administration compared with other routes.

INTRAVENOUS ADMINISTRATION

The direct administration of drugs into veins is the only route with which bioavailability considerations are not important. This route provides an almost instantaneous response with controllability of the rate of drug input into the body. This route is especially suitable for those drugs which cannot be absorbed adequately from the gastrointestinal tract or tissue depots (e.g., intramuscular administration) or where there is a significant first pass effect upon oral administration. The drugs which would be intolerably painful in the subcutaneous or muscle tissue by virtue of their irritant properties may be injected slowly into a vein without much difficulty, e.g., nitrogen mustard in cancer chemotherapy.

There are however, several disadvantages with the use of the intravenous route. A drug administered intravenously cannot be recalled, whereas some such measures can be taken with other routes. Rapid intravenous injections may evoke catastrophic effects in the circulatory and respiratory systems due to the transient wave of concentrated solute suddenly reaching the myocardium and the chemoreceptors in the aortic arch and carotid sinus. Intravenous injections should, therefore, be administered slowly, preferably over a period of one minute or more, during which time the blood completes its circulation. The possibility of anaphylactoid reactions is much greater than with any other route of administration.

Drugs can only be administered intravenously if they are present in a solution form or in an aqueous-miscible system. However, upon dilution with blood the drugs may precipitate, resulting in serious embolism which can be life-threatening. Drugs like sodium phenytoin have been shown to precipitate following intravenous administration—due care should be exercised with such drugs.

The tonicity of solutions is also important since hypotonic or hypertonic solutions can cause hemolysis or agglutination of erythrocytes. The damage of the vascular wall also leads to local venous thrombosis, especially after prolonged infusions.

The possibility of microbiologic contamination and pyrexia due to pyrogens is a serious concern in the use of intravenous administration.

The intravenous route is especially suitable when a rapid response is required, as in the treatment of epileptic seizures, acute asthmatic attacks, cardiac arrhythmias, etc. The fluctuation of plasma concentration is generally very small if a drug is administered by slow intravenous infusion, as is employed for lidocaine, theophylline, and many antibiotics.

INTRAARTERIAL ADMINISTRATION

This route is used only rarely—principally for the injection of substances used in diagnosis. A typical example is the injection of a radiopaque compound into the carotid artery to trace the circulation of the brain by roentgenography. In addition, certain specialized techniques in cancer chemotherapy call for regional infusion of drugs by arterial routes, which may provide a significant advantage over other routes.[12]

INTRAMUSCULAR ADMINISTRATION

More than 50 percent of hospitalized patients receive intramuscular drug administration.[3] The popularity of this route is due to the decreased hazard of administration when compared with the intravenous route. Large volumes of solution can be injected (2 to 10 ml) by this route, generally with less pain and irritation than is encountered with the subcutaneous route.

Aqueous solutions of drugs are usually absorbed from intramuscular administration sites within 10 to 30 minutes, but faster or slower absorption rates are possible depending on the vascularity of the site (blood flow rates range from 0.02 to 0.07 ml/min), the ionization and lipid solubility of the drug, the volume of injection, the osmolality of the solution, and other variables, including coadministered drugs and adjuvants in the formulation. The small molecules are absorbed directly into the capillaries from the intramuscular administration sites, whereas large molecules gain access to the circulation by way of the lymphatic channels.

Drugs which are poorly water-soluble, such as digoxin and diazepam, or those drugs which dissolve at pH values far above the physiologic range are often administered in nonaqueous media such as propylene glycol or in strongly acid or alkaline aqueous solutions. However, after intramuscular administration these may not stay in solution once the solutions either diffuse or are buffered to the physiologic pH, resulting in slow or incomplete absorption.[13] In some instances the total bioavailability may be less than that from oral administration, as is demonstrated with phenytoin,[14] diazepam,[15] and a new antibiotic, cefamandol.

Highly lipid-soluble molecules are quickly absorbed from intramuscular administration sites, whereas lipid-insoluble molecules diffuse between interstitial fluid and plasma only through the pores in the capillary membrane,—this is usually not, however, the rate-limiting step in the absorption. Only very large lipid-insoluble molecules which must be absorbed through the lymphatic system have a rate limitation in their absorption, due to the slow rate of lymph flow (0.1 percent of the plasma flow).

The concentration of the injected solution can also affect the rate of absorption. For example, atropine is absorbed more rapidly when administered in a smaller volume of more concentrated solution.[16] The absorption rates can be accelerated by spreading the solution over large tissue areas, e.g., by massaging or using high pressure injection devices.

The blood flow to the administration site is often the rate-limiting step in the absorption of drugs. Absorption is more rapid after injection into the deltoid than into the vastus lateralis, and is slowest after gluteal muscle injection. The drugs can be absorbed faster after administration into a buttock in males than in females due to greater adipose tissue in females.[17,18] Absorption rates increase during exercise regardless of the site of intramuscular administration, since this results in increased blood flow to skeletal muscles. Conversely, absorption rates decrease in circulatory shocks, hypotension,

CHF, myxedema, and other disturbances of the circulatory system.

Absorption rates can often be quite erratic upon intramuscular administration of drugs. This is due to increased membrane contact as the solution spreads, change in drug concentration as a result of absorption, a possible hypertonic effect drawing water to the site, or to the precipitation of the drugs. The precipitation can lead to incomplete absorption due to extremely slow redissolution or to phagocytosis of the drug particles. Examples of these incompletely absorbed drugs are ampicillin, cephaloridin, cephradine, chlordiazepoxide, diazepam, dicloxacillin, digoxin, phenylbutazone, phenytoin, and quinidine. Conversely, the slow absorption of drugs can itself be exploited to produce prolonged administration. The slow absorption can be accomplished by the use of injection vehicles of high viscosity, such as glycerin, cottonseed oil, sesame oil, or polyethylene glycols. Another technique involves preparation of fatty acid ester derivatives, such as decanoate derivative of fluphenazine,[19] which hydrolyzes slowly and provides gradual release. Benzathine penicillin and procaine penicillin are injected as water-insoluble suspensions for the same purpose. Slowly released preparations of antipsychotic agents have been useful in the maintenance therapy of schizophrenia.

The side effects of intramuscular administration include pain, elevation of serum creatine phosphokinase as a result of trauma, and often sciatic nerve damage following gluteal injections. Other complications include skin pigmentation, hemorrhage, septic or sterile abscesses, cellulitis, muscular fibrosis, tissue necrosis, and gangrene.

SUBCUTANEOUS ADMINISTRATION

The factors affecting intramuscular drug absorption also determine subcutaneous drug availability. The blood flow rates are poorer than in muscles and so are the rates of absorption. Yet some drugs are absorbed as rapidly from a subcutaneous site as from intramuscular administration, e.g., anionic dye, phenosulfonphthalein, and insulin.[20]

A prime determinant of the absorption rate of a subcutaneous depot is the total surface area over which the absorption can occur. Although the subcutaneous tissues are somewhat loose, and moderate amounts of fluids can be administered, the normal connective tissue prevents indefinite lateral spread of the injected solutions. These barriers can be bypassed with the aid of hyaluronidase, an enzyme that breaks down mucopolysaccharides of the connective tissue matrix and results in wider spreading of solutions and faster absorption rates. The absorption rates can also be increased by massage or by application of heat to increase blood flow. Quite frequently drugs affect their own rates of absorption if they alter the blood supply or capillary permeability. For example, methacholine, a cholinergic drug, causes vasodilation, which results in an immediate systemic response following subcutaneous administration.

The absorption of drugs from the depots formed following subcutaneous administration can be retarded to provide prolonged effect by such techniques as immobilization of the limb, local cooling to cause vasoconstriction, and the application of a tourniquet proximal to the injection site to block the superficial venous drainage and lymphatic flow. Inclusion of minute amounts of epinephrine (1:100,000 or 1:200,000) in the subcutaneous injection may retard absorption by constricting the local vessels. This technique is also of special value if local rather than systemic effects are desired, e.g., in the administration of local anesthetics.

The subcutaneous route of administration has frequently been used to provide prolonged release of drugs by incorporating the drugs into compressed pellets that can be implanted under the skin. The drug must be present in a relatively insoluble form and the pellet must resist disintegration by the subcutaneous fluid environment and mechanical stress. These conditions have been achieved with certain steroid hormones. For example, cylindrical pellets of testosterone, about 5 mm in thickness and diameter and weighing about 100 mg, are implanted subcutaneously in humans. The absorption rates can be determined by weight loss of the pellet upon removal. An absorption of about 1 percent per day is generally obtained during the steady state for up to two months.

An ideal shape for achieving constant rates of absorption is a flat disc. A change in the weight of the disc due to absorption results in very little change in the total surface area exposed since the release of the drug takes place from the flat surfaces.

For spherical pellets the ratio of surface area to volume increases with decreasing diameter. Thus, when drugs are prepared as spheres of known diameter the rate of absorption can be predicted: the larger the sphere the slower the rate of absorption. This principle has been used in the design of long-acting insulin preparations. Prompt insulin consists of small particles, and extended insulin is made up of relatively large particles.

Two examples of pellets used for subcutaneous implantation are Oreton pellets (75 mg testosterone) and Pyrogynon pellets (25 mg estradiol).

Some drugs produce severe pain when injected subcutaneously. Local necrosis and sterile abscesses may also occur. Such drugs may have to be administered intravenously because no solution concentrated enough to be useful can be given subcutaneously or intramuscularly.

PERCUTANEOUS ADMINISTRATION

The absorption of drugs through the skin should be a difficult matter since the function of the skin is to act as a barrier between the outside environment and the vulnerable tissues under the skin. Yet drugs are absorbed, sometimes quite efficiently, from the skin.

A major function of skin is to retard the diffusion and evaporation of water from within the body, except at the sweat glands. The stratum corneum, also known as the horny layer, which is densely packed with keratin, is probably responsible for this retardation. Beneath the horny layer, separating it from the underlying granular layer of epithelial cells, is the so-called barrier area, a clear dense region which is quite different from the horny layer both in microscopic appearance and in chemical properties. If the horny layer is stripped but the barrier area is left intact, little change in permeability occurs although water loss increases. However, removal of the barrier area leads to an abrupt increase in permeability for all kinds of molecules, large or small, lipid- or water-soluble. The dermis is generally freely permeable to all types of molecules.

The penetration rates of drugs through the skin are determined largely by their lipid/water partition coefficients excluding significant absorption of ions or water-soluble structures, except for very small molecules. Highly lipid-soluble molecules also penetrate skin slowly compared to their penetration through other membranes.

Drugs may be applied to the skin for a local effect, especially on the superficial layers of the epidermis. The drugs are incorporated into vehicles which adhere to the skin, allowing diffusion of drug molecules out of the vehicle and into the epidermis. If a pathologic condition exists in the deeper layers of the skin the systemic administration may be more desirable, especially if the drug is water-soluble. For example, antifungal and antibacterial agents are often much more effective in skin infections when given orally or by injection than when applied to the skin. Highly lipid-soluble drugs, such as griseofulvin, are also effective in systemic administration for local skin infections.

Some recent studies suggest the use of pharmacologically inactive solvents, such as dimethyl sulfoxide, to facilitate the absorption of drugs through the skin. Examples of drugs whose absorption has been increased are corticosteroids, antineoplastics, antibiotics, carcinogens, and insulin. There is, however, great controversy on the toxicity of these solvents in topical formulations and it is difficult to justify their use at the present time.

As discussed earlier, a recent approach to utilizing Therapeutic Systems consists of a multilaminate structure of small size which is worn in the postauricular region, providing optimum drug permeability. Scopolamine is the first drug applied in this way for prevention or treatment of motion-induced nausea with reduced parasympatholytic effects.[21]

The ionic drugs can often be administered through the skin by applying electrical gradients to the skin. This method of iontophoresis involves applying galvanic current to electrodes placed at the absorption site and at other parts of the body.

The fast absorption of lipid-soluble molecules through the skin indicates an environmental hazard which continues to grow with increasing pollution in the atmosphere. For example, carbon tetrachloride and other organic solvents penetrate the body through the skin and cause serious toxic effects.

Organic phosphates (DFP, parathion, malathion) and nicotine insecticides have caused deaths in agricultural workers as a result of percutaneous absorption upon in-field contact. Chlorvinyl arsine dichloride (lewisite), a mustard gas, is readily absorbed through the skin and has been used in chemical warfare. Most of the carcinogens in the atmosphere can be efficiently absorbed through the skin and it is no wonder that there is a higher cancer incidence rate in people living around the industrial centers, even though these people may not be directly exposed to these chemicals.

PULMONARY ADMINISTRATION

Drugs can be introduced into the pulmonary system as gases or in aerosol forms. An almost instantaneous absorption can be expected due to the extremely large surface area available for absorption. The primary mechanism of absorption is passive diffusion but the lipid solubility tends to play a smaller role than in gastrointestinal absorption. For example, Table 4.5 lists the partition coefficient between blood and air for various gases and the time required for 99 percent equilibration between pulmonary blood and alveoli.[22] It should be noted that an almost 120-fold difference in the solubility in the blood has little effect on the rate of equilibration.

Table 4.5. COMPARISON OF SOLUBILITY AND TIME REQUIRED FOR EQUILIBRATION BETWEEN PULMONARY BLOOD AND ALVEOLI FOR VARIOUS GASES

COMPOUND	PARTITION COEFFICIENT*	$t^{\dagger}_{0.99}$
Ether	15.00	14.43
Chloroform	7.30	18.28
Ethyl chloride	2.50	13.42
Halothane	2.30	23.55
Vinyl ether	1.50	14.02
Fluorocarbon 11	0.94	19.66
Acetylene	0.82	8.55
Cyclopropane	0.47	10.86
Nitrous oxide	0.47	11.13
Fluorocarbon 12	0.28	18.46
Fluorocarbon 114	0.26	21.95
Propylene	0.22	10.86
Ethylene	0.14	8.89

(From Niazi: Math Biosci 27:169, 1975)
* Partition coefficient between blood and air
† Time required for 99% equilibration between pulmonary blood and alveoli, μseconds.

The main limiting step in the utilization of this route has been the need to design dosage forms which accurately deliver the drugs. Most of these drugs are administered as aerosols, and their delivery is to a great extent dependent on the particle size distribution. Particles greater than 10 μ are almost completely removed by impaction in the nasal passages. IMPACTION refers to the deposition of particles in the respiratory tract. The precipitation of particles arises from the tendency of a particle moving in a stream of air to continue in its original direction when the air current changes direction at bronchial branch points and at curves in the bronchial tree. Impaction due to diffusion is negligible except for very small particles.

Particles below 10 μ in diameter are of great significance since these include bacteria, viruses, smoke, industrial fumes, dust laden with fission product, pollens, insecticide dusts or sprays, and inhalant sprays used in the therapy of pulmonary diseases.

In order for a drug to be absorbed from an aerosol its particles must impact, preferably in the alveolar sacs, and dissolve in the available fluids. Larger particles are retained in the upper respiratory tract and as they become smaller they penetrate deeper into the pulmonary tree. Particles larger than 2 μ in diameter probably do not reach the alveolar sacs. Particle sizes approximating 1 μ are most desirable, but there is a greater tendency for these particles to be exhaled without being impacted. Thus many formulations include hygroscopic substances in the formulation to increase the size of particles as they penetrate deeper into the trachea. The tidal volume is also an important consideration. At a given respiratory rate, the air stream velocity is greater at high tidal volumes and thus particles of all sizes tend to be driven deeper into the pulmonary tree before impaction.

Pulmonary administration has been used mainly for local therapy. For example, aerosols of epinephrine, isoproterenol, and dexamethasone are commonly used for acute asthmatic attacks, and antibiotics are sometimes incorporated for the treatment of complicated bronchopulmonary infections. In some instances, the systemic absorption of drugs meant for local action may be appreciable. For example, isoproterenol in a 0.5 percent aerosol is an effective bronchodilator, but a 1 percent aerosol is apt to cause undesirable cardioaccelerator and hypertensive actions after only a few inhalations. The quick responses can, however, be beneficial in the treatment of anaphylactic episodes, as in the use of epinephrine inhalations rather than parenteral administration of the drug for allergic reactions to insect or other venoms.

Although the pulmonary route is used mainly for local effects, several drugs have been successfully administered in this way for systemic effect, including penicillin, digitalis glycosides, diuretics, and tranquilizers.

The problem of accurate dosing in pulmonary administration remains a serious obstacle to greater use of this route. The use of metered dose devices is certainly an improvement and a recent product uses an inhalation device designed to deliver the drug as a powder aerosol into the lungs when actuated by the inspiratory effort. The powder particle sizes range primarily between 2 μ and 6 μ. This device, currently used for disodium cromoglycate (Cromolyn Sodium:Aarane Inhaler) provides a greater and more consistent absorption than can be obtained from metered dose devices.

OPHTHALMIC ADMINISTRATION

As with permeability in most other routes of administration, the permeability of drugs into and through the cornea is a function of their lipoid and aqueous solubility. The cornea is composed of three distinct layers: the outer epithelium, an inner stroma, and the endothelium. The epithelium and endothelium are much more lipoidal than the stroma. Therefore drugs must possess biphasic solubility characteristics in order to be absorbed through this route.

Weakly basic drugs, such as tropicamide, epinephrine, pilocarpine, atropine, homatropine, or cyclopentolate, freely penetrate the cornea because of rapid equilibration between their lipid-soluble un-ionized forms and their water-soluble ionized forms. The penetration of quarternary ammonium compounds, such as carbachol, echothiophate iodide, and demecarium bromide, which are charged and water-soluble at all pH values, is postulated on a binding mechanism which permits a small but sufficient quantity of these potent antiglaucomic agents to reach aqueous humor and evoke a therapeutic response.[23] Tetracycline, gentamicin, carbenicillin, and methicillin do not penetrate the cornea because of their low lipid solubility, but chloramphenicol shows good penetration.[24]

Fluorescein is used for diagnostic purposes because of its high lipid solubility, which prevents its entry into the stroma unless there is an abrasion. If there is an abrasion, fluorescein enters the stroma and possibly the aqueous humor, giving a brilliant green color due to its alkaline pH. In the precorneal film fluorescein exists in a yellow or orange form.

A variety of physiologic factors influence corneal drug absorption. Lacrimal drainage of an instilled drug solution competes for drug with corneal penetration and can account for a considerable loss of drug. When a drop of solution is applied to the eye two processes occur simultaneously: the solution is diluted by reflex tearing and the added volume in excess of the normal lacrimal volume is drained from the eye, which is partly facilitated by reflex blinking. In humans, administration of 25 μliters of solution to the eye at three-minute intervals will minimize volume build-up, dilution, excess drainage, and overflow. Shorter intervals of administration would reduce ophthalmic bioavailability. The normal lacrimal volume in humans is about 7 μliters, and if blinking does not occur the human eye can hold approximately 10 μliters. Since the size of commercial ophthalmic drops is between 50 and 75 μliters, the loss of drug due to spillage out of the eye can be considered a significant factor in the reduction of bioavailability.[25]

Ophthalmic dosage forms include solutions, ointments, suspensions, lyophilized powders, and oily solutions. Two new dosage forms have recently been introduced to the market. One is an ophthalmic insert, an elliptical device consisting of a drug-containing core surrounded by a flexible copolymer membrane through which pilocarpine diffuses while the ocular delivery system remains in contact with conjuctiva (Occusert). A spray device has also been designed for accurate delivery of drugs.

Polymers such as methylcellulose, hydroxypropyl methylcellulose, and polyvinyl alcohol decrease the surface tension and increase the viscosity of solutions, thus enhancing bioavailability. Soft contact lenses soaked in pilocarpine have also been used.[26] Biodegradable polymers have been employed for the controlled delivery of hydrocortisone and tetracycline.

MISCELLANEOUS ROUTES OF ADMINISTRATION

Drugs are also administered through such routes as the urethra, vagina, and spinal cord. For example, urethral suppositories and vaginal inserts or suppositories are frequently used for treatment of localized infections. Anesthetics are often administered in the spinal fluid, as are other drugs on occasion for localized effect.

Recent studies suggest that vaginal administration of drugs for systemic effect may be a valid alternative to rectal or even oral routes of administration because of fast and complete absorption from this site. Direct controlled delivery of fertility-controlling hormones has also been successfully made.

References

1. Forbes LS: Antihelminthic toxicity, administration technique and routes of absorption. Aust Vet J 47:601, 1971
2. Needleham P, Lang S, Johnson EM: Organic nitrates: relationship between biotransformation and rational angina pectoris therapy. J Pharmacol Exp Ther 181:489, 1972
3. Greenblatt DJ, Koch-Weser J: Intramuscular injection of drugs. N Engl J Med 294:562, 1976
4. Dollery CT, Davies DS, Connolly ME: Differences in the metabolism of drugs depending upon their routes of administration. Ann NY Acad Sci 179:108, 1971
5. Rowland M: Influence of route of administration on drug bioavailability. J Pharm Sci 61:70, 1972
6. Riegelman S, Rowland M: Effect of route of administration on drug disposition. J Pharmacokinet Biopharm 1:419, 1973
7. Goldman P: Therapeutic implications of the intestinal microflora. N Engl J Med 289:623, 1973
8. Alkalay D, Khemani L, Wagner WE: Sublingual and oral administration of methyltestosterone. A comparison of drug bioavailability. J Clin Pharmacol 13:142, 1973
9. Beckett AH, Triggs EJ: Buccal absorption of basic drugs and its application as an in vivo model of passive drug transfer through lipid membranes. J Pharm Pharmacol 19:315, 1967
10. Bickel MH, Weder HJ: Buccal absorption and other properties of pharmacokinetic importance of imipramine and its metabolites. J Pharm Pharmacol 21:160, 1969
11. Beckett AH, Moffat AC: The buccal absorption of some barbiturates. J Pharm Pharmacol 23:15, 1971
12. Eckman WW, Patlack CS, Fenstermacher JD: A critical evaluation of the principles governing the advantages of intra-arterial infusions. J Pharmacokinet Biopharm 2:257, 1974
13. Greenblatt DJ, Koch-Weser J: Clinical pharmacokinetics. N Engl J Med 293:702, 1975

14. Wilensky AJ, Lowden JA: Inadequate serum levels after intramuscular administration of diphenylhydantoin. Neurology (Minneap) 23:318, 1973
15. Assaf RAE, Dundee JW, Gambler JAS: The influence of route of administration on the clinical activity of diazepam. Anesthesia 30:152, 1975
16. Schriftman H, Kondritzer AA: Absorption of atropine from muscles. Am J Physiol 191:591, 1957
17. Vukovich RA, Brannick LJ, Sugerman AA: Sex differences in the intramuscular absorption and bioavailability of cephradine. Clin Pharmacol Ther 18:215, 1975
18. Evan EF, Proctor JD, Fratkin MJ: Blood in muscle groups and drug absorption. Clin Pharmacol Ther 17:44, 1975
19. Dreyfuss J, Ross, Jr JJ, Shaw JM: Release and elimination of ^{14}C-fluphenazine enanthate and decanoate esters administered in sesame oil to dogs. J Pharm Sci 65:502, 1976
20. Nora JJ, Smith DW, Cameron JR: The route of insulin administration in the management of diabetes mellitus. J Pediatr 64:547, 1964
21. Shaw JE, Chandrasekaran SK, Campbell P: Percutaneous absorption: controlled drug delivery for topical or systemic therapy. J Invest Dermatol 67:677, 1976
22. Niazi S: Evaluation of pulmonary clearance and accumulation model for gases. Math Biosci 27:169, 1975
23. Smolen VF, Clevenger JM, William EJ, Bergoldt MW: Biophasic availability of ophthalmic carbachol I: Mechanisms of cationic polymer and surfactant-promoted miotic activity. J Pharm Sci 62:958, 1973
24. Beasley H, Boltralik JJ, Baldwin HA: Chloramphenicol in aqueous humor after topical application. Arch Ophthalmol 93:184, 1975
25. Akers MJ, Schoenwald RD, McGinity JW: Practical aspects of ophthalmic drug delivery. Drug Dev Ind Pharm 3:185, 1977
26. Podos SM, Becker B, Assiff C, Harstlin J: Pilocarpine therapy with soft contact lenses. Am J Ophthalmol 73:336, 1972

Questions

1. Cite ten examples, with explanations, of route-dependent dosing of drugs.
2. What are the possible first pass effects in the delivery of drugs through various routes of administration?
3. What is the driving force in the absorption of drugs by passive diffusion?
4. Under what circumstances may the chemical potential of a drug not be same as the total concentration in the gastrointestinal tract?
5. What are the characteristics of an active transport process? Why can this process be more significant at lower drug concentrations in the gastrointestinal tract?
6. What are the characteristics and roles of solvent drag, facilitated transport, ion-pair transport, and pinocytosis in the absorption of drugs?
7. What are the anatomic and physiologic reasons for variations in the absorption of drugs from different regions of the gastrointestinal tract?
8. How does the consistency of the gastrointestinal contents affect absorption rates?
9. Discuss the factors which affect gastric emptying.

10. Cite examples of drugs for which either luminal or hepatic first pass biotransformation has been reported.
11. In determining absolute bioavailability, how would one account for first pass biotransformation?
12. What are the major advantages of sublingual/buccal routes of administration over gastrointestinal administration? Cite examples of drugs which are commonly administered through these routes.
13. What are the dosage form and physiologic factors which alter rectal absorption of drugs? Why is it that some drugs cannot be administered by this route?
14. Why must some drugs be administered slowly during intravenous injection?
15. What is the potential problem with administration of sodium phenytoin either intramuscularly or intravenously? Cite two additional examples of drugs which can cause a similar problem.
16. What is the role of blood flow rates and anatomy in the absorption of drugs following intramuscular administration?
17. What are the methods that can be used to either increase or decrease the absorption of drugs from intramuscular or subcutaneous sites of administration?
18. What are the possible side effects of intramuscular administration?
19. Why does a disc-shaped subcutaneous implant provide almost constant rates of absorption?
20. What are the dosage form and physiologic factors which affect the absorption of drugs from subcutaneous sites?
21. Which part of the skin anatomy provides the rate-limiting factors in the percutaneous absorption of drugs?
22. Under what conditions is it preferable to use systemic drug delivery for skin infections?
23. Why does the Therapeutic System of scopolamine provide a reduced incidence of parasympatholytic effects?
24. What is meant by iontophoresis?
25. What are the environmental hazards of percutaneous absorption?
26. Which part of the pulmonary tract is most suitable for systemic absorption of drugs?
27. What is the effect of the partition coefficient between blood and air and the total uptake of gases on pulmonary intake? How would you correlate the rate of equilibration with the extent of absorption in each cycle?
28. Why must aerosols be present at below the 10 μ range to be effectively absorbed? What is the role of impaction on drug absorption from the lungs?
29. How does the tidal volume affect pulmonary absorption?
30. What are the disadvantages of pulmonary administration?
31. How does the delivery of drug from the Aarane Inhaler differ from that of other inhalers of the conventional type?

32. Why is biphasic solubility most important for the ophthalmic delivery of drugs?
33. What is the rationale behind using fluorescein for diagnostic purposes? What is it used to diagnose?
34. How does lacrimal drainage affect the bioavailability of drugs after ophthalmic administration?
35. What is the role of viscosity in drug absorption from the eye?

CHAPTER 5

The Distribution of Drugs

Once drug molecules enter the blood stream, either by direct vascular administration or by absorption through any of the extravascular routes, these molecules start to mix with the body fluids, eventually distributing to the site of action. Therefore, the distribution process initiates the drug action.

The human body is composed of a variety of tissue structures with different lipophilic characteristics, blood supply, and abilities to interact with foreign molecules. These differences in the tissue properties make the drug concentration nonuniform throughout the body. The rates with which these nonuniform concentrations are attained depend mainly on the blood supply to the tissues and are therefore highly variable.

Since the drug action is dependent on the access of drug molecules to a site of action, the distribution process plays a very important role in the onset and intensity of a pharmacologic response. For example, the short duration of action of thiopental is due to its fast distribution to body muscles and fat, removing the drug from its site of action, the central nervous system. Conversely, drugs which can be stored in the body tissues may provide prolonged duration of action if the site is embedded in these tissues.

This chapter examines the distribution characteristics of drugs and their effects on clinical response.

DISTRIBUTION IN THE BLOOD

The distribution of drugs in the blood is important since further distribution to body tissues takes place from this compartment (blood). It is the free drug concentration which equilibrates with the body tissues (Scheme 5.1).

The first interaction between drugs and body tissues occurs in the blood compartment, where the drugs can distribute unevenly between red cells, white cells, plasma proteins, and plasma water. Although in many instances the plasma concentration is equal to the total blood concentration, there can be significant differences between the whole blood and plasma concentrations due to specific binding of the drugs to plasma proteins, hemoglobin, blood cell walls, and other components.

Several drugs enter the erythrocyte quite rapidly and others at a slower

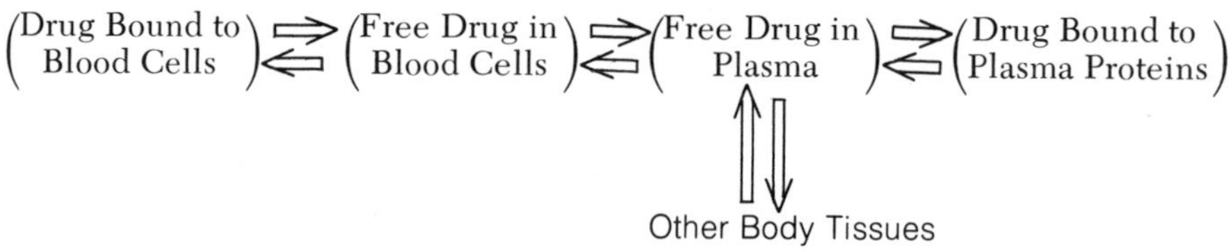

$$\begin{pmatrix}\text{Drug Bound to}\\ \text{Blood Cells}\end{pmatrix} \rightleftarrows \begin{pmatrix}\text{Free Drug in}\\ \text{Blood Cells}\end{pmatrix} \rightleftarrows \begin{pmatrix}\text{Free Drug in}\\ \text{Plasma}\end{pmatrix} \rightleftarrows \begin{pmatrix}\text{Drug Bound to}\\ \text{Plasma Proteins}\end{pmatrix}$$

$$\Updownarrow$$

Other Body Tissues

Scheme 5.1. Equilibration and distribution from blood compartment.

rate. These rates and the extent of penetration can often be correlated to the partition coefficients of these drugs—highly lipid-soluble drugs show faster accumulation in the erythrocytes. Hemoglobin does not bind many drugs which are bound by albumin. The concentration of drugs in erythrocyte water is generally equal to the concentration in plasma water after equilibration. Erythrocyte membranes have a carbonic acid mechanism which permits rapid

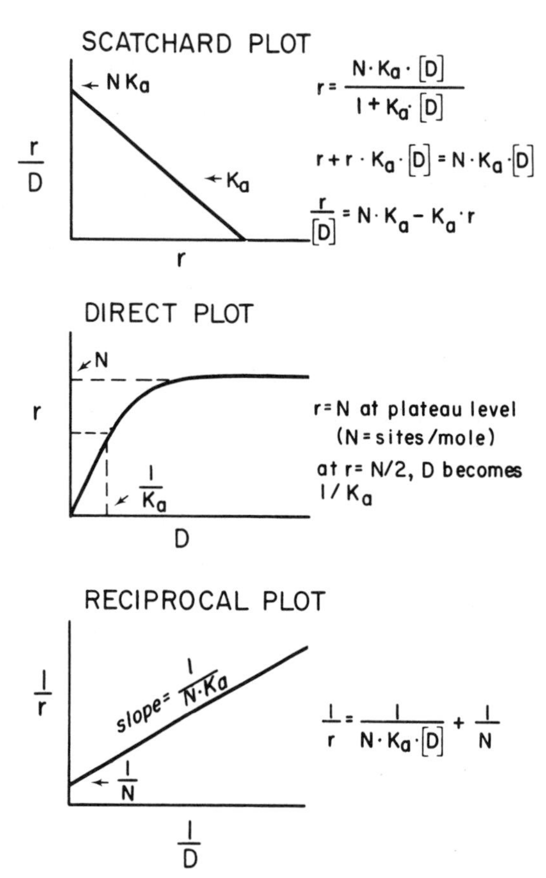

Figure 5.1. Calculations of protein-binding parameters.

penetration of anions without energy expenditure, whereas only highly lipophilic cations penetrate the membrane.

The main interaction in the blood compartment is due to the presence of a variety of plasma proteins which can bind the drug molecules, mainly by reversible physical forces. Stronger binding of a chemical nature is rare.

The binding of drugs to plasma proteins is a dynamic process:

$$[P] + [D] \underset{K_d}{\overset{K_a}{\rightleftharpoons}} [PD] \qquad \text{(Eq. 5.1)}$$

$$K_a = \frac{[PD]}{[P][D]} \qquad \text{(Eq. 5.2)}$$

where $[P]$ = free protein concentration
$[D]$ = free drug concentration
$[PD]$ = concentration of bound complex
K_a = association constant; K_d = dissociation constant

The ratio of bound drug and total protein is:

$$[PD]/[P_t] = \frac{N\,K_a[D]}{1 + K_a\,[D]} = r \qquad \text{(Eq. 5.3)}$$

where N = number of binding sites
$[P_t] = [PD] + [P]$
r = number of moles of drug bound per mole of protein

If r is plotted against the free drug concentration (Fig. 5.1), the binding parameters can be easily calculated, including the number of binding sites on each protein molecule. Many mathematical transformations of Equation 5.3 are made for calculation purposes. The protein binding parameter which is most widely used is percent binding, which can be totally meaningless if the total concentration at which this determination is made is not provided. This is because of the nonlinear nature of the percent binding and the total concentration as a result of saturation of protein binding sites. Table 5.1 lists the percent binding characteristics of some commonly used drugs at therapeutic levels. This list only represents average values and a large variation is possible for various reasons. The primary reason is that for many drugs a well-defined therapeutic concentration is difficult to assign. Even if such concentrations can be assigned, they will vary between individuals as a result of physiologic or pathologic factors. Therefore, this list should only be used for illustration purposes and as a general guide for comparison of drugs.

The effect of percent binding on the capacity of the plasma to contain a drug can be easily understood through a hypothetical example of a drug which distributes uniformly throughout the total body water (which is generally 60 percent of the total body weight). The plasma water is assumed to be 4 percent of the body weight. The percentage of total drug in the body retained in the plasma can be calculated with the following equation:

$$\text{Percent in plasma} = \frac{4}{(F_u \cdot 56) + 4} \times 100 \qquad \text{(Eq. 5.4)}$$

where F_u = fraction unbound in plasma

The following tabulation shows a solution of the equation above:

% Bound in Plasma	% of Total Drug Retained in Plasma
0	6.7
25	8.7
50	12.2
60	15.2
70	19.2
75	22.2
80	26.3
90	41.7
95	58.8
98	78.1
99	87.7

Therefore, it is only when the percent binding is high that the plasma can serve as a significant storage compartment for drugs.

The binding strength, K_a, is an important parameter in determining the effect of binding on the distribution properties. Generally, if the binding constant is less than $10^4 M^{-1}$, the effect on distribution is minimal. Figure 5.2 shows the binding constants of some drugs.[1] These drugs have two binding sites and thus two binding constants. The values in parentheses show the span of binding constants obtained by a ratio of these constants. A recent review article on the effect of protein binding on the drug distribution and pharmacokinetics summarizes the binding parameters.[2]

The most important component of the blood compartment responsible for binding is albumin, which carries a net negative charge at the physiologic pH and interacts with both anions and cations (Fig. 5.3). The binding site for acids is generally the N-terminal amino acid, whereas bases seem to be bound non-specifically. Thus the capacity of albumin to bind acids is limited and the capacity for bases is apparently large.

The consequences of drug binding to blood components can be summarized as follows:

1. Insoluble drugs and endogenous chemicals are carried in the blood in the bound form. In the absence of such binding the body may be deprived of lipophilic steroids and vitamins, and the water-insoluble drugs may not show a pharmacologic response due to lack of their distribution to the sites of action.
2. Plasma protein binding can increase the absorption rates of drugs, especially those which remain ionized in the gastrointestinal tract. The binding

Table 5.1. PERCENT PROTEIN BINDING OF DRUGS AT THERAPEUTIC LEVELS

DRUG	PERCENT BOUND	DRUG	PERCENT BOUND
Acetaminophen	25	Heparin	0
Acetazolamide	90	Imipramine	85
Allopurinol	0	Indomethacin	90
Aminopyrine	18	Isoniazid	0
Aminosalicylic acid	65	Kanamycin	10
Amitriptyline	96	Lincomycin	85
Ampicillin	25	Mepacrine	90
Antipyrine	4	Meperidine	40
Atropine	50	Methadone	40
Barbital	10	Methicillin	45
Bishydroxycoumarin	97	Methotrexate	45
Carbamazepine	72	Nitrofurantoin	70
Carbenicillin	47	Nortriptyline	94
Cephalexin	22	Novobiocin	96
Chloramphenicol	25	Oxacillin	94
Chloroquine	55	Oxyphenbutazone	90
Chlorpheniramine	70	Oxytetracycline	28
Chlorpromazine	95	Phenacetin	30
Chlorpropamide	80	Phenylbutazone	95
Chlortetracycline	47	Phenytoin	87
Desipramine	80	Prednisolone	90
Diazepam	96	Probenecid	80
Diazoxide	99	Procainamide	15
Dicoumarol	97	Promethazine	8
Digitoxin	95	Quinidine	70
Digoxin	23	Rifampicin	85
Doxycycline	93	Streptomycin	34
Erythromycin	18	Sulfadimethoxine	95
Ethambutol	8	Sulfamethizole	90
Fenfluramine	32	Sulfathiazole	70
Furesamide	75	Theophylline	15
Glutethimide	54	Thiopental	75

of the un-ionized form of a drug in the blood decreases the free concentration and therefore the driving force for absorption, the concentration gradient, is increased.

3. Binding of drugs to blood components leads to an even distribution throughout the body tissues. In the absence of binding, drugs may quickly accumulate only in the specific parts of the body which show affinity for the particular drug. The total body exposure will therefore be decreased, resulting in possibly infinite half-lives which make every drug extremely toxic.
4. Binding also serves a storage function. Although this is only significant when the extent of binding is high, small binding coupled with strong binding forces sometimes makes it important. An excellent example of

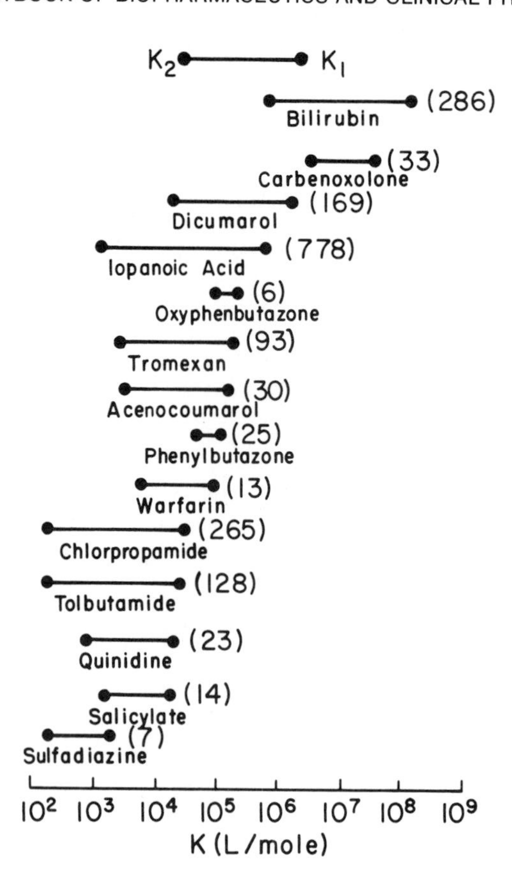

Figure 5.2. Association constants of selected drugs reported to be bound by two sets of sites on human serum albumin. Values in parentheses indicate span of the two binding constants. (From Wosilait: Gen Pharm 8:7, 1977)

storage in the plasma is that of bilirubin, which is contained in the blood through its high binding with plasma proteins. Several drugs, including sulfonamides, compete with bilirubin for the same binding site, resulting in the decreased binding of bilirubin. The increased levels of free bilirubin are extremely toxic, especially in children, and often cause a syndrome termed kernicterus. Conversely, diazoxide extensively binds to plasma proteins (approximately 79 percent) and if it is administered slowly by intravenous route the drug becomes trapped in the plasma, indicating a need for fast intravenous injection to assure quick body exposure.[3]

5. Only the unbound fraction of any drug in the plasma is available for biotransformation or excretion and a high degree of binding can result in delayed elimination. For example, at therapeutic plasma levels, only 5 percent of digitoxin is unbound to plasma proteins, compared with 77 percent of digoxin, and as a result digitoxin is less available for elimination and shows a longer biological half-life when compared with digoxin.[4] However, the extent of binding alone should not be extrapolated to include elimination rates. The strength of binding is in fact the primary determi-

nant of the elimination rates. For example, penicillin binds extensively to plasma proteins but shows very fast elimination from the body due to rapid equilibration between the unbound and bound fractions. For some drugs, a high degree of binding may even be responsible for high elimination rates if the eliminating organs can overcome the binding strength by fast removal. In these situations higher binding results in higher concentrations of drugs being transported to these organs, resulting in shorter half-lives of the drugs.

6. Slight change in the binding of highly bound drugs can result in significant change in clinical response or cause a toxic response. Since the free drug in the plasma equilibrates with the site of pharmacologic or toxic response, a slight change in the extent of binding, such as from 99 to 98 percent, results in an almost 100 percent increase in the free concentration. A large

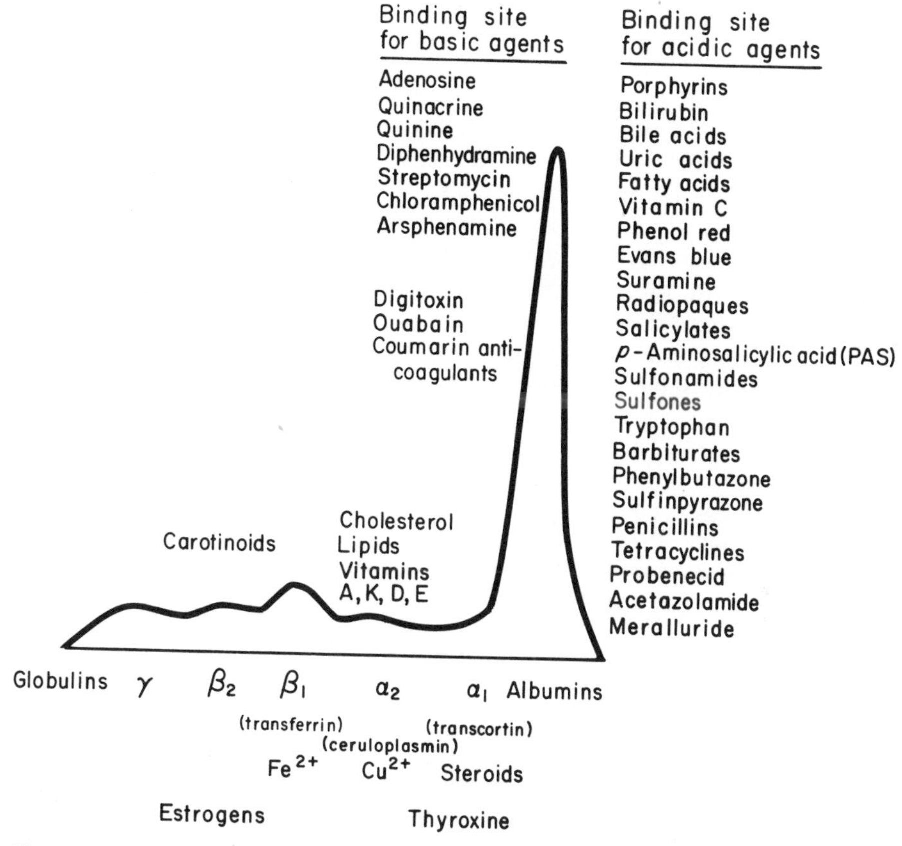

Figure 5.3. Schematic representation of the binding of various drugs to plasma proteins. (From Ariens and Simonis: in Cluff and Petrie, eds., *Clinical Effects of Interaction Between Drugs*, 1975. Courtesy of Excerpta Medica)

number of drugs, including coumarin anticoagulants, phenytoin, sulfonamides, and corticosteroids, show their actions to be dependent on the free drug concentration, and any changes brought about by drug interactions and physiologic and pathologic changes will seriously modify the actions of these drugs.

Briefly, any variation in plasma protein binding will have a significant effect on the therapeutic response if protein binding plays an important role in the distribution process. One of the major sources of protein binding variation is the alteration in the plasma albumin concentration as a result of a number of diseases and physiologic states. This is due to changes in the synthesis rate, the catabolic rate, or in the distribution between extravascular and intravascular spaces. In acute injury or disease, reduced plasma levels of albumin and increased γ-globulin levels are due to normal host reaction. Burns result in a significant decrease in plasma protein levels due to direct loss of proteins from the body and also to increased capillary permeability decreasing plasma protein levels. A decrease in plasma albumin levels is also noted in pregnancy, neoplastic diseases, and acute viral hepatitis (due to diminished synthesis of albumin). In severe renal diseases excessive proteins may be filtered—nephrosis can be associated with increase in the degradation of albumin. Hypoalbuminemia in cystic fibrosis is usually caused by dilution of the normal circulating albumin mass in an enlarged plasma volume. Often cigarette smokers have lowered albumin levels.

It should, therefore, be clear that any disease state or physiologic change which alters the level of plasma proteins can significantly affect the dose-response curve through variation in the free drug concentration. However, few studies have reported the clinical effects of alteration in the protein binding.

Another important source of the variation of free drug concentration in the plasma is the competitive nature of binding between drugs and endogenous substances. For example, free fatty acids are extensively bound to albumin and represent the major transport form of fat in plasma. Normal free fatty acid concentration in the plasma ranges between 200 and 500 μEq/liter but levels of more than 2500 μEq/liter have been observed in humans in various conditions of fasting, exercise, sympathetic stimulation, bacterial infection, diabetes mellitus, and hyperthyroidism.[5] These increased levels of free fatty acids can displace several drugs, as is demonstrated with phenytoin and warfarin, where significant change in the unbound fraction of drug in the plasma was noted at free fatty acid levels of above 800 μEq/liter.[5] In vitro displacement by free fatty acids has been demonstrated for bilirubin, hydroxyphenylazobenzoic acid, salicylates, phenylbutazone, thiopental, and tryptophan.[6–9] In some instances an increase in the lipid contents of the plasma, such as an increase in cholesterol and triglycerides, increases the binding of the total drug in the plasma. A recent study showed that a hyperlipidemic serum sample can retain about 50 percent more minocycline as a result of interaction with these lipids.[10]

The variation in the free drug concentration also occurs as a result of drug interaction. For example, the binding of phenytoin is significantly affected by salicylates, sulfafurazole, phenylbutazone, and acetazolamide. It has been suggested, therefore, that in correlating plasma concentration profiles of phenytoin, free drug should be calculated instead of the total amount of drug in the plasma. Corticosteroids, anticoagulants, sulfonamides, and various other drugs also show therapeutic activity dependent on free drug concentration.

Although in most instances competitive binding increases the free concentration of one drug alone, it is also possible to observe a decrease in the free drug concentration as a result of drug interaction. For example, pempidine, which is not normally bound to plasma proteins, becomes extensively bound in the presence of chlorothiazide,[11,12] and tetracycline increase the binding of both promazine and chlorpromazine.[13]

Various disease states result in a change in the electrolyte balance in the blood, which changes the binding of drugs since the activity coefficients of drugs also change. There is also a possibility of changes in the three-dimensional structures of proteins when the electrolyte balance is disturbed. The effect of age on the binding of drugs can also be significant, since the plasma volume and its composition change with age.

Table 5.2. RELATIONSHIP BETWEEN PHYSICOCHEMICAL AND PHARMACOKINETIC PROPERTIES OF TETRACYCLINE ANALOGS

DRUG	K_p*	*PERCENT BOUND*	*PERCENT EXCRETED UNCHANGED*	*HALF-LIFE*
Oxytetracycline	25	34	70	9.5
Tetracycline	36	55	60	6.8
Demethylchlortetracycline	50	68	46	12.7
Methacycline	430	78	34	14.3
Doxycycline	600	93	45	15.1

* K_p = octanol/water (pH 7.5) partition coefficient × 10^3
After Jusko and Gretch;[2] Colaizzi and Klink;[14] Wittenau and Yeary;[15] Kunin;[16] Rosenblatt *et al.*;[17] Kunin *et al.*;[18] Steigbigel *et al.*;[19] and Ritschel[20]

Finally, the physicochemical properties of drugs also play an important role in their binding to various blood components. For example, the analogs of tetracycline show an interesting correlation between their physical properties and disposition characteristics (Table 5.2). As the molecules become more lipid-soluble, their interaction with proteins increases and their elimination from the body decreases.[14–20]

BEYOND THE BLOOD COMPARTMENT

The foreign drug molecules permeate the capillary walls freely to reach the interstitial spaces unless they are bound in the blood compartment or have molecular weights in excess of 500 to 600. Lipid-soluble molecules might show a slight advantage over water-soluble compounds in their rates of penetration, but these can generally be regarded as equal. An ultrafiltrate of plasma, almost protein-free, passes back and forth across the capillary wall due to hydrostatic pressure, carrying with it any non-protein-bound ingredients.

The process of capillary permeation is so rapid that the overall rate of delivery of a drug from the circulation to other regions of the body, except where extraordinary barriers exist, is generally set by the blood flow rate rather than by the barrier at the capillaries.

The human body is almost 60 percent fluids, a breakdown of which is shown in Figure 5.4. This high hydrophilicity of the body explains why highly lipophilic drugs may not reach their sites of action. Almost invariably a balance of hydrophobic/hydrophilic properties is needed to adequately supply the tissues with active drug molecules.

The tissue membranes behave as typical lipid barriers. An equilibration between the inside and the outside of the tissues is maintained by diffusion of lipid-soluble moieties. Most tissue membranes are not selective in their uptake, except for such special barriers as those around the central nervous system (CNS). For most weakly acidic or basic drugs, the penetration through these tissue membranes depends on their pK_a and hence their ratios of ionized to un-ionized species. The pH of the extracellular fluids is kept

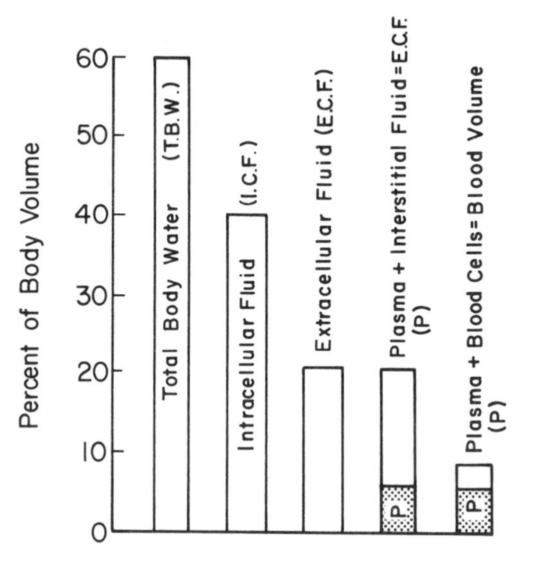

Figure 5.4. Body fluid compartments.

constant by the several buffer systems and is not easily changed. A change in the pH of extracellular fluids will result in a significant change in drug activity. For example, an acidosis will potentiate the effect of phenobarbital, and alkalosis will decrease the anesthetic effect due to change in the ionization of phenobarbital. The effect of systemic acidosis or alkalosis is minimal, however, in changing the fluid pH. The major effect lies in the change in urinary excretion rates. Active transport processes for the transport of drug molecules across tissue membranes are rare.

Most of the central nervous system is surrounded by a specialized barrier which behaves like an extreme form of a lipid membrane. This membrane is highly selective for lipid-soluble foreign molecules, which cross it by simple diffusion. Most natural substrates, such as amino acids and sugars, are apparently transported to the brain by active processes and thus structurally similar foreign molecules can also be transported by these mechanisms. There are, however, some parts of the CNS, such as the pituitary, which are less protected by the lipid membrane and thus allow permeation of relatively polar compounds.

The loss of drugs from the brain and other parts of the nervous system generally takes place by back diffusion of lipid-soluble forms. There is little possibility of biotransformation in the brain. An additional mechanism of loss of drug molecules is by filtration across arachnoid villi. Some organic anions and quarternary compounds, such as hexamethonium, may also leave the cerebrospinal fluids by means of active transport in the choroidal epithelium. This is, however, rare in the CNS and the cerebrospinal fluid (CSF).

Drug permeability sufficient to allow crossing of the cerebrospinal fluid barrier can be correlated with the drugs' lipid/water partition coefficients. For example, thiopental has a chloroform/water partition coefficient which is 50 times higher than that of barbital. The permeability coefficient of thiopental is therefore 22 times higher than that of barbital.[21] In many instances the onset of action can be explained by the lipid solubility and the consequent ability of drug molecules to cross CNS or CSF barriers.

Most antibiotics which are water-soluble and ionized at plasma pH (such as penicillin G) do not enter the central nervous system. However, in the presence of an infection (such as meningeal infections) the permeability of the membranes may change, allowing greater penetration of antibiotics. This effect has been noted for ampicillin, penicillin G, lincomycin, and cephalothin.[22,23]

The specificity of barrier transport makes appropriate selection of a drug form an integral part of therapy. For example, in parkinsonism, where depletion of dopamine is found, levodopa should be administered instead of dopamine since the latter does not enter the brain whereas levodopa, a precursor of dopamine, enters the brain readily and is biotransformed to dopamine.[24]

Another barrier in drug transport is the placenta. Most drug molecules are transported across the placenta by simple diffusion, with an upper limit of molecular weight of 1,000. Lipid-soluble compounds such as ethanol, chloral hydrate, paraldehyde, chlorpromazine, gaseous anesthetics, sulfonamides, some antibiotics, barbiturates, morphine, heroin, etc. readily cross the membrane. Compounds with low lipid solubility, such as highly ionized quaternary ammonium muscle relaxants, do not cross the placental membrane.

The fetal liver and kidney are immature and therefore no biotransformation or excretion takes place—the molecules entering the fetus simply circulate and return to the maternal side. The concentration of drugs which cross the placenta is generally the same on either side of the membrane and no specific accumulation can be expected. However, as pregnancy progresses the placental surface area increases and the membrane becomes thinner, facilitating the placental transfer of drugs.

The fetus is invariably exposed to the drugs taken during pregnancy. Teratogenic effects occur largely in the first trimester when the fetus is forming organs and systems. Well known teratogenic effectors are thalidomide, testosterone, methotrexate, and aminopterin. It has recently been suggested that alcohol intake during pregnancy, even at moderate levels, can cause serious birth defects. All drugs should be restricted during pregnancy due to the uncertainty of their possible effects.

TISSUE LOCALIZATION

Drugs localize extensively in tissues. The extent of localization depends on the physicochemical properties of drugs and ratios between plasma and tissue concentrations of up to 100 are not uncommon. The rate of tissue localization is dependent on the rate of blood flow to a particular tissue. The tissue localization process is generally reversible, like any other equilibration process. Thus no drug residue can exist in a tissue without a finite amount, however small, existing in the plasma water. Some drugs are non-reversibly bound, such as mepacrine and such chlorinated insecticides as DDT. Some antimalarial drugs also show very slow rates of de-equilibration and thus persist in the tissues for prolonged periods of time. An interesting example of tissue localization is the binding of 3-hydroxy-2,4,6-triiodo-α-ethylhydrocinnamic acid (Fig. 5.5) with plasma proteins. The drug persists in the plasma for years and possibly indefinitely.[25]

In most instances tissue localization serves only as a storage for drugs. If the storage sites are also the sites of pharmacologic response, prolonged effects can be expected.

The driving force for tissue localization includes pH differences, binding, lipid solubility, active transport, and some special processes. The pH difference between the plasma and tissue milieu will determine the variation in the drug concentration due to ionization effects and pH partition behavior of drug transport. The protein binding to tissues has the same effect as was discussed for plasma protein binding. The partitioning of drugs to lipid

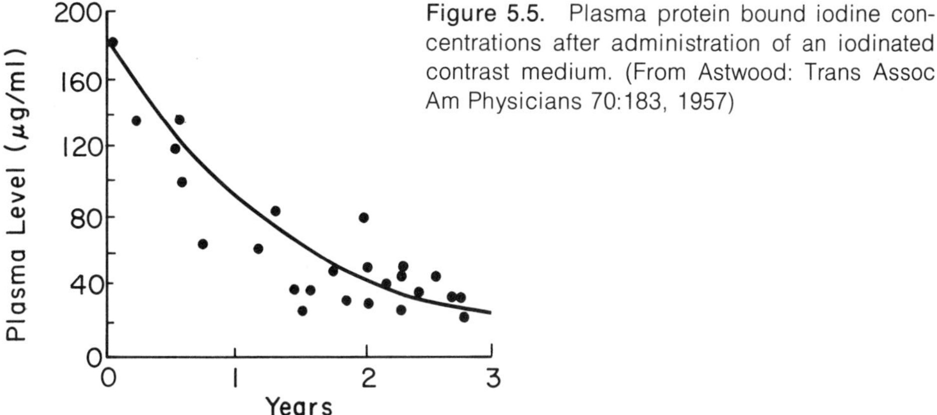

Figure 5.5. Plasma protein bound iodine concentrations after administration of an iodinated contrast medium. (From Astwood: Trans Assoc Am Physicians 70:183, 1957)

contents of the body can be correlated to the partition coefficients of these compounds.

As mentioned earlier, active transport processes are not common in tissue localization, but one exception is that of guanethidine reaching the cardiac muscles.[26]

Many drug-tissue interactions are not easily reversible, as with the deposition of phenothiazine, chloroquine, and arsenicals in hair shafts and of tetracycline in the bones and teeth, for which the half-life may be several months. This deposition of tetracycline results in a serious depression of bone growth. A number of polycyclic aromatic compounds, such as chloroquine and phenothiazines, interact with melanin, resulting in retinopathies. The hepato-toxicity of carbon tetrachloride, bromobenzene, and phenacetin can be attributed to the covalent bonding of their epoxides to macromolecules in the liver, causing necrosis. A variety of chemicals interact with DNA, involving intercalations between adjacent pairs of the double helix. Examples are chloroquine and quinacrine.

The degree and rate of tissue localization is controlled by blood flow rates to the tissues. The vascularity of tissues determines the equilibration rate. Almost instantaneous equilibration can be achieved with such highly perfused tissues as kidney, liver, lung, and brain (Table 5.3). Muscles have intermediate perfusion, and fat depots, which are poorly perfused, often require hours for equilibration. It is noteworthy that about three-fourths of the cardiac output is delivered to tissues that comprise less than one-tenth of the body weight.

The onset of action is often determined by the rate of equilibration with tissues. The fast action of thiopental is caused by rapid equilibration with the brain. The short duration is explained on the basis of removal of thiopental from the brain and localization in muscles and tissues. This phenomenon of

Table 5.3. BLOOD FLOW TO VARIOUS ORGANS AND TISSUES

TISSUE	*BLOOD FLOW (LITERS/MIN)*	*TISSUE MASS (% TOTAL BODY WEIGHT)*
Blood	5.4	8.0
Rapidly perfused		
Brain	0.75	2.0
Liver	1.55	3.5
Kidney	1.2	0.5
Heart musculature	0.25	0.5
Less rapidly perfused		
Muscle	0.8	48.0
Skin	0.4	6.5
Poorly perfused		
Fat	0.25	14.0
Skeleton	0.2	17.0

redistribution is of great importance since most vital organs receive a high concentration of drug as soon as it enters the blood stream.

It is interesting to note that the addictive drugs are generally administered by the intravenous route, whereby they reach the nervous system rapidly and elicit a rapid response. For example, heroin addicts can easily tell heroin from morphine upon intravenous injection even though heroin rapidly deacetylates to morphine in the body. Intravenous injection of cocaine or amphetamines produces an orgasmic physiologic effect of a sort not experienced upon oral administration.

The importance of tissue perfusion is well demonstrated in the elimination of drugs through the lungs. Since the vascularity of the lungs is extremely high, the elimination rates of drugs which are removed by this route are also very high.

Tissue localization is sometimes intended for specific therapy. Regional perfusion is used for carcinolytic agents too toxic to be administered systemically. Intraarterial injection can be used to localize the action of respiratory stimulants acting on the chemoreceptors of the carotid body. A much smaller dose is required to stimulate respiration when injected into the carotid artery than when injected into the vertebral artery or any other vessel, when the site of action lies in these chemoreceptors.

VOLUMES OF DISTRIBUTION

Volume of distribution is a mathematical parameter which estimates the extent and intensity of the exposure of the body to a given dose of a drug. Upon intravenous administration the drug distributes to and equilibrates with all body tissues simultaneously. This results in a change in the ratio between the tissue content and the amount contained in the plasma. Once an equilibrium is reached, this ratio does not change. The plasma concentration fall is initially attributed to the loss of drug from the blood compartment to the tissues as well as to less significant elimination. The volume of distribution relates the plasma concentration to the total amount of drug in the body:[27]

$$\text{Volume of distribution} = \frac{\text{Amount of drug in the body}}{\text{Plasma concentration}} \qquad \text{(Eq. 5.5)}$$

For those drugs with which the equilibrium is quickly reached in the body (e.g., a few minutes), only one volume of distribution term is defined, but if the distribution is a time-dependent process, as with perfusion to bones, muscles, and other poorly perfused organs, then the volume is a function of time. It is customary to define one volume term after the drug reaches equilibrium with highly perfused tissues and another term for the overall equilibration.

The volume of distribution has the following physiologic significance:

1. If a drug is restricted to a certain segment of the body water, the volume of distribution will be equal to the volume of that segment (Fig. 5.4). For example, Evans blue dye remains restricted to plasma whereas antipyrine distributes throughout the total body water. The drug molecules which are restricted to certain body water compartments can be used to calculate the volumes of these compartments. For example, the volume of distribution for D_2O or antipyrine should be about 60 percent of the body weight (assuming 1 liter = 1 kg). The retention of water in the body or, conversely, dehydration can be ascertained by these methods. As suggested above, the volume of distribution of Evans blue dye will provide a direct estimate of the total blood volume.
2. If a drug specifically binds to tissue components or localizes due to its physicochemical properties, then the volume of distribution is going to be a large number and will have no tangible physiologic meaning. For example, the volume of distribution of digoxin is about 600 liters, due to its localization in the deep body tissues. A high volume of distribution generally means greater exposure of the body to the drug or a high degree of selective exposure. Both of these aspects can be used to define the general toxic nature of drugs. Drugs with high volumes of distribution leave the body slowly and are generally more toxic than the drugs which do not distribute deeply into body tissues.

The applications of the term "volume of distribution" are many, including the dosage regimen calculation of drugs based on the characteristics of an individual patient. It is possible to observe different plasma concentration values in different individuals after the administration of the same dose because of the differences in the volume of distribution. Therefore a rationale can be formed for the calculation of dosage adjustment according to body weight, body surface area, lean body weight, body density, etc. This is especially important in pediatric therapy, because infants and children have higher body water content than adults due to incomplete calcification of the bones. Lower plasma concentrations in infants and children may be observed if the dose is only adjusted in proportion to the body weight.

Volume of distribution measurements provide an excellent tool to correlate the physicochemical properties with the duration and intensity of action, based on their distribution. Table 5.4 lists the volumes of distribution of some commonly used drugs to illustrate the wide range of differences possible between drugs.

Table 5.4. VOLUMES OF DISTRIBUTION OF SOME DRUGS IN NORMAL SUBJECTS

DRUG	*VOLUME OF DISTRIBUTION RANGE (LITERS/KG)*	*DRUG*	*VOLUME OF DISTRIBUTION RANGE (LITERS/KG)*
Antipyrine	0.48–0.70	Nalidixic acid[33]	0.26–0.45
Amylobarbital[28]	0.50–1.11	Nortriptyline[34]	22.50–56.90
Diazepam[29]	0.18–1.30	Phenylbutazone	0.04–0.15
Growth hormone[30]	0.071–0.093	Procainamide[35]	1.74–2.22
Heparin[31]	0.055–0.059	Theophylline[36]	0.33–0.74
Insulin[30]	0.054–0.112	Warfarin	0.09–0.24
Lignocaine[32]	0.58–1.91		

References

1. Wosilait WD: Aspects of multiple binding of anticoagulants and other drugs. Gen Pharm 8:7, 1977
2. Jusko WJ, Gretch M: Plasma and tissue protein binding of drugs in pharmacokinetics. Drug Metab Rev 5:43, 1976
3. Mroczek WJ, Leibel BA, Davidov M, Finnerty FA: The importance of rapid administration of diazoxide in accelerated hypertension. N Engl J Med 285:603, 1971
4. Lukas DS, DeMartino AG: Binding of digitoxin and some related cardenolides to human plasma proteins. J Clin Invest 48:1041, 1969
5. Gugler R, Shoeman DW, Azarnoff DL: Effect of in vivo elevation of free fatty acids on protein binding of drugs. Pharmacology 12:160, 1974
6. Jacobsen J, Thiessen H, Brodersen R: Effect of fatty acids on the binding of bilirubin to albumin. Biochem J 126:7, 1971
7. Chan G, Schiff D, Stern L: Competetive binding of free fatty acids and bilirubin to albumin: differences in HBABA dye versus Sephadex G-25 interpretation of results. Clin Biochem 4:208, 1971
8. Spector AA, Imig B: Effect of free fatty acid concentration on the transport and utilization of other albumin-bound compounds; hydroxyphenylazobenzoic acid. Mol Pharmacol 7:511, 1971
9. Lipsett D, Madras BK, Wurtman RJ, Munro HN: Serum tryptophan level after carbohydrate ingestion: selective decline in non-albumin bound tryptophan coincident with reduction in serum free fatty acids. Life Sci 12:57, 1973

10. Raff MJ, Summersgill JT, Fontana FJ, Barnwell PA, Waterman NG, Scharfenberger L: Effect of serum lipid content on the binding of minocycline. J Antibiot Tokyo 30:593, 1977
11. Breckenridge A, Rosen A: The binding of chlorothiazide to plasma proteins. Biochem. Biophys Acta 229:610, 1971
12. Dollery CT, Emslie-Smith D, Muggleton DF: Action of chlorothiazide in hypertension. Proc R Soc Med 53:592, 1960
13. Franz JW, Jahnchen E, Krieglstein J: Der Einflus verschiedener Pharmaka auf das Bindungsrermogen einer Albuminlosung fur Promazin und Chlorpromzain. Naunyn-Schmiedebergs Arch Exp Path Pharmak 264:462, 1969
14. Colaizzi JL, Klink PR: pH-partition behavior of tetracycline. J Pharm Sci 58:1184, 1969
15. Wittenau AM, Yeary R: The excretion and distribution in body fluids of tetracycline after IV administration to dogs. J. Pharmacol Exp Ther 140:258, 1963
16. Kunin CW: Comparative serum binding, distribution, and excretion of tetracycline and a new analog, methacycline. Proc R Soc Exp Biol Med 110:311, 1962
17. Rosenblatt JE, Barrett JE, Brodie JL, Kirby WMM: Comparison of in vitro activity and clinical pharmacology of doxycycline with other tetracyclines. Antimicrob Agents Chemother 6:134, 1966
18. Kunin CW, Dornbash AC, Finland M: Distribution and excretion of 4 tetracycline analogues in normal young men. J Clin Invest 38:1950, 1959
19. Steigbigel NH, Reed CW, Finland M: Absorption and excretion of 5 tetracycline analogues in normal young men. Am J Med Sci 255:296, 1968
20. Ritschel WA: Biological half-lives and their clinical applications. In Francke DE, Whitney HAK (eds): Perspectives in Clinical Pharmacy. Hamilton, Ill., Drug Intelligence Publications, 1972, Chapter 16
21. Brodie BB, Kurz H, Schanker LS: The importance of dissociation constant and lipid-solubility in influencing the passage of drugs into the cerebrospinal fluid. J Pharmacol Exp Ther 130:20, 1960
22. Thrupp LD, Leedon JM, Ivler D: Ampicillin levels in the cerebrospinal fluid during treatment of bacterial meningitis. Antimicrob Agents Chemother 5:206, 1965
23. Lerner PI: Penetration of cephalothin and lincomycin into the cerebrospinal fluid. Am J Med Sci 257:125, 1969
24. Gibaldi M: Biopharmaceutics and Clinical Pharmacokinetics. Philadelphia, Lea and Febiger, 1977, p. 84
25. Astwood EB: Occurrence in the sera of certain patients of large amounts of a newly isolated iodine compound. Trans Assoc Am Physicians 70:183, 1957
26. Schanker LS, Morrison AS: Physiological disposition of guanethidine in the rat and its uptake by heart slices. Int J Pharmacol 4:27, 1965
27. Niazi S: Volume of distribution as a function of time. J Pharm Sci 65:452, 1976
28. Mawer GE, Miller NE, Turnberg LA: Metabolism of amylobarbitone in patients with chronic liver diseases. Br J Pharmacol 44:549, 1972
29. van der Kleijn E: Pharmacokinetics of distribution and metabolism of ataractic drugs and an evaluation of the site of antianxiety activity. Ann NY Acad Sci 179:115, 1971
30. Sonksen PH, Srivastava MC, Tompkins CV: Antibiotic levels on continuous intravenous infusion. Lancet 2:491, 1971
31. Estes JW, Pelikan EW, Kruger-Thiemer E: A retrospective study of the pharmacokinetics of heparin. Clin Pharmacol Ther 10:329, 1969
32. Rowland M, Thompson PD, Guichard A, Melmon KL: Disposition kinetics of lidocaine in normal subjects. Ann NY Acad Sci 179:383, 1971
33. Portmann GA, McChesney EW, Stander H, Moore WE: Pharmacokinetic model for nalidixic acid in man II: Parameters

for absorption, metabolism, and elimination. J Pharm Sci 55:72, 1969
34. Alexanderson B: On interindividual variability in plasma level of nortriptyline and desmethylimipramine in man: a pharmacokinetic and gentic study. Linkoping Univ Med Dissert 6, Linkoping, Sweden
35. Koch-Weser J: Pharmacokinetics of procainamide in man. Ann NY Acad Sci 179:370, 1971
36. Jenne JW, Wyze W, Rood FS, MacDonald FM: Pharmacokinetics of theophylline. Clin Pharmacol Ther 13:349, 1972

Questions

1. What is the driving force for the diffusion of drug molecules to the body tissues?
2. Explain why drug concentrations may not be uniform throughout the body tissues upon equilibration.
3. What form of a drug freely equilibrates with the body tissues? Relate the importance of chemical potential in the distribution of drugs.
4. Under what conditions may the plasma concentration of a drug not be same as the concentration in the whole blood? Why should one be concerned with this difference?
5. What drug properties are necessary for erythrocyte penetration? Is there any active transport across erythrocyte membranes?
6. What is the nature of the interaction between drug molecules and plasma proteins? What are the different kinds of plasma proteins? Specify their concentrations and functions in healthy individuals.
7. Derive Equation 5.3 and its modifications in Figure 5.1.
8. What is the meaning of percent protein binding? Under what condition does this parameter have kinetic or physiologic meaning?
9. Derive Equation 5.4 and calculate the percentage of total drug retained in the plasma when the percentages bound in the plasma are 10, 30, and 85 percent.
10. What is the role of binding strength in the distribution of drugs? What is approximate threshold value of K_a necessary to affect distribution?
11. How is it possible that two binding constants are as widely different as shown in Figure 5.2?
12. What is the component of plasma proteins most important in binding drugs? What kind of drugs bind extensively to albumin?
13. How does plasma protein binding supply essential nutrients to the body, increase the absorption rates of drugs, and provide a homogeneous concentration of drugs throughout the body?
14. Why must diazoxide be administered rapidly when given intravenously?
15. Elaborate on the statement, "Plasma protein binding can either increase or decrease the elimination rates of drugs."

16. Under what conditions can a slight change in the degree of plasma protein binding precipitate serious toxicologic responses?
17. List physiologic and disease states which affect the concentration of plasma proteins.
18. What is the role of the free fatty acids in plasma in the binding of drugs?
19. Why is it necessary during multiple drug therapy to frequently correlate the free drug concentration rather than the total concentration with a pharmacologic response?
20. Cite examples of increased binding of drugs by other drugs.
21. What is the possible effect of electrolyte balance on plasma protein binding?
22. Explain the correlation between lipid solubility and the binding of tetracyclines.
23. Is the permeation of capillary walls a rate limiting step in the distribution of drugs?
24. What is the fluid composition of the human body?
25. How do the characteristics of tissue membranes differ from those of the CNS/CSF barrier?
26. Explain the reason for the therapeutic effectiveness of antibiotics in meningeal infections.
27. Why must levodopa be administered in parkinsonism instead of dopamine?
28. What are the characteristics of the placental transfer of drugs?
29. Should alcohol intake be restricted during pregnancy?
30. Is it possible to have tissue concentrations 100 times higher than plasma concentrations at equilibrium?
31. Are all tissue distribution processes reversible?
32. What is the driving force in tissue accumulation?
33. Cite examples of toxic drug-tissue interactions.
34. How does the blood flow rate affect the rate of tissue equilibration?
35. Which organs can be classified as highly perfused or poorly perfused?
36. Discuss the reasons for the immediate onset of thiopental's action. Why is the effect of thiopental so short-lived?
37. How is tissue distribution responsible for some drug addictions?
38. Under what conditions is a tissue localization desirable for direct therapeutic purposes?
39. What is the meaning of the term "volume of distribution?" Is it possible to have two volumes of distribution for one drug?
40. How would you estimate blood volume and total body water through the respective distribution characteristics of inulin and D_2O?
41. What is the physical meaning of a volume of distribution of 600 liters for digoxin?
42. What is the importance of volume of distribution in drug therapy, especially in pediatric and geriatric care?

CHAPTER 6

The Termination of Drug Action

Absorption and distribution processes determine the onset of action. The duration of action is determined by:

1. Excretion of intact active molecules.
2. Biotransformation of active molecule to inactive structures.
3. Tissue redistribution.

An in-depth understanding of the processes above is necessary to understand the rationale behind dosing interval, loading dose, and maintenance dose selection and the effect of disease states on dosage regimen adjustments.

EXCRETION

Active drug molecules are removed from the body through various body fluids. The transfer of drug molecules from the general circulation to these fluids is governed by the principles applicable to the distribution of drugs. The excretion process involves de-equilibration of tissues, or back diffusion of drug molecules from tissues to the general circulation, and therefore the rates of drug excretion are also dependent on the hemodynamic properties. As a result, the excretion profiles of drugs can significantly change in disease states and show marked variation between and often within individuals.

Renal Excretion

The volume and composition of body fluids are kept constant despite fluctuations imposed by fluid and food intake and by the metabolic products of the body. The function of homeostatis is to conserve the essential substances and remove undesirable components. The kidneys play a very important strategic role in the regulation of homeostatis: they eliminate the majority of unnecessary nonvolatile, water-soluble substances while maintaining the constant volume and osmotic pressure of the blood. Kidneys also adjust the relative and absolute concentrations of normal constituents of the plasma (Table 6.1).

Table 6.1. ELECTROLYTE COMPOSITION OF NORMAL BLOOD PLASMA MAINTAINED CONSTANT BY KIDNEY FUNCTIONS

CONSTITUENT	CONCENTRATION (μEq/liter)
Cations	
Sodium	142
Potassium	4
Calcium	5
Magnesium	2
Anions	
Chloride	101
Bicarbonate	27
Phosphate	2
Sulfate	1
Organic acids	6
Proteins	16

The functional unit of the kidney is the nephron (Fig. 6.1). The endothelium of the glomerular capillaries (a tuft of capillaries in Bowman's capsule) contains large pores which readily permit passage of all plasma constituents except macromolecules (such as protein), thus acting as a kind of sieve. This porous nature coupled with hydrostatic pressure provides a rapid movement of plasma constituents into the capsule. This filtrate of plasma, lacking only the plasma proteins, is called ultrafiltrate. About one-fifth of the blood passing through the capsule is filtered—that is, about 130 ml of ultrafiltrate in an average healthy adult, where about 1200 ml of blood or about 650 ml of plasma flow through the glomeruli each minute. Therefore, about 180 liters of fluid are presented to the kidneys each day for processing.

The filtrate contains such indispensable substances as water, ions, glucose, and other nutrients, as well as such waste materials as phosphate, sulfate, and urea, the end-products of protein metabolism. More than 99 percent of the original filtrate is reabsorbed. The total quantity of water and solutes lost daily in the urine is equal to that acquired by the body, minus only the amounts excreted through other routes. The movement of solutes out of the tubular urine and back into the circulation involves physicochemical processes. The driving force in this transport is generally either the concentration gradient or the energy expenditure in an active transport system. The selectivity principles are those common to other biologic membranes. The drugs diffuse passively back into the circulation in accordance with their respective lipid/water partition coefficients, degrees of ionization, and molecular sizes.

The pH of the filtrate is the same as that of plasma (7.4), but the pH of the urine may vary from 4.5 to 8.0, depending on the secretion of protons and absorption of bicarbonate. An increased urine acidity generally favors the excretion or clearance of weakly basic drugs, which will tend to exist in an

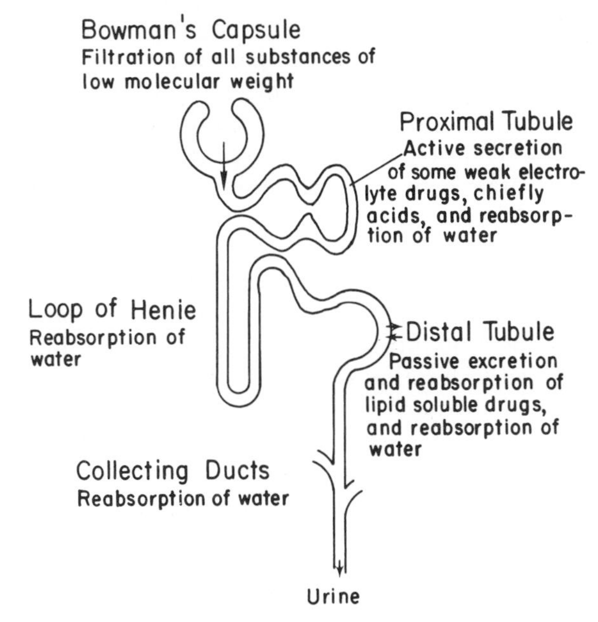

Figure 6.1. Principal processes involved in the renal excretion of drugs.

ionized lipid-insoluble form. However, weakly acidic drugs are also removed, depending on their ionization constants. A change in the urinary pH can be brought about by sodium bicarbonate, ammonium chloride, or other compounds changing the rate of acid secretion. Such a change alters the excretion of weakly acidic and basic drugs (Table 6.2). The alkalinization of urine is often necessary to hasten elimination of an overdose of phenobarbital.

Table 6.2. pH-DEPENDENT CLEARANCE OF DRUGS

CLEARANCE GREATER IN ACIDIC URINE	*CLEARANCE GREATER IN ALKALINE URINE*
Amphetamine	Acetazolamide
Chloroquine	Amino acids
Codeine	Barbiturates
Imipramine	Nalidixic acid
Levorphanol	Nitrofurantoin
Mecamylamine	Phenylbutazone
Mepacrine	Probenecid
Meperidine	Salicylic acid
Morphine	Sulfonamides
Nicotine	
Procaine	
Quinine	

If urinary pH is maintained at about 5, approximately 55 percent of a dose of dextroamphetamine is excreted in 16 hours, as compared to about 3 percent excretion when the urinary pH is increased and maintained at 8. Erythromycin is regarded as being primarily active against gram-positive organisms. However, when the urine is alkalinized, erythromycin becomes effective against certain gram-negative organisms that commonly cause urinary tract infections.

Although active reabsorption of essential components of the filtrate is quite common, only a few drugs are reabsorbed by a carrier-mediated process. On the contrary, the active secretion of drugs (Fig. 6.1) is more significant than it is for essential organic components (Table 6.3). It appears that only two transport processes are responsible for almost all of the secretion. This may at times restrict the transport of substances sharing the same system. Such restrictions can often be exploited for therapeutic advantage by decreasing or increasing the excretion rates of various compounds. For example, salicylic acid inhibits the excretion of uric acid, especially when uric acid is present in high quantities—e.g., in the state of gout. Probenecid acts as a uricosuric agent by inhibiting the tubular reabsorption of urate. It also inhibits the tubular secretion of penicillin and usually increases penicillin plasma levels regardless of the route of administration. A two- to four-fold elevation has been demonstrated for various penicillins. Probenecid also decreases the urinary excretion of aminosalicylic acid, aminohippuric acid, phenolsulfonphthalein, pantothenic acid, 17-ketosteroids, and sodium iodomethamate. It decreases both renal and hepatic excretion of sulfobromophthalein. Probenecid does not affect plasma concentrations of salicylates, nor the excretion of streptomycin, chloramphenicol, chlortetracycline, oxytetracycline, or neomycin. Organic cations compete with each other but not with anions, and the process of active secretion is generally blocked by metabolic inhibitors.

The extent to which the organic acids or bases are eliminated in the urine following their secretion is dependent on their degree of ionization within

Table 6.3. DRUGS SECRETED BY ACTIVE TRANSPORT INTO THE TUBULI

ACIDS	*BASES*
Amino acids	Choline
p-Aminohippuric acid	Dihydromorphine
Acetazolamide	Dopamine
Ethacrynic acid	Histamine
Furosemide	Mepiperphenidol
Indomethacin	Methylnicotinamide
Mersalyl	Neostigmine
Penicillins	Quinine
Phenol red	Tetraethylammonium
Phenylbutazone	
Salicylates	
Thiazide	

the tubular urine. For example, quarternary ammonium compounds are fully ionized at all pH values and are therefore completely excreted. Organic acids such as penicillin are also highly ionized and thus undergo little reabsorption.

Since the elimination of drugs and other compounds involves an elaborate balance of the various kidney functions, an estimation of the excretion rates of drugs provides a useful procedure for diagosing the functional status of this organ. For example, the glomerulus filtration rate can be calculated by the rate of excretion of a compound if the compound is filterable, unbound to plasma proteins, and nontoxic, remains chemically unaltered during passage through the kidneys, and can be easily quantitated. The polymeric carbohydrate inulin meets all of these requirements. Its excretion takes place only by glomerular filtration. Since excretion is proportional to the plasma concentration:

$$\begin{array}{c}\text{Amount of inulin}\\ \text{excreted in}\\ \text{urine per minute}\end{array} = \begin{array}{c}\text{Concentration of inulin in plasma} \times\\ \text{volume of plasma filtered per minute}\end{array} \qquad \text{(Eq. 6.1)}$$

The volume of plasma filtered per minute is the glomerular filtration rate and can be easily calculated. Quantitative data on kidney function obtained in this manner are termed as the results of a plasma renal clearance study RENAL CLEARANCE is defined as the volume of plasma needed to supply the amount of a specific substance excreted in the urine in one minute. For example, in the case of inulin the value will be around 125 ml. Any physiologic or pathologic condition which affects the glomerular filtration of the kidneys can be easily detected through such plasma clearance studies, which are an integral part of the dosage regimen adjustments of a large number of drugs in disease states.

Some organic acids, such as paraaminohippuric acid, are secreted so rapidly that they are almost entirely removed from the plasma in a single passage through the kidneys. The plasma clearance of these substances will therefore be equal to the total plasma flow to the kidneys and forms an additional tool for the evaluation of total blood flow to the kidneys. The average plasma flow in normal healthy individuals is about 650 ml/min. Drugs such as penicillin show complete clearance in each cycle through the kidneys despite their extensive plasma protein binding. This is due to the rapid equilibration between bound and free drug in the plasma.

The determination of the renal clearance of a drug provides an insight into the mechanism by which it is excreted when its clearance is compared with the normal glomerular filtration rate—i.e., 125 to 130 ml/min, as obtained using inulin as an index. If the unbound plasma concentration is used to calculate the renal clearance, an expression of excretion ratio can be obtained:

$$\text{Excretion ratio} = \frac{\text{Renal clearance of drug (ml/min)}}{\text{Normal inulin clearance (ml/min)}} \qquad \text{(Eq. 6.2)}$$

An excretion ratio of less than one indicates that the drug is filtered, perhaps secreted, and then perhaps partially reabsorbed. A value greater than

Table 6.4. EXAMPLES OF DRUGS WHICH ARE EXTENSIVELY (>90%) CLEARED INTACT FROM THE KIDNEYS AND WHICH ARE LITTLE CLEARED (<5%).

EXTENSIVE CLEARANCE	*LITTLE CLEARANCE*	
Acetazolamide	Acetaminophen	Hydrallazine
Amantadine	Acetohexamide	Imipramine
Amiloride	Acetophenetidin	Indomethacin
Amphetamine	Aminopyrine	Levodopa
Barbital	Antipyrine	Lorazepam
Disodium cromoglycate	Aspirin	Nalidixic acid
Gentamicin	Bupivacaine	Novobiocin
Lithium	Carbenoxolone	Pentazacine
Methotrexate	Chlormethiazole	Phenacetin
Penicillins	Chlorpromazine	Phenytoin
Pentolinium	Desipramine	Prednisolone
Practolol	Diamorphine	Succinylcholine
Tranexamic acid	Diazepam	Suxamethonium
Vancomycin	Diazoxide	Thiopental
	Diphenoxylate	Thyroxine
	Griseofulvin	Vinblastine
	Heroin	Vincristine

one indicates secretion in addition to filtration and perhaps reabsorption. The highest excretion ratio value that can be obtained is about five, at which value the renal clearance of the drug becomes equal to the total volume of plasma flowing per minute. An excretion ratio of zero will be expected for such compounds as glucose, which are not expected to be excreted through the kidneys because of complete reabsorption. Table 6.4 lists the drugs which are extensively cleared from the body by renal clearance as well as those which are cleared insignificantly.

Table 6.5. RENAL FUNCTION TESTS

FUNCTION	*SPECIFIC TEST*	*CLINICAL TEST*
Glomerular filtration	Inulin clearance	Creatinine clearance
	^{125}I-Iothalamate	Plasma creatinine
	^{169}Yb-Diethylenetriaminepentoacetic acid	Plasma urea
	^{51}Cr-Ethylenediaminetetraacetic acid	
Renal plasma flow	Paraaminohippuric acid clearance	Phenolsulfonphthalein excretion
	^{125}I-Hippuran	
Proximal tubular transport	Maximum glucose transport	Plasma phosphate, urate
	Maximum paraaminohippuric acid secretion	Urinary amino acids
		Phenolsulfonphthalein excretion
Distal tubular transport	Maximal urine/plasma osmolarity	Maximum urinary osmolarity
	Acidifying capacity	Acid and bicarbonate loading

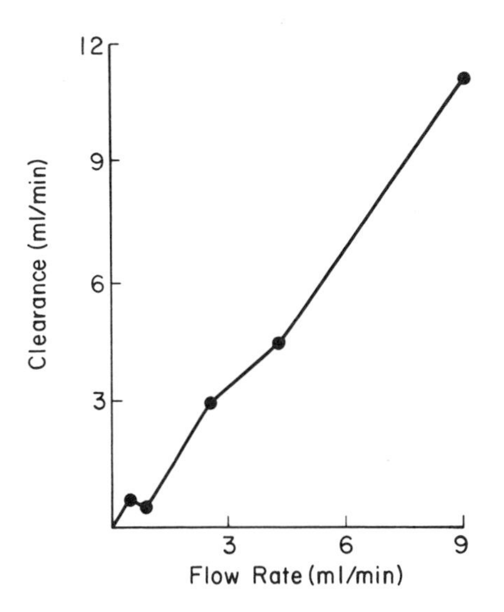

Figure 6.2. Urinary clearance of ethanol as a function of urine flow rate. (From Curry: Drug Disposition and Pharmaco-kinetics, 1977. Courtesy of Blackwell Scientific Publications)

In clinical practice, the glomerular filtration rate is often determined by the clearance of creatinine, an endogenous product of body metabolism. Creatinine is eliminated entirely by the kidneys, mainly by glomerular filtration. There is some indication of both its active transport and its metabolism in the body. The normal range of serum creatinine concentration is about 1 to 2 mg/100 ml, and its clearance, which is dependent on the body mass, is usually adjusted on the basis of the adult body surface area of 1.72 m^2. Although inulin clearance provides the most accurate measurement of glomerular filtration, the advantage in using the creatinine clearance is that it can be correlated to the steady state concentration of creatinine in the blood, often eliminating the need for collection of urine. For example:

$$\text{Creatinine clearance} = \frac{100 - 0.5\,(\text{age}-20)}{\text{serum creatinine (mg/100 ml)}} \qquad \text{(Eq. 6.3)}$$

$$= \frac{(140-\text{age, yr})\,(\text{body wt, kg})}{72 \times \text{serum creatinine (mg/100 ml)}} \qquad \text{(Eq. 6.4)}$$

A large number of equations and nomograms are available for direct calculation of creatinine clearance based on the serum concentration. A variety of renal function tests are also available to ascertain the effect of renal clearance on the elimination of drugs from the body (Table 6.5).

It is generally agreed that the clearance of drugs is not affected by the urine flow rate, since the reabsorption from tubuli takes place proximal to the most distal sites of water transfer so that dilution or concentration of the urine will not affect the rate of loss of drug from the body. However, if the reabsorption of drugs takes place after the concentration of the urine, or if the transfer takes place from the bladder, then the urine flow rate will affect the clearance significantly, as is shown for ethanol, glutethimide, amylobarbital, butobarbital, cyclobarbital, and phenobarbital (Fig. 6.2).[1] Increasing the flow rates of urine will increase the excretion rates of these drugs.

Biliary Excretion

Liver secretes about 0.5 to 1 liter of bile into the duodenum through the common bile duct every day (Fig. 6.3). Large quantities of bile are discharged into the duodenum in response to food intake. The major portion of bile is, however, conserved by reabsorption. The bile acids are functionally important for the digestion and absorption of fats. Almost 90 percent of secreted bile acids are reabsorbed from the intestine and transported through the portal blood back to the liver to be available again for secretion. The transport of bile acids is an active process—passive diffusion would not allow sufficient transport because of the ionized nature of these compounds and the diffusion against the concentration gradient.

Many drugs are also excreted by the liver into the bile. However, most are completely reabsorbed by passive diffusion across the intestinal barrier and are resecreted until they are excreted in the urine. This phenomenon is referred to as enterohepatic cycling (Fig. 6.3).

The transport of organic acidic and basic drugs from the liver into the bile takes place through mechanisms very similar to those effecting the secretion of these same substances into the tubular urine. The protein-bound component in the plasma is fully available for biliary secretion. The enterohepatic cycling of drugs takes place only if the drugs are passively reabsorbed or if there are specific absorption sites available—this is not the case for many drugs. For example, quarternary ammonium compounds ionize completely upon biliary secretion and are not reabsorbed, thus providing an effective means for their elimination from the body. A large number of drugs are eliminated into an enteric pool as a consequence of biliary secretion. For some drugs, such as stilbesterol, rifamycin, and lysergic acid diethylamide, this process can be so rapid and extensive that from 50 to 100 percent of the available dose is localized into an enteric pool within a few hours after administration.

The biochemical mechanism of drug excretion into the bile is not very well-known, but a large number of factors such as molecular weight, chemical structure, polarity, species, sex, and the nature of the biotransformation processes affect the rate and extent of biliary elimination. The molecular weight of the drug or its biotransformation product seems to affect the biliary secretion most significantly. For example, it is only when the molecular weight is around 500 that a significant biliary excretion (>5 to 10 percent) can be expected in humans.[2]

The molecular weight threshold is also dependent on the chemical nature of the drug. For example, the threshold is much lower for monoquarternary ammonium cations than is required for aromatic anions. Such animal species as the rat, guinea pig, and rabbit require a threshold of 200 for quarternary

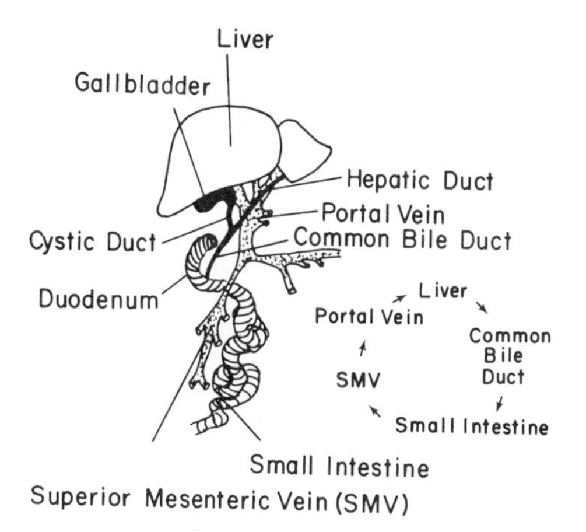

Figure 6.3. The enterohepatic cycle.

compounds and above 300 (325 to 475) for aromatic anions. For humans, a threshold of about 300 is required for quarternary structures.

A drug may be eliminated in the bile in three forms: unchanged; as a conjugate with glucuronic acid; or as a glutathione conjugate, provided the combined molecular weight meets the threshold requirement. Such biotransformations as methyl, acetyl, or amino acid conjugates are of limited value in biliary excretion. This is because these transformations do not increase the molecular weight of the drug significantly when compared with glucuronide conjugation (where the molecular weight increases by 176) or with glutathione conjugation (where the molecular weight increases by about 300) (Table 6.6). The latter conjugation reactions are more possible in the body because of the large number of different drug structures which can accept or be activated to accept these conjugating agents. Drugs which are excreted unchanged must meet an additional requirement—they must have a polar functional group, as with carboxyl, sulfonate, or ammonium, before they can be eliminated in the bile.

Table 6.6. EXAMPLES OF CHEMICAL FORMS IN WHICH SOME DRUGS ARE ELIMINATED IN THE BILE

CHEMICAL FORM	*EXAMPLE*
Unchanged	
Carboxylic acid	Cromoglycate
Sulfonic acid	Amaranth
Quarternary ammonium	Tribenzylmethylammonium
Glycoside	Ouabain
Glutathione	Chlorobenzene
Glucuronic acid	Morphine, naphthalene, indomethacin, chloramphenicol

A possible rate limiting step in the biliary excretion of drugs is delay due to storage in the gall bladder. A drug can be stored in the gall bladder for hours before it is intermittently released into the intestine. This phenomenon is responsible for the secondary plasma peaks of some drugs, which often coincide with the food intake. This is demonstrated for digitalis glycosides, diazepam, and practolol, which may undergo hepatic circulation. However, such drugs as phthalylsulfathiazole, succinylsulfathiazole, cromoglycate, fluoresceins, tribenzylmethylammonium, and dibenzyldimethylammonium are not reabsorbed from the enteric pool during the enterophepatic circulation and are excreted extensively. An interesting example of biliary elimination is that of clomiphene citrate, an ovulation-inducing agent. This drug is almost completely removed from the body via biliary secretion, regardless of the route of administration.

A number of drugs, including dichloromethotrexate, indomethacin, carbenoxolone, and the antibiotics rifamide and erythromycin, undergo biliary excretion in man, but the extent to which these and other drugs undergo hepatic recirculation is not known and is difficult to establish. This is due to the fact that most of the biliary excretion data have been collected during gall bladder surgery and the information remains scanty for most compounds.

The efficiency of biliary system excretion of a drug can often be assessed by administering a drug which is exclusively and totally eliminated in the bile, such as sulfobromophthalein, a synthetic organic acid dye. In an average individual with normal liver function most of the dye is excreted into the intestine within 30 minutes. The extent of liver function depression can be monitored by the concentration of the dye in the blood 30 minutes after the beginning of the test.

Pulmonary Excretion

The lung is the major organ of excretion for gaseous and volatile substances. The breath test given to drunken drivers is the best example of the pulmonary excretion of drugs such as alcohol. The use of paraldehyde has been limited as a sleep-inducing agent due to the foul smell that appears in the breath upon its excretion.

A large number of gaseous anesthetics are eliminated through the pulmonary route.

Salivary Excretion

The excretion of drugs in saliva is determined largely by their pH-partition properties. The salivary glands behave as lipoid membranes and lipid-soluble forms of drugs are excreted by simple diffusion since there is little possibility of any filtration process. The salivary pH varies from 5.5 to about 8.4 and therefore, based on the pH-partition theory, the concentrations of drugs can be higher or lower than in the plasma (pH = 7.4). The plasma protein binding, however, makes the total concentration of drugs on the plasma side generally higher than on the saliva side. There is some indication that basic drugs tend to accumulate in the salivary glands preferentially, often reaching concentrations as high as 0.1 percent by weight. This localization is often responsible for dryness of the mouth due to the inhibition of salivary flow by these basic drugs.[1]

There are also indications of the active transport of various drugs in the saliva. For example, weak acids such as penicillins are generally transported by an active process which can be partially blocked by probenecid. The excretion of lithium also appears to involve an active process, whereby salivary concentrations as high as two to three times the plasma concentrations have been found.[3]

A variety of drugs have been reported excreted in saliva, including sulfonamides, phenobarbital, clonidine, antipyrine, rifampin, phenytoin, theophylline, salicylates, primidone, quinidine, acetaminophen, tolbutamide, procainamide, and digoxin.

In some instances salivary excretion is responsible for localized side effects. For example, excretion of antibiotics may cause black hairy tongue, and gingival hyperplasia can be a side effect of phenytoin.[4,5]

The salivary excretion of drugs offers an excellent means of monitoring the plasma concentration if the partition coefficients between plasma and saliva remain constant (which is often the case). For example, the half-life of antipyrine can be determined by monitoring its salivary level (Fig. 6.4).[6] The effectiveness of rifampin against meningococcal infection can be determined

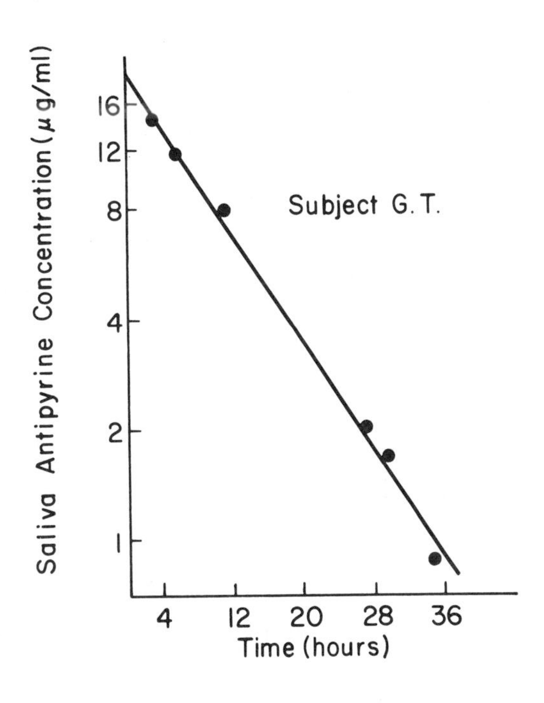

Figure 6.4. Antiyprine elimination in saliva after oral administration of a 10 mg/kg dose. (From van Boxtel et al.: Eur J Clin Pharmacol 9:327, 1976)

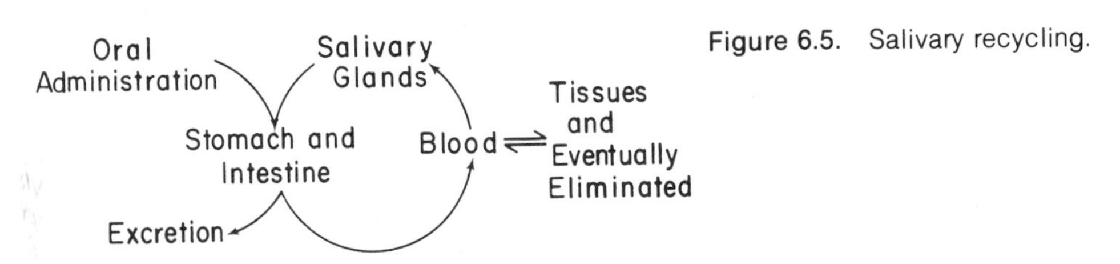

Figure 6.5. Salivary recycling.

by the salivary level of the antibiotic. A salivary concentration of rifampin exceeding the minimum inhibitory concentrations for the infecting organism appears to provide adequate results.[7]

The use of saliva levels in monitoring drug therapy is an attractive proposition which is subject to significant error if improperly utilized. A common error is inappropriate cleansing of the mouth following oral administration of drugs before saliva samples are collected. There also seems to be little need for stimulating the salivary output by mastication. The salivary concentrations are generally independent of the volume of saliva. Despite all precautions, some drugs show extreme variations in their plasma:saliva concentration ratios, making salivary monitoring of plasma concentrations relatively useless. One such example is that of procainamide, where ratios ranging from 0.3 to 8.8 have been reported.[8]

Drugs which are excreted in the saliva can undergo cycling similar to biliary cycling, as is noted for sulfonamides, antibiotics, and the hypotensive agent clonidine (Fig. 6.5). The pharmacologic significance of this recycling will depend on the extent of salivary excretion and the degree of oral bioavailability of drugs. Few studies have noted this aspect in relation to the disposition of drugs from the body.

Mammary Excretion

The extent of drug excretion from the plasma into breast milk is independent of the milk volume. It is almost entirely determined by the pH-partition behavior, molecular weight, and lipid solubility of the drug molecules. The pH of human milk ranges from 6.8 to 7.3 and therefore variable amounts of a drug can be transferred to the milk, depending upon the ionization constant of the drug. Most drugs enter the mammary alveolar cells in the un-ionized, non-protein-bound form. Under normal pH conditions, the concentration of the weakly acidic component in milk is lower than it is in the plasma, while the concentration in weakly alkaline drugs in some instances equals or exceeds the plasma level (Table 6.7).[9–11] Some drugs in free solution may pass into the alveolar milk directly via spaces between mammary alveolar cells. Nonelectrolytes such as ethanol and tetracycline enter the milk by diffusion through membrane pores and may reach the same concentration as in the plasma.

The total amount of drug excreted is generally less than 1 percent and the fraction of it ingested by the infant is generally too small to achieve thera-

Table 6.7. CONCENTRATION OF VARIOUS DRUGS IN MATERNAL BLOOD AND BREAST MILK UNDER NORMAL pH CONDITIONS

DRUG ADMINISTERED (THERAPEUTIC DOSAGE)	DRUG LEVELS (UNITS/100 ML)		ADMINISTERED DRUG APPEARING IN MILK (%/DAY)
	Plasma or Serum (pH 7.4)	Milk (pH 7.0)	
Aspirin	1-5 mg	1-3 mg	0.5
Bishydroxycoumarin	11-16.5 mg	0.2 mg	0.5
Chloral hydrate	0-3 mg	0-1.5 mg	0.6
Chloramphenicol	2.5-5 mg	1.5-2.5 mg	1.3
Chlorpromazine	0.1 mg	0.03 mg	0.07
Colistin sulfate	0.3-0.5 mg	0.05-0.09 mg	0.07
Cycloserine	1.5-2 mg	1-1.5 mg	0.6
Diphenylhydantoin	0.3-4.5 mg	0.6-1.8 mg	1.4
Erythromycin	0.1-0.2 mg	0.3-0.5 mg	0.1
Ethanol	50-80 mg	50-80 mg	0.25
Ethyl biscoumacetate	2.7-14.5 mg	0-0.17 mg	0.1
Folic acid	3 μg	0.07 μg	0.1
Imipramine hydrochloride	0.2-1.3 mg	0.1 mg	0.1
Iodine 131	0.002 μc	0.13 μc	2-5
Isoniazid	0.6-1.2 mg	0.6-1.2 mg	0.75
Kanamycin sulfate	0.5-3.5 mg	0.2 mg	0.05
Lincomycin	0.3-1.5 mg	0.05-0.2 mg	0.025
Lithium carbonate	0.2-1.1 mg	0.07-0.4 mg	0.12
Meperidine hydrochloride	0.07-0.1 mg	trace (<0.1 mg)	<0.1
Methotrexate	3 μg	0.3 μg	0.01
Nalidixic acid	3-5 mg	0.4 mg	0.05
Novobiocin	1.2-5.2 mg	0.3-0.5 mg	0.15
Penicillin	6-120 μg	1.2-3.6 μg	0.03
Phenobarbital	0.6-1.8 mg	0.1-0.5 mg	1.5
Phenylbutazone	2-5 mg	0.2-0.6 mg	0.4
Pyrilamine maleate	—	0.2 mg	0.6
Pyrimethamine	0.7-1.5 mg	0.3 mg	0.3
Quinine sulfate	0.7 mg	0.1 mg	0.05
Rifampin	0.5 mg	0.1-0.3 mg	0.05
Streptomycin sulfate	2-3 mg	1-3 mg	0.5
Sulfapyridine	3-13 mg	3-13 mg	0.12
Tetracycline hydrochloride	80-320 μg	50-260 μg	0.03
Thiouracil	3-4 mg	9-12 mg	5

Adapted from Vorherr: The Breast: Morphology, Physiology, and Lactation, 1974. Courtesy of Academic Press.

peutic or toxic levels in the plasma. Moreover, drugs may be excreted in milk in inactive forms—for example, almost 50 percent of the chloramphenicol excreted in the milk is antimicrobially ineffective. Some drugs are even inactivated after being excreted into the milk.

Nevertheless, some drugs passing into the milk are very potent and are able to induce toxic responses in the infant, due primarily to the infant's insufficient hepatic and renal detoxification potentials. Specific effects include neonatal jaundice as a result of sulfonamide interaction with bilirubin, superinfection from antibiotics, and dental mottling upon tetracycline ingestion in infants.

An interesting but relatively unnoticed interaction is due to cigarette smoking. Mothers smoking more than 20 to 30 cigarettes a day have a decreased milk yield along with the possibility of inducing nausea, vomiting, abdominal cramps, and diarrhea in the infant.[12]

In short, whereas there is no need for alarm, lactating mothers should be encouraged to bottle-feed if prolonged drug therapy of any kind is recommended.

Skin Excretion

The excretion of drugs into sweat is a well-established fact. Iodine, bromine, benzoic acid, salicylic acid, lead, arsenic, mercury, iron, alcohol, and antipyrine are some of the compounds that are excreted in the sweat (Table 6.8).[13] Drug excretion in sweat is primarily dependent on the diffusion of nonionized species, and the sweat:plasma ratios are generally independent of the plasma concentration, indicating passive diffusion.

Table 6.8. EXCRETION OF VARIOUS COMPOUNDS DURING SWEATING IN MAN

COMPOUND	*SWEAT/PLASMA RATIO*	pK_a
Urea	1.84	13.8
Sulfanilamide	0.69	10.4
Sulfapyridine	0.58	8.4
Sulfathiazole	0.13	7.1
Sulfadiazine	0.11	6.5
p-Aminohippuric acid	0.02	3.8

From Thayson and Schwartz: J Exp Med 98:261, 1953.

Gastrointestinal Excretion

A number of basic drugs can be excreted into the stomach after intravenous administration due to partitioning equilibration with the ionized species in the stomach (as with quinine and nicotine). Water-soluble drugs and weak electrolytes can be excreted into the intestine on the basis of the pH-partition theory. The excreted drug in the stomach is generally reabsorbed when it reaches the intestine, where it may convert into a nonionized form (making this route of excretion of little importance). However, if the excreted drug in the stomach undergoes a decomposition reaction or binds with the components of the gastrointestinal tract, the total excretion may become very important in the overall picture of drug disposition.

Genital Excretion

Some drugs are known to be excreted in prostate secretions and are detectable in semen. No systemic investigation has ever been made of this route of excretion. Whereas this route seems irrelevant in the overall disposition of drugs, there remains a possibility of drug-induced abnormalities during conception.

BIOTRANSFORMATION

The term biotransformation is preferable to the term metabolism for describing the chemical aspects of the fate of foreign compounds in the body. Metabolism refers to the processes in living cells by which energy is provided for the vital systems of the body.

Biotransformation reactions take place in the presence of enzymes, which act as catalysts in the process and are not themselves consumed. The activity of enzymes is often dependent on the presence of dissociable coenzymes, nondissociable prosthetic groups, and such small ions as magnesium.

The major purpose of biotransformation reactions is to deactivate the drug and promote its elimination. The majority of biotransformation reactions lead to conversion of relatively lipid-soluble compounds into relatively water-soluble forms which are easily excreted in the kidneys. Biotransformation reactions generally inactivate the drug molecules, but there are several exceptions to this rule (Table 6.9).

The importance of biotransformation can be learned from such drug examples as mepacrine and thiopental, which will remain in the body for several years unless biotransformed. However, not all biotransformations produce readily excretable forms. Methylation and acetylation lead to less water-soluble forms and are excreted slowly when compared with the parent drug (Table 6.10).

Acetylation reactions are undesirable since they lead to relatively water-insoluble forms which can precipitate in the kidneys, as is demonstrated by sulfonamides. One alternative to this problem is the use of trisulfapyrimidines, in which a mixture of sulfonamides are used instead of a single entity. The acetylated derivatives of each sulfonamide are prevented from precipitating in the kidneys because there are smaller concentrations of each, due to their independent precipitation processes. However, the combined activity of the sulfonamides is similar to the activity obtained from one form of the drug.

The various pathways of biotransformation can be divided into two phases. Phase I involves exposure or addition of functionally reactive groups by oxidation, reduction, or hydrolysis. Phase II consists of conjugation of these reactive groups. For example, phenacetin is first dealkylated to expose a reactive hydroxyl group, which subsequently conjugates with glucuronic acid

Table 6.9. BIOTRANSFORMATION OF DRUGS TO ACTIVE FORMS

PARENT DRUG	ACTIVE FORM
Acetohexamide	Hydrohexamide
Acetylsalicylic acid	Salicylic acid
Allopurinol	Alloxanthine
Amitriptyline	Nortriptyline
Chloral hydrate	Trichloroethanol
Chlordiazepoxide	Desmethylchlordiazepoxide
Codeine	Morphine
Diazepam	Desmethyldiazepam
Digitoxin	Digoxin
Flurazepam	Desalkylflurazepam
Glutethimide	4-Hydroxyglutethimide
Imipramine	Desipramine
Meperidine	Normeperidine
Mephobarbital	Phenobarbital
Methabarbital	Barbital
Methamphetamine	Amphetamine
Nitroprusside	Thiocyanate
Phenacetin	Acetaminophen
Phenylbutazone	Oxyphenbutazone
Prednisone	Prednisolone
Primidone	Phenobarbital
Procainamide	N-Acetylprocainamide
Propranolol	4-Hydroxypropranolol
Spironolactone	Canrenone
Sulindac	Sulfapyridine
Sulfasalazine	Sulindac sulfide
Trimethadione	Dimethadione

to yield a water-soluble compound. In those instances where reactive groups are already exposed, phase I reactions are not required. Examples are provided by chloramphenicol and morphine, which have hydroxyl groups for direct conjugation with glucuronide or sulfonamide, which in turn have amine groups for direct acetylation (Table 6.10).

The major site of biotransformation is the liver. Most of the specific enzymes are found in smooth surface endoplasmic reticula. Microsomal enzymes are nonspecific in nature and are able to biotransform various structures. The first exposure of drug molecules to hepatic biotransformation comes during the first pass in oral or intraperitoneal administration, when a large number of drugs can be significantly removed from the blood (Tables 4.3; 4.4).

The biotransformation in the gastrointestinal tract affects the bioavailability as well as the activity of some drugs. For example, peptidases of the intestinal secretions destroy orally administered polypeptides, reducing their bioavailabilities to almost zero. The drug conjugates excreted in the bile are also decomposed by intestinal enzymes, which promote the enterohepatic cycling of these drugs, as is shown with morphine and chloramphenicol. Hydrolysis reactions in the intestine release the active component emodin from cascara and senna. Isoprenaline and chlorpromazine undergo sulfate conjugation in the intestine, resulting in their inactivation. As a result, a 10 μg intravenous dose is about equal to a 10 mg oral dose of isoprenaline.

An interesting example of manipulating intestinal biotransformation is that of levodopa, which is decarboxylated to dopamine in the gastrointestinal tract. If the enzymes responsible for this reaction are inhibited, an almost ten-fold increase in the activity of levodopa is obtained.

Some amines also undergo biotransformation due to the presence of amine oxidases in the intestine.

Other sites of biotransformation include the plasma and the lungs. Plasma cholinesterases are responsible for the inactivation of hexamethonium, procaine, and propanidid, resulting in the extremely short durations of action of hexamethonium.

Some drugs are biotransformed in the lungs by amine oxidase and other enzymes. Few studies have addressed the role of the lungs in the overall biotransformation of drugs. Prostaglandins seem to be biotransformed during the first passage through the lungs.

The products and rates of biotransformation are generally dependent on the site of biotransformation, due to the high degree of specificity associated

Table 6.10. PATHWAYS OF BIOTRANSFORMATION

REACTION	*EXAMPLES*	*REACTION*	*EXAMPLES*
Oxidation		*Reduction*	
Aromatic hydroxylation	Antipyrine	Azoreduction	Protonsil
Aliphatic hydroxylation	Pentobarbital	Nitroreduction	Chloramphenicol
Deamination	Amphetamine	Carbonylreduction	Chloral hydrate
N-dealkylation	Imipramine		
N-hydroxylation	Trimethylamine		
O-dealkylation	Codeine, phenacetin	*Conjugation*	
Desulfuration	Thiopental		
S-dealkylation	6-Thiopental	Glucuronidation	Morphine
Sulfoxidation	Chlorpromazine	N-methylation	Noradrenaline
Dehalogenation	Dicophane-DDS	S-methylation	Dimercaprol-BAL
Dehydrogenation	Ethanol	Acetylation	Sulfadimidine
Oxidative Deamination	Thiamine	Amino Acid (e.g. glycine)	Salicylate
		Mercapturic Acid Formation	Napthalene
Hydrolysis		Sulfate Formation	Salicylamide
De-esterification	Procaine Suxamethonium		
Deamidation	Procainamide		

with the function of enzymes. This specificity of enzyme action is also responsible for differences in the biotransformation of enantiomers in the racemic mixture of a compound. For example, S(−) warfarin is a more potent anticoagulant than R(+) warfarin, but the former is more rapidly removed from the body.[14] Similar differences in the rates of biotransformation have also been noted for hexobarbital,[15] propranolol,[16] and ibuprofen.[17]

In most instances the extent of biotransformation of a drug is a constant just like other parameters related to the disposition. However, disease states and inter- and intra-subject variations can lead to significant changes in the degree of biotransformation. These variations are partly responsible for the inherent variability in human drug response and will be discussed in a later chapter.

The kinetics of biotransformation reactions is somewhat similar to the kinetics of protein binding or of any other reaction described by the Michaelis-Menton equation:

$$\text{Rate of reaction} = \frac{V_m C}{K_m + C} \qquad \text{(Eq. 6.5)}$$

where V_m is the maximum reaction velocity, C is the drug concentration, and K_m is the Michaelis constant as well as the drug concentration at which the rate of reaction is one-half of the maximum rate.

Depending on the drug concentration, C, Equation 6.5 can be a first order or a zero order equation. At low drug concentrations, when the concentration is much smaller than K_m:

$$\text{Rate of reaction} = \frac{V_m C}{K_m} = k' \cdot C \qquad \text{(Eq. 6.6)}$$

where the rate of biotransformation is proportional to the drug concentration. At higher drug concentrations, the rate becomes equal to V_m, or the maximum rate.

The Michaelis-Menton behavior of drug biotransformation can lead to nonlinear rates of drug disposition if the concentration exceeds the saturation level, i.e., if it is above K_m. For example, K_m for phenytoin is about 4 μg/ml, and therefore at the desired plasma concentrations of 10 to 20 μg/ml the rate of biotransformation is not proportional to the plasma concentration.[18] This will result in a net decrease in the rate of elimination of phenytoin at higher dose levels. Another example where nonlinear biotransformation kinetics is operative is in the case of aspirin, where the elimination is considerably slowed down in arthritic patients taking large doses.[19] It is understandable that the effect of nonlinear biotransformation will be noted only for those drugs which undergo extensive biotransformation. If the extent of biotransformation is not significant, little effect will be noticed in the elimination rates.

Another source of variation in biotransformation is the change in the enzyme content or its activity as a result of drug interaction. The enzyme

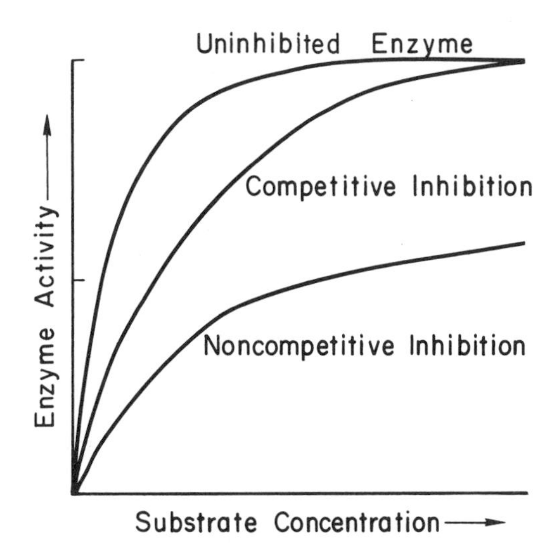

Figure 6.6. Effect of enzyme inhibitors on enzyme activity.

inhibition can be of a competitive or noncompetitive type. In competitive inhibition structurally similar compounds compete for the same site on an enzyme (Fig. 6.6), whereas in noncompetitive inhibition an agent unrelated in structure to the drug combines with the enzyme in such a way as to prevent the formation of an enzyme:drug complex.

The competitive inhibition is reversible and the effect can be overcome at higher substrate (drug) concentrations, whereas in noncompetitive inhibition the maximum capacity of the enzymes to affect a reaction is decreased. An example of competitive inhibition is the action of amphetamine, which inhibits the biotransformation of tyramine by the monoamine oxidase enzyme. Methacholine inhibits the biotransformation of acetylcholine by competing with it for the reaction site on cholinesterase enzymes. Noncompetitive inhibition is brought about by many heavy metals, such as mercury, lead, and arsenic, and organic phosphate insecticides.

The net effect of enzyme inhibition is a decrease in the biotransformation rate and prolongation of the drug effect. Table 6.11 lists examples of drug interactions leading to inhibited biotransformation. Many of these interactions can lead to serious toxic manifestations if left unchecked.

The drug interactions are also responsible for the stimulation of enzymes and for thereby increasing the rate of drug action termination. Prior treatment with a large number of drugs results in either an increased rate of enzyme synthesis or the induction of new enzyme production. Drugs often stimulate their own biotransformations. Table 6.12 lists some examples of enzyme stimulations. It is interesting to note that there may not be any structural or pharmacologic similarity between the drug molecules which are stimulated by a given compound. For example, phenobarbital increases the rate of biotransformation of a wide range of compounds. An interesting example is

Table 6.11. EXAMPLES OF BIOTRANSFORMATION INHIBITION

INHIBITOR	*BIOTRANSFORMATION-INHIBITED DRUG*
Allopurinol	Mercaptopurine
Chloramphenicol	Hexobarbital
Desipramine	Amphetamine
Methandrostenolone	Oxyphenylbutazone
MAO inhibitors	Tyramine, barbiturates
p-Aminosalicylic acid	Hexobarbital, phenytoin
Oxyphenylbutazone	Coumarins
Dicumarol/disulfiram/isoniazid/warfarin	Phenytoin
Warfarin	Tolbutamide

the effect of phenobarbital on the activity of bishydroxycoumarin (Fig. 6.7).[20]

The enzyme stimulation is also exploited as a means of treating disease states. For example, newborn infants suffering from congenital nonhemolytic jaundice have high levels of bilirubin. These can be lowered by the administration of phenobarbital, which stimulates the enzymatic biotransformation of bilirubin.

In summary, the function of biotransformation is to terminate the drug action. However, the rate of biotransformation can be highly dependent on the patient history of concurrent medication, disease states, nutritional state, and a large number of physiologic factors, all of which add up to inherent uncertainty in the extent of biotransformation.

Table 6.12. EXAMPLES OF BIOTRANSFORMATION STIMULATION

STIMULATOR	*BIOTRANSFORMATION-STIMULATED DRUG*
Alcohol	Pentobarbital/tolbutamide
Aminopyrine	Estradiol/pentobarbital/hydrocortisone
Antihistamines	Phenobarbital/progesterone/testosterone
Phenobarbital	Phenobarbital/bishydroxycoumarin/phenytoin/ warfarin/griseofulvin/digitoxin/bilirubin/ cortisol/testosterone/phenylbutazone/etc.
Phenylbutazone	Phenylbutazone/aminopyrine/cortisol
Phenytoin	Cortisol
Chloral hydrate	Bishydroxycoumarin
Glutethimide	Glutethimide/warfarin/dipyrone
Meprobamate	Meprobamate
Griseofulvin	Warfarin
Cigarette smoke	3,4-Benzpyrine/nicotine

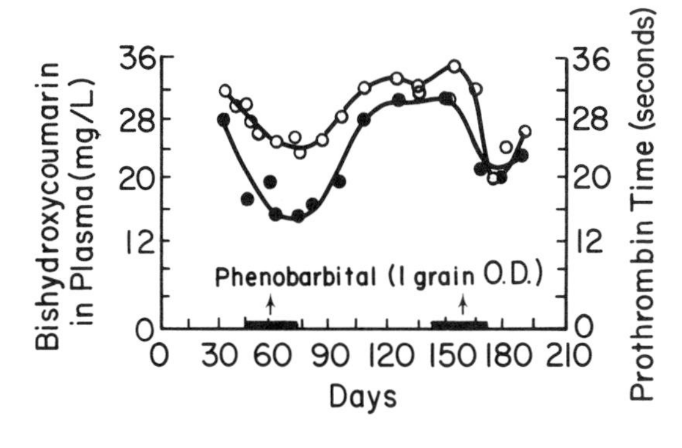

Figure 6.7. Effect of phenobarbital on plasma levels of bishydroxycoumarin (●) and on prothrombin time (○) in a human subject (dose of bishydroxycoumarin: 75 mg per day). Phenobarbital was administered (60 mg once daily) during the period indicated by heavy marks on the abscissa. (Adapted from Cucinell et al.: Clin Pharmacol Ther 6:420, 1965)

TISSUE REDISTRIBUTION

In addition to excretion and biotransformation, drug actions are also terminated by tissue distribution through a dilution of the drug concentration at the site of action as a result of equilibration with other body tissues. A classic example is that of thiopental, whose short duration of action is explained in terms of its extensive distribution to body muscles and fat. The quick onset of action is attributed to the high lipid solubility of thiopental allowing rapid equilibration with the brain.

Tissue distribution is also responsible for the prolonged half-lives of several compounds. For example, extensive tissue distribution of DDT allows the compound to remain in the body, from whence it is slowly eliminated mainly by biotransformation.

References

1. Curry SH: Drug Disposition and Pharmacokinetics. London, Blackwell, 1977 pp 60 and 189
2. Hiron PC, Millburn P, Smith RL: Some physiologic factors influencing the concentration of drugs at body sites. J Mond Pharm 1:15, 1972
3. Groth U, Prellwitz W, Jahnchen E: Estimation of pharmacokinetic parameters of lithium from saliva. Clin Pharmacol Ther 16:490, 1974
4. Gibaldi M: Biopharmaceutics and Clinical Pharmacokinetics. Philadelphia, Lea and Febiger, 1977, p 98
5. Ellinger P, Shattock FM: Black tongue and oral penicillin. Br Med J 2:208, 1946
6. van Boxtel CJ, Wilson JT, Lindgren S, Sjoqvist F: Comparison of antipyrine half-life in plasma, whole blood and saliva in man. Eur J Clin Pharmacol 9:327, 1976
7. Devine LF, Johnson DR, Rhodes SL: Rifampin: effect of two-day treatment on the meningococcal carrier state and the relationship to the levels of drug in sera and saliva. Am J Med Sci 261:79, 1971
8. Koup J, Jusko WJ, Goldfarb AL: pH-dependent secretion of procainamide into saliva. J Pharm Sci 64:2008, 1975

9. Vorherr H: The Breast: Morphology, Physiology, and Lactation. New York, Academic, 1974
10. Vorherr H: Drug excretion in breast milk. Postgrad Med 56:97, 1974
11. O'Brien TE: Excretion of drugs in human milk. Am J Hosp Pharm 31:844, 1974
12. Perlman HH, Dannenberg AM, Sokoloff N: The excretion of nicotine in breast milk and urine from cigarette smoking. JAMA 120:1003, 1942
13. Thayson JH, Schwartz IL: The permeability of human sweat glands to a series of sulfonamide compounds. J Exp Med 98:261, 1953
14. Hewick DS, McEwen J: Plasma half-lives, plasma metabolites and anticoagulant efficacies of the enantiomers of warfarin in man. J Pharm Pharmacol 25:458, 1973
15. Breimer DD, van Rossum JM: Pharmacokinetics of (+), (−), and (±) hexobarbitone in man after oral administration. J Pharm Pharmacol 25:762, 1973
16. George CF, Fenyvesi T, Connolly ME: Pharmacokinetics of dextro-, laevo-, and racemic propanolol in man. Eur J Clin Pharmacol 4:74, 1972
17. Vangiessen GJ, Kaiser DG: GLC determination of ibuprofen [dl-2-(p-isobutylphenyl) propionic acid] enantiomers in biological specimens. J Pharm Sci 64:798, 1975
18. Mawer GE, Mullen PW, Rodgers M: Phenytoin dose adjustment in epileptic patients. Br J Clin Pharmacol 1:163, 1974
19. Levy G, Tsuchiya T: Salicylate accumulation in man. N Engl J Med 287:430, 1972
20. Cucinell SA, Conney AH, Sansur M, Burns JJ: Drug interactions in man. I: Lowering effect of phenobarbital on plasma levels of bishydroxycoumarin (Dicumarol) and diphenylhydantoin (Dilantin). Clin Pharmacol Ther 6:420, 1965

Questions

1. What are the three mechanisms of the termination of drug action?
2. What is the role of hemodynamics in the excretion of drugs?
3. What is the function of homeostatis?
4. Describe the functions of the various parts of a nephron.
5. What are the characteristics of an ultrafiltrate?
6. What is the total volume of plasma filtered each day in the kidneys?
7. Describe the role of tubular reabsorption in the disposition of drugs.
8. Why does the clearance of dextroamphetamine decrease in alkaline pH? Draw a chemical structure to justify your response.
9. Identify the ionizable groups in the drugs listed in Table 6.2.
10. Cite examples of drugs secreted actively into the renal tubule.
11. Cite examples in which the competitive inhibition of tubular secretion is utilized for therapeutic purposes.
12. How would you estimate the glomerular filtration rate with inulin as an index? Why is inulin an ideal compound for this purpose?
13. How would you estimate the total blood flow to the kidneys?
14. How is the excretion ratio related to the mechanism(s) of the renal excretion of drugs?

15. What is the advantage in using creatinine clearance rather than inulin clearance to estimate kidney function?
16. What is the normal range of serum creatinine concentration and how can this be used to obtain the creatinine clearance?
17. Why must the creatinine clearance be adjusted for body surface area when there is no such need for inulin?
18. Under what conditions can urine flow rate affect the clearance of drugs? Cite examples in which it does.
19. What is the average daily output of bile? What is its composition?
20. What is the mechanism of the conservation of bile salts?
21. What is enterohepatic cycling? How does it affect the duration of action?
22. Can plasma-protein binding affect the biliary secretion?
23. Cite examples of drugs which are very rapidly secreted into the bile.
24. What are the physical and chemical characteristics required for biliary secretion? How are these acquired?
25. What is the molecular weight threshold for biliary secretion and how does it vary with the chemical nature of the drug?
26. What are the three forms in which a drug can be eliminated in the bile?
27. What is the rate limiting step in biliary excretion?
28. How would biliary excretion affect bioavailability studies?
29. How would you study the efficiency of biliary excretion?
30. What kinds of compounds are most likely to be eliminated through the lungs?
31. What is the mechanism of the salivary excretion of drugs?
32. Comment on the statement, "Some basic drugs inhibit the salivary flow by accumulation."
33. Cite examples of local effects of the salivary excretion of drugs.
34. What requirements must be met before salivary levels can be used to monitor drug therapy?
35. What are the sources of error in salivary drug concentration monitoring and how can these be overcome?
36. What are the mechanisms of the mammary excretion of drugs?
37. Cite examples of the possible effects on nurslings of the mother taking a drug while nursing.
38. What types of drugs are known to be excreted in the sweat?
39. Why are basic drugs preferentially excreted into the gastrointestinal tract? What is the effect of gastrointestinal excretion on the bioavailability of drugs?
40. What is the difference between biotransformation and metabolism?
41. What is the role of enzymes in biotransformation reactions?
42. What is the major function of a biotransformation reaction?
43. Cite examples of biotransformation reactions which lead to the active forms of the respective drugs.
44. Which biotransformation reaction leads to slower drug elimination from the body? Cite examples.

45. What is the rationale behind using sulfonamide combinations instead of a single sulfonamide for the treatment of infections?
46. What are Phase I and Phase II reactions?
47. Draw the chemical structures of some drugs and their products as suggested in Table 6.10.
48. What are the various sites of drug biotransformation in the body? Give examples of some site-specific biotransformation reactions.
49. S(−) warfarin is a more potent anticoagulant than R(+) warfarin but in practice there may not be an apparent difference in their activities. Why?
50. Derive Equation 6.5.
51. Show how a Michaelis-Menton equation yields first order or zero order responses.
52. Explain why the duration of action of phenytoin is prolonged at concentrations above 4 μg/ml.

53. Differentiate between competitive and noncompetitive inhibition of enzymes. Cite examples of each.
54. Cite examples of enzyme inhibition due to drug interaction and its effect on the duration of action.
55. Discuss the nature of the drug interactions shown in Figure 6.7.
56. What are the possible mechanisms of enzyme stimulation?
57. What is the rationale behind the use of phenobarbital in the treatment of congenital nonhemolytic jaundice?
58. How does the tissue distribution affect the duration of action? Give examples.
59. Give examples whereby tissue distribution leads to prolongation of the disposition phase.

CHAPTER 7

Pharmacokinetic Principles

PHARMACOKINETICS is the study of the time course of absorption, distribution, biotransformation, and excretion of drugs and their metabolites in the body. Unlike a variety of behavioral, physiologic, and pharmacologic responses, the principles of pharmacokinetics are described in terms of mathematical equations which are used to quantitatively predict the nature of these processes.

The underlying principle of pharmacokinetics is that the movement of molecules in the body across various barriers and their conversion to other chemical forms is dependent on the concentration of the drug molecules themselves. That is, it is a first order process:

$$\frac{dX}{dt} = -KX \qquad \text{(Eq. 7.1)}$$

where K = rate constant characterizing the decreasing amount of drug, X.

An integrated form of the differential equation above is (see appendix A for details):

$$X = X_0 e^{-Kt} \qquad \text{(Eq. 7.2)}$$

where X_0 is the amount of drug present at the initiation of the kinetic process.

INSTANTANEOUS DISTRIBUTION MODELS

The principle described in Equation 7.1 states simply that the rate of reaction decreases as the amount of drug decreases. This principle is applicable to the absorption of drugs, where the absorption rates can be related to the amount of drug remaining at the site of absorption. The rate of drug distribution from the blood compartment can be described in terms of Equation 7.1 and similarly the rates of biotransformation and excretion can also be predicted in terms of Equation 7.1.

The simplest example of applying this basic principle to predict the time course of drug action is the intravenous administration of drugs. Let us assume that a drug can be instantaneously placed into the veins (well, within two or three minutes) so that the amount of drug present in the body at time $t = 0$ is X_0. Knowing that the blood circulation completes its cycle in one

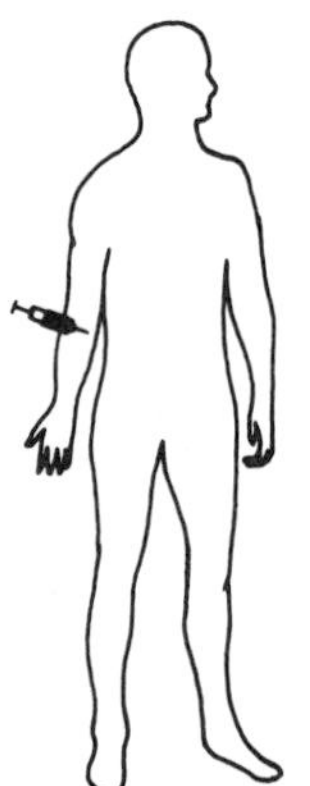
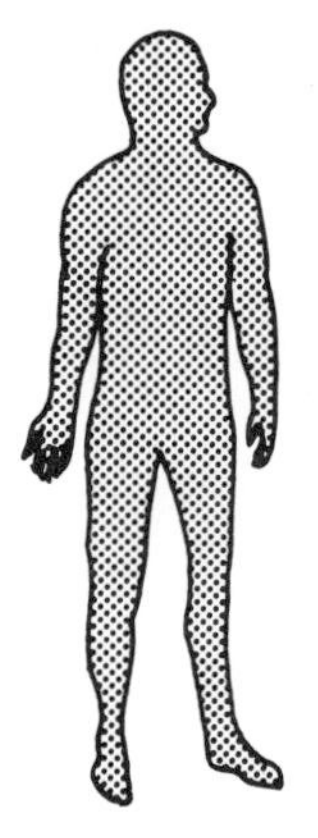
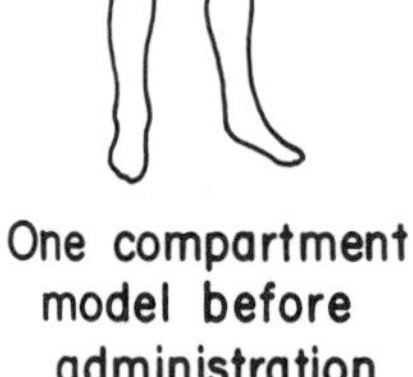

Figure 7.1. Single compartment visualization.

minute, it is reasonable to expect that by the time the injection is completed all those tissues which are highly perfused will have equilibrated with the concentration in the plasma (Fig. 7.1). Not surprisingly, a large number of drugs do show almost instantaneous equilibration with the body tissues and are distributed homogeneously throughout the body. Note that the word homogeneous does not mean equal concentration in this discussion. It simply means that an equilibration is reached between the plasma and the various tissues and fluids in the body and that any change in the plasma concentration can be attributed to the elimination of drug from the body rather than uptake by the tissues. Of course, as the drug is removed from the plasma, body tissues de-equilibrate and achieve a new equilibrium. All of these processes are very quick in nature, so the plasma concentration can be used to predict the amount of drug remaining in the body:

$$X = V\,C \qquad \text{(Eq. 7.3)}$$

where V = volume of distribution or a proportionality constant between the amount of drug, X, in the body and the plasma concentration, C. A substitution of Equation 7.3 into Equation 7.2 and conversion to $\log_{10}$ base yields (see Appendix A):

$$\log C = \log C_0 - \frac{Kt}{2.303} \qquad \text{(Eq. 7.4)}$$

where C_0 is the plasma concentration immediately after intravenous injection or the concentration extrapolated to time zero on a plasma concentration : time profile (Fig. 7.2). Thus if a known amount of dose, X_0, is introduced into the body and the plasma concentration is measured periodically the unknown pharmacokinetic parameters can be easily obtained:

$$V = X_0/C_0 \qquad \text{(Eq. 7.5)}$$

$$K = -\text{ slope} \times 2.303 \qquad \text{(Eq. 7.6)}$$

Once we have calculated these two parameters by sampling the plasma at various times, the complete profile can be predicted using Equation 7.2. The calculated pharmacokinetic parameter can be used to describe the general nature of the drug elimination, i.e., biotransformation and excretion. A quick understanding of the rate of drug removal from the body can be obtained by calculating the time needed to decrease the body level by one-half, or the HALF-LIFE, which is calculated from:

$$\log X = \log X_0 - \frac{Kt}{2.303} \qquad \text{(Eq. 7.7)}$$

Thus if $X = \frac{1}{2}X_0$, $t = t_{0.5}$, or the half-life:

$$\log \tfrac{1}{2}X_0 = \log X_0 - \frac{kt_{0.5}}{2.303} \qquad \text{(Eq. 7.8)}$$

$$\log \tfrac{1}{2} = -\frac{Kt_{0.5}}{2.303} \qquad \text{(Eq. 7.9)}$$

$$t_{0.5} = 0.693/K \qquad \text{(Eq. 7.10)}$$

It should be noted from Equation 7.10 that the half-life is independent of the dose administered, a characteristic of first order processes. The half-life can also be calculated directly from a plasma concentration profile (Fig. 7.2) by reading the time needed for the concentration to decrease by one-half from any arbitrary point on a log concentration:time plot. Since the calculation of

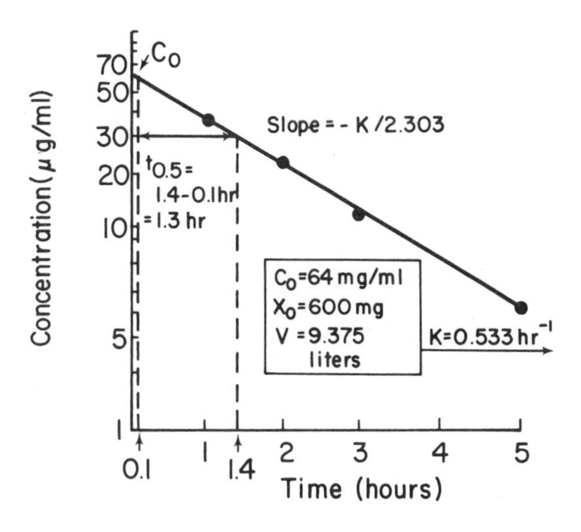

Figure 7.2. Plasma concentration profile following intravenous administration in a single compartment open model. Data from Example 7.1. See text for details.

half-life is simpler than calculating the rate constant, it is suggested that the rate constant be calculated from the half-life values.

The model presented in Figure 7.1 is called a single compartment model since the body can be assumed to have homogeneous characteristics as far as the elimination of the drug is concerned. The term COMPARTMENT refers to the homogeneity of the rate process rather than a physical part of the body within which the drug has distributed, though the latter use is always possible.[1]

The following example illustrates calculations of the pharmacokinetic parameter in a single compartment open model.

Example 7.1:

A single 600 mg dose of ampicillin was administered intravenously to an adult and the following plasma concentration profile was obtained:

t (hr)	*C (μg/ml)*
1	37
2	21.5
3	12.5
5	4.5

The pharmacokinetic parameters can be calculated by first plotting the plasma concentration data on a semilogarithmic scale (Fig. 7.2) to obtain a straight line drawn through these points (best fit line).

C_0 (64 μg/ml) is directly read as the intercept of the extrapolation on the ordinate.

V (9.375 liters) is obtained by dividing C_0 into the dose administered.

$t_{0.5}$ (1.3 hours) is obtained from the time needed for the plasma concentration to decrease from 60 μg/ml to 30 μg/ml. You should try using other ranges as well.

K (0.533 hr^{-1}) is obtained by dividing the half-life into 0.693 or by determining the slope of the best fit line and multiplying that by −2.303. Note that the slope can be calculated by:

$$\text{slope} = \frac{\log A - \log B}{t_A - t_B} \qquad \text{(Eq. 7.11)}$$

where A and B are any two concentrations on the profile and t_A and t_B are the corresponding times at which these concentrations occur. A common error made in the calculation of slopes is the failure to convert the concentration on the ordinate scale to log scale. Remember that the values read from the plasma concentration axis are not log C but simply C.

All of these parameters can be summed up in a compartmental representation of the model, as shown in the inset of Fig. 7.2.

The pharmacokinetic parameters have great physiologic and clinical meaning and the following discussion is hoped to shed some light on it. The full potential of the pharmacokinetic modeling excercises can only become apparent after all applications are discussed (as presented in later chapters).

The Half-life

As defined above, the half-life is the time required for the amount of drug in the body to decrease to one-half of its initial value. Thus if the initial amount is 100 mg and if the half-life is one-hour, only 50 mg will remain in the body at the end of a one hour period:

Time (hr)	*Amount (mg)*
0	100 (dose)
1	50
2	25
3	12.5
4	6.25

Note that after each hour the amount is decreased to one-half of the amount in the previous hour. Thus the half-life can be calculated at any time on the amount/concentration profile. The half-life denotes how quickly a drug is removed from the body by biotransformation or excretion. Since most drugs require a minimum effective concentration in the plasma, a drug which is eliminated quickly requires more frequent dosing than a drug with a long half-life. Half-lives vary greatly between drugs. For instance, penicillin has a half-life of about 30 minutes, phenobarbital has a half-life of about 5 days and, as shown in Figure 5.5, a drug can persist in the blood for years. The half-lives can also vary between structurally similar drugs. For example, sulfamethylthiazide has a half-life of about two hours, whereas sulfamethoxypyridazine requires 34 hours for 50 percent elimination.

The half-lives also vary between species and even within a given species, depending on various physiologic and pathologic functions. In normal subjects the half-lives remain fairly constant, but subtle variations due to such factors as diurnal functions are always possible.

In bioavailability studies it is recommended that the urinary excretion be monitored for at least seven half-lives to assure complete elimination. The rationale for this comes from the following calculations:

Cumulative $t_{0.5}$	*Fraction Remaining*	*Percent Remaining*
0	1	100
1	1/2	50
2	1/4	25
3	1/8	12.5
4	1/16	6.25
5	1/32	3.12
6	1/64	1.56
7	1/132	0.78

Thus at the end of seven half-lives less than 1 percent of the administered dose remains in the body. Note that at least in theory infinite half-lives will be required to reach a 0 percent value in the body.

Since the half-life is an indication of the efficiency of the elimination processes of the body, any change in the half-life will reflect changes in these elimination organ functions, such as liver biotransformation or excretion in the kidneys. Thus such reference compounds as creatinine or inulin can be used to estimate the kidney function or bromosulfonphthalein can be used to study the liver function in terms of the half-lives of these compounds.

The changes in the half-lives of drugs are used as a prime measure for dosage adjustment in disease states. Table 7.1 lists half-lives of some commonly used drugs. Note that most commonly used drugs can be classified into four broad categories based on their half-lives. The group of drugs undergoing ultra-fast disposition (UFD) must be administered more frequently than other drugs in order to maintain a desirable plasma concentration in the blood. In some instances a continuous intravenous infusion is needed to provide effective drug concentration due to these drugs' rapid disposition. On the other hand, with the drugs which are slowly or very slowly eliminated from the body (SD and VSD), administration of a dose even once a day is often sufficient, since the fluctuation in the plasma concentration during the dosing interval is not quite as large as that observed for UFDs or FDs.

In clinical pharmacokinetic studies, the half-lives of drugs in the patients are individually determined rather than simply assuming the reported values in normal individuals and the dosage regimens are adjusted accordingly.

Table 7.1. DRUG CLASSIFICATION BASED ON THE HALF-LIVES OF DISPOSITION

DRUG	*HALF-LIFE*	*DRUG*	*HALF-LIFE*
Ultra-Fast Disposition ($t_{0.5} \leq 1$ hour)		*Moderate Disposition ($t_{0.5}$ 4 to 8 hours)*	
Acetylsalicylic acid[2]	0.25	Amiloride[47]	6.0
Para-aminosalicylic acid[3]	0.90	Chlortetracycline[48]	5.5
Amoxacillin[4]	1.00	Lincomycin[49]	2.5–11.5
Carbenicillin[5]	1.00	Sulfisoxazole[50]	6.0
Cephalexin[6]	1.00	Tetracycline[51]	7.0–9.0
Cephalothin[7]	0.50	Theophylline[52]	4.0–7.0
Cloxacillin[8]	0.40	Tolbutamide[53]	6.0–9.0
Cortisone[9]	0.50	Trimethoprim[54]	9.0
Dicloxacillin[10]	0.90		
Furosemide[11]	0.50	*Slow Disposition ($t_{0.5}$ 8 to 24 hours)*	
Hetacillin[12]	0.30		
Insulin[13]	0.10	Amphetamine[55]	7.0–14.0
Methicillin[10]	0.40	Antipyrine[56]	7.0–35.0
Nafcillin[14]	0.50	Chlordiazepoxide[57]	6.0–15.0
Nalidixic acid[15]	1.00	Cycloserine[58]	3.0–15.0
Oxacillin[14]	0.40	Dapsone[59]	17.0–21.0

Table 7.1. (*CONTINUED*)

DRUG	*HALF-LIFE*	*DRUG*	*HALF-LIFE*
Ultra-Fast Disposition (cont.)		*Slow Disposition (cont.)*	
Penicillin G[16]	0.70	Daunorubicin[60]	12.0–27.0
Pivampicillin[17]	0.10	Demeclocycline[61]	15.0
Propylthiouracil[18]	1.00	Desimipramine[62]	14.0–25.0
		Doxycycline[51]	12.0
Fast Disposition ($t_{0.5}$ 1 to 4 hours)		Glutethimide[63]	5.0–22.0
		Griseofulvin[64]	9.0–22.0
Acetaminophen[19]	1.0–3.0	Iodochlorhydroxyquin[65]	11.0–14.0
Alprenolol[20]	2.0	Lithium[66]	14.0–24.0
Amikacin[21]	2.5	Meprobamate	6.0–16.0
Ampicillin[5]	1.0–1.5	Methacycline[61]	12.0
Bupivacaine[22]	2.5	Methadone[68]	15.0
Cefazolin[7]	2.0	Minocycline[69]	16.0
Cephaloridine[7]	1.5	Practolol[70]	12.0
Chloramphenicol[23]	1.7–2.8	Sulfadiazine[71]	13.0–25.0
Clindamicin[24]	3.0		
Colistimethate[25]	3.0	*Very Slow Disposition ($t_{0.5} > 24$ hours)*	
Cyclophosphamide[26]	3.0–6.0		
Cytarabine[27]	0.4–3.5	Acetylsulfisoxazole[71]	9.0–32.0
Disodium cromoglycate[28]	1.0–1.8	Amobarbital[56]	14.0–42.0
Ethambutol[29]	4.0	Apobarbital[72]	12.0–36.0
Gentamicin[30]	2.0	Atropine[73]	12.0–38.0
Heparin[31]	0.7–2.5	Barbital[72]	60.0–78.0
Hydrocortisone[32]	2.0	Carbamezapine[74]	36.0
Indomethacin[33]	2.0	Chlorpromazine[75]	30.0
Isoniazid[34]	3.5	Chlorpropamide[76]	25.0–42.0
Kanamycin[21]	2.0	Diazepam[77]	55.0
Lidocaine[35]	2.0	Dicumarol[78]	8.0–74.0
Meperidine[36]	3.0	Digitoxin[79]	200.0
Methyltestosterone[37]	3.5	Digoxin[80]	12.0–132.0
Morphine[38]	2.0	Ethosuximide[81]	52.0
Phenacetin[19]	0.75–1.5	Haloperidol[82]	13.0–35.0
Prednisolone[39]	3.5	Methaqualone[83]	10.0–40.0
Procainamide[40]	3.0	Nortriptyline[62]	18.0–35.0
Propranolol[41]	4.5	Pentobarbital[84]	48.0
Rifampicin[34]	3.0	Phenobarbital[72]	48.0–120.0
Salicylamide[42]	1.2	Phenylbutazone[85]	84.0
Salicylic acid[43]	4.0	Sulfadimethoxine[71]	40.0
Streptomycin[44]	2.5	Sulfamerazine[71]	15.0–45.0
Testosterone[45]	1.8		
Tobramycin[30]	2.0		
Warfarin[46]	2.0		

The Elimination Rate Constant

The first order rate constant of elimination, K, has units of reciprocal time and represents the fraction of drug removed per unit time. Thus if $K = 0.1$ min^{-1}, it means that 10 percent of the remaining amount is removed per minute. Note the term **REMAINING AMOUNT** in the definition. Since the drug is continuously removed from the body, the remaining amount is continuously changing and thus accurate calculation of the amount remaining in the body can only be made by using Equation 7.2. However, if the time interval for which the calculation is made is very small (about one-seventh or one-eighth of the half-life) compared to the half-life of the drug, a direct calculation can be made with little error without using Equation 7.2. For example, if K is equal to 0.1 day^{-1} (half-life = 6.93 days) and the amount remaining in the body at the end of each day is calculated using the three approaches described above, then there will be little error if the half-life is much longer than the duration of calculation. However, as the half-life decreases, significant errors can be introduced into the direct calculation of the amounts remaining in the body, based on the rate constant (Table 7.2).

Table 7.2. COMPARISON OF ELIMINATION RATE CALCULATIONS

TIME (day)	*AMOUNT REMAINING IF 10% OF DOSE REMOVED PER DAY (mg)*	*AMOUNT REMAINING IF 10% OF REMAINING DOSE REMOVED PER DAY (mg)*	$X = X_0 e^{-0.1t}$ *(mg)*
0	100	100	100
1	90	90	90.48
2	80	81	81.87
3	70	72.9	74.08
4	60	65.6	67.03
TIME (day)	*AMOUNT REMAINING IF 50% OF DOSE REMOVED PER DAY (mg)*	*AMOUNT REMAINING IF 50% OF REMAINING DOSE REMOVED PER DAY (mg)*	$X = X_0 e^{-0.5t}$ *(mg)*
0	100	100	100
1	50	50	60.65
2	0	25	36.79
3	—	12.5	22.31
4	—	6.25	13.53

Thus direct calculation of the amount remaining as a function of time is possible for MDs, SDs, and VSDs on an hourly basis, or on a daily basis if the half-lives are six or seven times longer than the time interval chosen.

The elimination rate constant represents the overall drug elimination from the body, which includes urinary excretion, biliary secretion, biotransformation, and all other mechanisms possible for the removal of the drug from the body. All of these individual processes are described by individual first order reaction constants and K is simply the sum of all of these rate constants:

$$K = k_e + k_b + k_b' + k_{bi} + k_{lu} + \cdots \quad \text{(Eq. 7.12)}$$

where k_e is the urinary excretion constant, k_b and k_b' are the biotransformation constants for two routes of structural modification, k_{bi} is the biliary excretion constant, and k_{lu} is the fraction of drug removed in the lungs. These are apparent first order constants describing the apparent nature of the order of reaction. It is possible that as the drug concentration changes in the body these constants might also change, as in the saturation of enzymes responsible for the biotransformation of a drug to a specific product.

The additive property of the rate constants is of great importance since it allows calculation of unknown rate constants and the total fractions of the drugs removed from the body by a specific route. For example, the total fraction of drug biotransformed in the body is given by:

$$F_b = \frac{k_b + k_b'}{K} \quad \text{(Eq. 7.13)}$$

where the numerator represents the sum of all rate constants representing the biotransformation of drug. One can similarly calculate the fraction of the available dose excreted in the kidneys or eliminated through the bile.

Example 7.2:

The half-life of oxacillin is 0.5 hours and 30 percent of the available dose is excreted unchanged in the urine; the rest undergoes biotransformation. What is the overall rate constant for biotransformation?

$$k_b = K - k_e$$

$$= (0.693/0.5) - k_e$$

$$\frac{k_e}{K} = 0.3$$

Thus

$$k_b = (0.693/0.5) - 0.3(0.693/0.5)$$

$$= 0.97 \text{ hr}^{-1}$$

Example 7.3:

In the example given above calculate the disposition half-life if the renal function decreases by one-half.

$$K = k_b + k_e$$

$$= k_b + 0.5k_e$$

$$= 0.97 + 0.5 \times 0.3 \times (0.693/0.5)$$

$$= 1.178$$

$$t_{0.5} = 0.963/1.178 = 0.59 \text{ hours}$$

The calculation above can be repeated with a 50 percent reduction in the biotransformation function to yield a half-life of 0.77 hours, a much greater increase since biotransformation is a much more important factor for this drug than renal excretion.

The Volume of Distribution

As discussed earlier, the volume of distribution is given by:

$$V = \frac{X_0}{C_0} \qquad \text{(Eq. 7.14)}$$

This is simply a proportionality constant between the amount of drug in the body and the plasma concentration, but when the drug distributes to specific fluids of the body, as occurs with Evans blue dye (which remains restricted to the blood compartment), the volume of distribution represents that pool of fluid. If an instantaneous equilibration is reached, then Equation 7.14 describes the volumes of distribution throughout the course of drug disposition. But when tissues which might equilibrate slowly are involved the volume of distribution changes with time.[1] This concept will be discussed later.

An important application of the volume of distribution comes in the calculation of clearance of the drug from the body. The clearance, Q, is defined as:

$$Q = K\,V \qquad \text{(Eq. 7.15)}$$

The units for clearance are volume/time and represent the part of the volume of distribution which is cleared of the drug per unit of time. The total body clearance is the sum of individual clearances:

$$K\,V = k_r V + k_b V + k_b' V + \ldots \qquad \text{(Eq. 7.16)}$$

Thus each of the component clearance values represents the fraction of volume of distribution cleared by the given elimination mechanism of excretion, biotransformation, biliary excretion, etc. The concept of clearance can be easily understood from the examples of the excretion of inulin and penicillin in the kidneys. Inulin distributes only in the blood and in each cycle through the kidneys about 125 ml of plasma is filtered per minute as ultrafiltrate containing inulin. If all of the drug contained in the filtered volume is removed, the renal clearance is equal to the volume of filtrate. In the case of penicillin almost 100 percent of the total drug circulating through the kidneys is removed in each cycle and thus the renal clearance of penicillin is equal to the total blood flow rate.

Example 7.4:

Calculate the half-life of a drug if it is completely removed in each cycle through the kidneys and this represents the only route of drug elimination. The volume of distribution is 50 liters.

$$V k_e = Q = K V = 650 \text{ ml/min}$$

$$K = 650/50{,}000 = 0.013 \text{ min}^{-1}$$

$$t_{0.5} = 0.693/0.013 = 53.3 \text{ min}$$

Example 7.5:
What is the shortest half-life possible for a drug if it is completely removed in each cycle through the kidneys and this represents the only mechanism of elimination?

The smallest volume of distribution is about 5 liters, or the blood volume:

$$K = 650/5{,}000 = 0.13 \text{ min}^{-1}$$

$$t_{0.5} = 0.693/0.13 = 5.33 \text{ min}$$

Example 7.6:
What is the half-life of a drug if its excretion ratio is 0.45, and 50 percent of the blood flowing through the liver is cleared of the drug in each cycle? The volume of distribution is 50 liters and the normal hepatic flow rate is 1.25 liters/min.

$$V k_e = \text{Excretion ratio} \times \text{inulin clearance}$$

$$= 0.45 \times 125 = 56.25 \text{ ml/min}$$

$$k_e = 56.25/50{,}000 = 1.125 \times 10^{-3} \text{ min}^{-1}$$

$$V k_b = 0.5 \times \text{hepatic flow rate}$$

$$= 0.5 \times 1.25 = 625 \text{ ml/min}$$

$$k_b = 625/50{,}000 = 1.25 \times 10^{-2} \text{ min}^{-1}$$

$$K = k_e + k_b = 0.01363 \text{ min}^{-1}$$

$$t_{0.5} = 0.693/0.01363 = 50.9 \text{ min}$$

It is quite obvious that these calculations are required for dosage calculations in the event of kidney or liver malfunction, when the half-life in the patient can be calculated based on the component clearance values. However, in calculating the half-life in disease states the clearance component due to biotransformation in the blood and excretion in the breath, saliva, lungs, and other locations should also be considered if it contributes significantly to the overall disposition of the drug.

The discussion presented above is applied to the pharmacokinetic analysis following intravenous administration, when the plasma concentration is monitored. However, other concentrations (such as urinary levels) can also be monitored for similar pharmacokinetic analyses.

Urinary Excretion Rate Data

One advantage in using urinary excretion data to analyze a pharmacokinetic system is the noninvasive nature of such data. It is much more convenient to collect a urine sample than to draw the blood periodically. Another advantage in using urinary excretion data is that this allows direct measurements of bioavailability, both absolute and relative, without the necessity of fitting the data to a mathematical model.

The rate of urinary drug excretion is proportional to the amount of drug in the body:

$$\frac{dX_u}{dt} \propto X \qquad \text{(Eq. 7.17)}$$

The proportionality constant is simply the rate constant for the renal excretion of drug:

$$\frac{dX_u}{dt} = k_e X = k_e X_0 e^{-Kt} \qquad \text{(Eq. 7.18)}$$

$$= k_e V C_0 e^{-Kt} = Q_e C_0 e^{-Kt} \qquad \text{(Eq. 7.19)}$$

The urinary excretion rate is therefore the product of renal clearance and the plasma concentration with units of amount/time. It is experimentally determined by collecting the urine during a specified time interval and analyzing the drug concentration in an aliquot of the collected volume:

$$\frac{dX_u}{dt} = \frac{\text{(urinary concentration) (urine volume)}}{\text{(time interval for collection)}} \qquad \text{(Eq. 7.20)}$$

For example, if after an intravenous dose of 100 mg of a drug the volume of urine collected during the first three hours is 250 ml and the concentration of drug in this volume is 20 μg/ml, the urinary excretion rate is:

$$\frac{20 \times 250}{3} = 1666.67\ \mu\text{g/hr} \qquad \text{(Eq. 7.20A)}$$

Analogous to the treatment of plasma level data, the urinary excretion rate data can be used to calculate the half-life and excretion rate constant of the drug. Equation 7.18 can be converted to:

$$\log (dX_u/dt) = \log k_e X_0 - Kt/2.303 \qquad \text{(Eq. 7.21)}$$

Therefore, a plot of the log of the excretion rate of unchanged drug in the urine against time (the midpoint time of the duration of urine collection) will yield a straight line (Fig. 7.3) from which the following information can be obtained:

1. Half-life of the drug is simply the time required for the excretion rate to decrease to one-half on the descending part of the plot.
2. The elimination rate constant, K, is determined by: $0.693/t_{0.5}$
3. The urinary excretion rate constant is determined by: intercept on Y-axis/ X_0

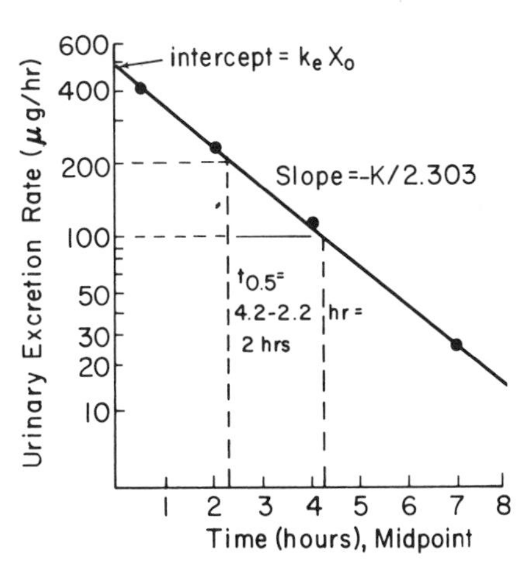

Figure 7.3. Calculation of pharmacokinetic parameters from urinary excretion rate data. See text for details of the data points.

4. The extrarenal excretion rate constant (mainly the biotransformation) is obtained by: $K - k_e$
5. If the volume of distribution is known—as may be obtained from blood level data—the total body clearance, renal clearance, extrarenal clearance, and the corresponding plasma concentration for a given urinary excretion rate can be easily obtained.

Another approach to handling urinary excretion data is by integrating Eq. 7.18 (Appendix A) to give the total urinary excretion as a function of time:

$$X_u = \frac{k_e X_0}{K}(1 - e^{-Kt}) \qquad \text{(Eq. 7.22)}$$

Figure 7.4 shows a characteristic plot for Equation 7.22. Note that as time approaches infinity (for our purpose, 6 to 7 half-lives):

$$X_u^{\infty} = \frac{k_e}{K} X_0 \qquad \text{(Eq. 7.23)}$$

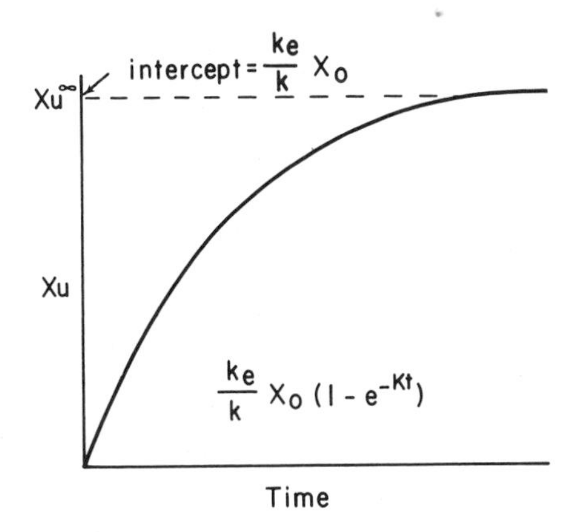

Figure 7.4. Cumulative urinary excretion.

If the drug is eliminated only by renal excretion, K will be equal to k_e and X_u^∞ will become equal to the dose administered (intravenously), X_o.

Example 7.7:

The following urinary excretion data are obtained following intravenous administration of a 100 mg dose:

Time (hr)	Urine Volume (ml)	Drug Concentration (μg/ml)
0-1	200	1.85
1-3	150	2.86
3-5	300	0.70
5-9	700	0.20

If the volume of distribution is 100 liters, calculate the half-life, K; k_e, renal clearance; the half-life in case of anuria, or in the case of 50 percent renal function; plasma concentration at $t = 3$; and percent biotransformation in normal subjects.

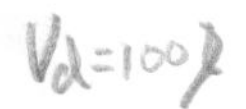

Midpoint (hr)	Excretion Rate (μg/hr)
0.5	370
2.0	214.5
4.0	105
7.0	35

From a plot of the excretion rate against time on a semi-logarithmic scale, the half-life is calculated to be around two hours:

$$K = 0.693/2 = 0.347 \text{ hr}^{-1}$$
$$k_e = \text{intercept}/X_0 = (450\ \mu\text{g/hr})/(100\ \text{mg} \times 1000\ \mu\text{g/mg}) = 0.0045 \text{ hr}^{-1}$$
$$Q_e = k_e V = 0.0045 \times 100 = 0.45 \text{ liters/hr}$$
$$k_b = K - k_e = 0.347 - 0.0045 = 0.3425 \text{ hr}^{-1}$$
$$\text{In anuria, } k_e = 0\text{: } k_b = K;\ t_{0.5} = 0.693/0.3425 = 2.02 \text{ hr}$$
$$\text{In 50\% renal function: } K = 0.5k_e + k_b = 0.34475 \text{ hr}^{-1}$$
$$t_{0.5} = 0.693/0.34475 = 2.01 \text{ hr}$$
$$\text{Plasma concentration at } t = 3\text{: } C_0 = X_0/V = 1\ \mu\text{g/ml}$$
$$C = C_0 e^{-Kt} = 1\, e^{-0.347 \times 3} = 0.353\ \mu\text{g/ml}$$
$$\text{Percent biotransformation} = (k_b/K)\, 100 = 98.7\%$$

First Order Input Data

The discussion presented above pertains to instantaneous input or intravenous administration, which shows an immediate distribution in the body. However, a large number of drugs are administered orally or by other routes from which the input or absorption is not instantaneous and instead follows a first order input:

$$\frac{dX}{dt} = k_a X_a - KX \qquad \text{(Eq. 7.24)}$$

where k_a is the absorption rate constant and X_a is the amount of drug remaining at the site of absorption, such as the stomach or intestine. Equation 7.24 can be integrated (Appendix A) to yield:

$$X = \frac{k_a F X_0}{(k_a - K)} (e^{-Kt} - e^{-k_a t}) \qquad \text{(Eq. 7.25)}$$

or

$$C = \frac{k_a F X_0}{(k_a - K)V} (e^{-Kt} - e^{-k_a t}) \qquad \text{(Eq. 7.26)}$$

where F is the fraction of the administered dose, X_0, which is absorbed following administration by the oral or other routes.

In most instances the absorption rate constant, k_a, is larger than the elimination rate constant, K, and the plasma concentration profiles (such as those shown in Figure 3.7) are obtained. The peak plasma concentration represents the time at which the absorption rate becomes equal to the rate of elimination:

$$k_a X_a = KX \qquad \text{(Eq. 7.27)}$$

As the amount of drug remaining at the site of absorption decreases, the rate of absorption also decreases until:

$$e^{-k_a t} \sim 0 \qquad \text{(Eq. 7.28)}$$

and the plasma concentration is described only by the elimination constant of the drug:

$$C' = \frac{k_a F X_0}{(k_a - K)V} e^{-Kt} \qquad \text{(Eq. 7.29)}$$

This is referred to as the post-absorptive phase, in which the half-life of the drug can be easily determined from the slope of the plasma concentration : time profile (Fig. 7.5). The absorption rate constant, k_a, is determined by a technique which is appropriately termed the **METHOD OF RESIDUALS**. This involves the subtraction of two rate processes, in this case the subtraction of the elimination rate process from the overall disposition of the drug, to yield the rate of absorption. Recalling Equation 7.26:

$$C = Ae^{-Kt} - Ae^{-k_a t} \qquad \text{(Eq. 7.30)}$$

where

$$A = \frac{k_a F X_0}{(k_a - K)V} \qquad \text{(Eq. 7.31)}$$

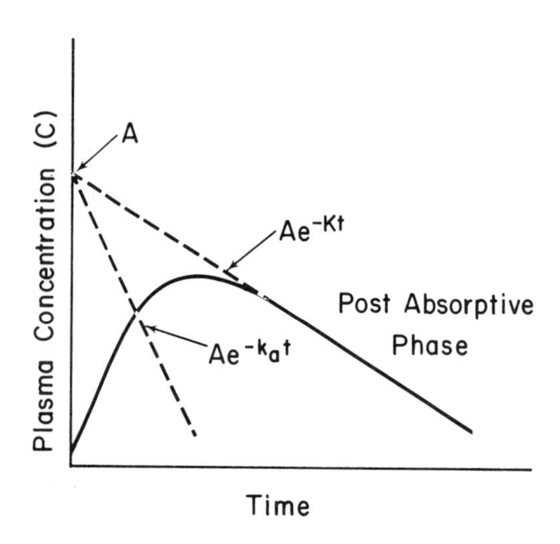

Figure 7.5. Semilogarithmic plot of plasma concentration against time. Note that it is composed of two exponential terms, one describing the absorption and the other the elimination. During the post-absorptive phase the plasma concentration is described by a single exponent.

Since the terminal portion of the plasma concentration (log scale): time plot represents only the elimination phase, or Equation 7.29, it can be extrapolated back to time zero on the plasma concentration axis and the absorption rate constant can be determined by the following subtraction:

$$Ae^{-k_a t} = Ae^{-Kt} - C \qquad \text{(Eq. 7.32)}$$

DRUG	t_p	DRUG	t_p
Acetaminophen	10 min to 1 hr	Hydroxyurea	0.5 to 2 hr
Acetazolamide	2 to 8 hr	Lidocaine	45 to 60 min
Acetohexamide	1.5 to 2 hr	Lithium carbonate	1.33 hr
Amitriptyline	2 to 4 hr	Methadone	4 hr
5-Azacytidine	30 min	Morphine	45 min
Cytosine	5 to 15 min	6-Mercaptopurine	2 hr
Cromolyn sodium	15 min	Methaqualone	2 hr
Chlorothiazide	1.5 to 2.5 hr	Nortriptyline	5.5 hr
Chlorodiazepoxide	2 hr	Pentazocine	2 hr
Chlorpropamide	2 to 4 hr	Propoxyphene	2 hr
Chlorphenesin carbamate	2 hr	Practolol	2 to 4 hr
Clofibrate	5 to 12 hr	Procainamide	1 hr
Chlorpromazine	3 to 4 hr	Propranolol	2 hr
Diazepam	1 hr	Procarbazine	1 hr
Ethchlorvinyl	1 hr	Phenformin	2 to 4 hr
Floxuridine	15 to 20 min	Protriptyline	24 to 30 hr
Furosemide	1 to 2 hr	Quinidine	0.5 to 4.5 hr
Glutethimide	2.2 hr	Salicylate	2 hr
Hexamethylamine	2 to 3 hr	Warfarin	2 to 9 hr

Equation 7.32 describes the residual line on the plot from which the absorption half-life can be calculated simply by reading the time required for a 50 percent decrease in the values on the plot. To obtain good estimates, at least three arbitrary points should be marked on the plasma concentration plot and the corresponding values on the extrapolated elimination phase line read from the graph. These values are then subtracted and the residual values are plotted at the same time values to obtain the line described by Equation 7.32 (see also Fig. 7.5). Note that the intercept for both lines on the graph is always the same and from this the volume of distribution can be calculated if the bioavailability is known:

$$V = \frac{k_a F X_0}{(k_a - K)\,(\text{intercept})} \qquad \text{(Eq. 7.33)}$$

The time at which the peak plasma concentration occurs can be determined easily since at this time:

$$k_a X_a = KX \qquad \text{(Eq. 7.34)}$$

or

$$\frac{dC}{dt} = 0 \qquad \text{(Eq. 7.35)}$$

$$= \frac{k_a^2 F X_0}{(k_a - K)V} e^{-k_a t_p} - \frac{k_a K F X_0}{(k_a - K)V} e^{-K t_p} \qquad \text{(Eq. 7.36)}$$

where t_p = time at which $C = C_{max}$, peak concentration. Equation 7.36 yields:

$$t_p = \frac{2.303}{(k_a - K)} \log \frac{k_a}{K} \qquad \text{(Eq. 7.37)}$$

C_{max} can be obtained simply by substituting the value of t_p in Equation 7.26. Table 7.3 reports the t_p values of some commonly used drugs.

It should be noted that as k_a becomes larger than K, t_p becomes smaller since the term $(k_a - K)$ increases much faster than log (k_a/K) in Equation 7.37. In some instances where the absorption is extremely fast, it is possible to miss the peak entirely and the plasma concentration profile resembles that obtained after intravenous or instantaneous input.

In many instances it is possible to calculate the absorption half-life (0.693/k_a) from the t_p and the elimination half-life:

$$t_{0.5}\,(\text{absorption}) = \frac{0.693\, e^{-k_a t_p}}{K\, e^{-Kt}} \qquad \text{(Eq. 7.38)}$$

It should be noted that Equation 7.38 is mathematically insoluble. However, with the aid of an iterating computer program, as described in Appendix

B, it can be easily solved for a given elimination half-life and a time for peak concentration.

The discussion presented above applies only to those situations where the absorption rate constant is larger than the elimination rate constant. However, in some situations these two constants may be of comparable dimensions, as with drugs listed as undergoing very fast or fast disposition (Table 7.1). The plasma concentration under these conditions is described as:[86]

$$C = \frac{k'FX_0}{V} te^{-k't} \quad \text{(Eq. 7.39)}$$

where k' is the absorption or elimination rate constant. The time at which the peak plasma concentration occurs is:

$$t_p = 1/k' \quad \text{(Eq. 7.40)}$$

and

$$C_{max} = \frac{0.37FX_0}{V} \quad \text{(Eq. 7.41)}$$

The term FX_0/V represents C_0 following intravenous administration of a dose equal to X_0. Thus the maximum plasma concentration obtained from oral, intramuscular, or other routes of administration which give first order absorption is only about 37 percent of the highest level achieved when a similar dose is given intravenously. If the bioavailability is less than 100 percent, then even lower concentrations will be obtained.

If the absorption rate constant is smaller than the elimination rate constant, then the terminal part of the plasma concentration plot will describe the absorption rather than the elimination. This is due to the relative decrease in the exponential terms, e^{-Kt} and e^{-k_at}. As t increases, each of these terms decreases—the larger the value of the rate constant the faster the decrease. Thus one exponent can vanish or approach zero before the other, at which time the plasma concentration is described only by the remaining exponent. This phenomenon is often termed the "flip-flop" model. The existence of such a model is quite possible for the drugs which are eliminated very fast or when a slow-release dosage form is utilized for sustained delivery of the drug.

Urinary Excretion Data in First Order Input

In a way similar to the use of urinary excretion rate data in the calculation of pharmacokinetic parameters after intravenous or instantaneous input, the excretion of drugs in urine can also be used to calculate absorption and elimination constants and other parameters. Since the urinary excretion rate is:

$$\frac{dX_u}{dt} = k_e X \qquad \text{(Eq. 7.42)}$$

substitution of Equation 7.25 into Equation 7.42 leads to:

$$\frac{dX_u}{dt} = \frac{k_e k_a F X_0}{(k_a - K)} (e^{-Kt} - e^{-k_a t}) \qquad \text{(Eq. 7.43)}$$

Therefore, a plot of the urinary excretion rate against time on a semilogarithmic scale will result in a similar biexponential plot, as seen with the plasma concentration profile. The only difference is that the intercept now includes the term k_e. The urinary excretion rate:time plots can be subjected to a residual method of isolation of the exponents, as described above, and the absorption, elimination, and renal excretion constants can be calculated from the slopes of the two lines generated and the intercept on the plot.

Example 7.8:
Oral administration of a 100 mg dose in a solution form to a 70 kg subject resulted in the following plasma levels:

Time (hr)	*C (μg/ml)*
0.2	1.65
0.4	2.33
0.6	2.55
0.8	2.51
1.0	2.40
1.5	2.00
2.5	1.27
4.0	0.66
5.0	0.39

Assuming that the administered dose is completely absorbed, calculate the absorption and elimination rate constants, the volume of distribution, and the amount of drug in the body at steady state.

The plasma concentration is first plotted on a semilogarithmic scale (Fig. 7.5) and the residuals are obtained as follows:

Time (hr)	*Residual* ($Ae^{-Kt} - C$)
0.2	3.60−1.65 = 1.95
0.4	3.30−2.33 = 0.97
0.6	3.00−2.55 = 0.45

The residual values are then plotted on the same scale corresponding to the times listed above. A straight line is forced through these points and the absorption half-life is determined from the slope.

$t_{0.5}$ for elimination $= 1.5$ hr

$K = 0.693/1.5 = 0.462\ \text{hr}^{-1}$

$t_{0.5}$ absorption $= 0.2$ hr

$k_a = 0.693/0.2 = 3.465\ \text{hr}^{-1}$

$$V = \frac{3.465\ \text{hr}^{-1} \times 100\ \text{mg} \times 1000\ \mu\text{g/mg} \times 1}{3.9\ \mu\text{g/ml} \times (3.465\ \text{hr}^{-1} - 0.462\ \text{hr}^{-1})}$$

$= 29.59$ liters

At steady state, $C_{\text{max}} = 2.55\ \mu$g/ml (peak)

The amount of drug in the body $= CV$

$= 2.55\ \mu\text{g/ml} \times 1000\ \text{ml/liter} \times 29.59$ liters

$= 74.45$ mg

Example 7.9:
Following intravenous administration of a 1 g dose of a drug dissolved in dilute alcohol to a patient with congestive heart condition, the following urinary excretion data were obtained:

Time (day)	*Urine Volume (ml)*	*Urine Concentration (mg%)*
0–1	990	0.0233
1–2	1190	0.01369
2–3	1350	0.00864
3–5	2100	0.00666

What is the half-life of this drug and the total amount of drug excreted unchanged in the urine? Assuming that the remaining portion is biotransformed, calculate the constants K, k_e, k_b and calculate the plasma concentration at 0.5, 1.5, 2.5, and 4 days if $V = 500$ liters. What is the half-life of the drug if the patient suffers from anuria?

The data are first converted to the following:

Time (day), midpoint	*Excretion Rate (µg/day)*
0.5	230.67
1.5	162.91
2.5	116.64
4.0	69.93

The excretion rate is then plotted against time on a semilogarithmic scale (Fig. 7.3) and the following calculations are made:

The half-life is 2.0 days

$$X_u = k_e X_0/K = \text{intercept}/K = (275\ \mu\text{g/day})/0.3465\ \text{day}^{-1}$$
$$= 793.65\ \mu\text{g} = 0.0794\%\ \text{of the administered dose}$$
$$k_b = 0.3465\ \text{day}^{-1} - (\text{intercept}/X_0)$$
$$= 0.3465 - 0.000275 = 0.346225\ \text{day}^{-1}$$

Since the excretion is almost negligible, anuria will result in little change in the half-life.

Plasma concentration at different time intervals is determined by using Equation 7.4:

$$C_{0.5} = 1.682\ \text{mg/liter},\ C_{1.5} = 1.189\ \text{mg/liter},$$
$$C_{2.5} = 0.84\ \text{mg/liter},\ C_4 = 0.5\ \text{mg/liter}$$

Example 7.10:
Following oral administration of a dose in solution form, an elimination half-life of 4 hours and an absorption half-life of 0.5 hours were obtained by plotting the urinary excretion rate against t_{mid} for urine collection. The intercept of the elimination phase extension on the ordinate was 38.5 mg/hr. Assuming complete absorption of the drug, what was the administered dose if 30 percent of the available dose is excreted intact in the urine?

The intercept on the excretion rate axis is:

$$\text{intercept} = \frac{k_e k_a F X_0}{(k_a - K)}$$

Since the half-life of elimination is 4 hours, $K = 0.693/4 = 0.1733\ \text{hr}^{-1}$. The absorption half-life is 0.5 hours and thus $k_a = 0.693/0.5 = 1.3860\ \text{hr}^{-1}$. The administered dose, X_0, can be determined easily since

$$F = 1 \text{ and } k_e = 0.3K$$
$$X_0 = \frac{38.5\ \text{mg/hr}\ (1.3860 - 0.1733)}{0.3 \times 0.1733 \times 1.386}$$
$$= 647.9\ \text{mg}$$

The Rate of Drug Absorption

The absorption half-life as determined by the methods described above will characterize the rate of drug absorption. The rates can also be estimated from the areas under the plasma concentration curves to yield plots of the amounts remaining to be absorbed.

The amount of drug absorbed up until time t is given by:

$$X_A = VC + KV \int_0^t C\,dt \qquad \text{(Eq. 7.44)}$$

where X_A = amount absorbed
VC = amount in the body

$KC \int_0^\infty Cdt$ = total amount eliminated up to time, t. This is obtained by integrating the rate of elimination equation which is equal to KX or KVC.

C = plasma concentration at time t

The total amount of drug absorbed is given by:

$$X_A^\infty = KV \int_0^\infty Cdt \qquad \text{(Eq. 7.45)}$$

where $\int_0^\infty Cdt$ is the area under the plasma concentration:time plot from time zero to infinity.

A ratio of Equation 7.44 versus Equation 7.45 will describe the fraction absorbed up until time t:

$$\frac{X_A}{X_A^\infty} = \frac{C + K \int_0^t Cdt}{K \int_0^\infty Cdt} \qquad \text{(Eq. 7.46)}$$

The percent of drug remaining to be absorbed is given by:

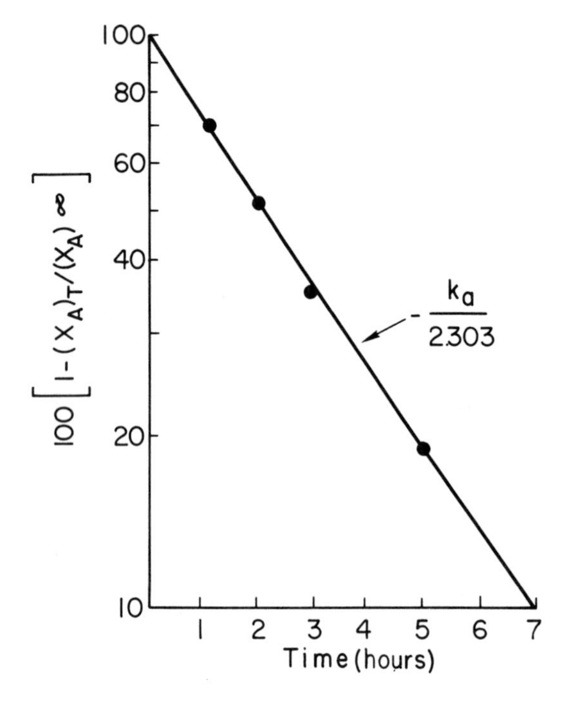

Figure 7.6. Semilogarithmic plot of the percentage of drug remaining to be absorbed versus time.

$$100\left(1-\frac{X_A}{X_A{}^{\infty}}\right) \qquad \text{(Eq. 7.47)}$$

A plot of the log of the percentage unabsorbed against time gives a straight line, from the slope of which the absorption rate constant can be determined (Fig. 7.6). A changing slope may indicate absorption profiles other than a true first order. One advantage of this method over the plasma concentration data fitting and applying residual method is that in some cases the plasma concentrations do not fit a model adequately. In such a case the trapezoidal rule may be applied to Equation 7.46 to calculate the amount unabsorbed.

Example 7.11:
Given the following plasma data, calculate the percent unabsorbed profile. The volume of distribution is 20 liters and the dose administered is 2 g. C_t is expressed in μg/ml.

t	C	$\int_0^t Cdt$*	$K\int_0^t Cdt$	$C+K\int_0^t Cdt$	$100\left(1-\frac{X_A}{X_A{}^{\infty}}\right)$
0					100
1	28.24	14.12	0.93	29.22	73.91
2	46.31	51.40	3.56	49.87	55.47
3	57.33	103.22	7.15	64.48	42.42
4	63.48	163.62	11.34	74.82	33.18
5	66.29	228.51	15.84	82.13	26.66
7	65.90	360.70	25.00	90.90	18.82
10	58.60	547.45	37.94	96.54	13.79
15	43.51	802.72	55.63	99.14	11.47
20	31.14	989.35	68.56	99.70	10.97
50	3.91	1515.10	105.00	108.91	2.74
100†	0.12	1615.85	111.98	112.10	0.00

* Calculated by trapezoidal rule. See Equation 3.2.
† Considered infinity.
The rate constant for elimination = 0.0693, calculated from a semilogarithmic plot of C_t against time.

It should be noted from exercise above that even though the peak plasma concentration occurs at around 5 hours, the absorption still continues. In theory the absorption will never be completed, since it is an exponential process, and the percent unabsorbed will be equal to zero only at time equal to infinity. However, for the purpose of calculation, at least 3 to 5 half-lives of elimination must pass after the peak plasma concentration before the contribution of the absorption phase can be considered to be negligible.

The area under the plasma curve is also obtained by integrating Equation 7.26 between the limits of "time equal to zero" and "time equal to t" or infinity:

$$\int_0^t Cdt = \frac{k_a F X_0}{(k_a - K)V}\left[\frac{(1-e^{-Kt})}{K} - \frac{(1-e^{-k_a t})}{k_a}\right] \qquad \text{(Eq. 7.48)}$$

$$\int_0^{\infty} Cdt = \frac{k_a F X_0}{(k_a - K)V}\left(\frac{1}{K} - \frac{1}{k_a}\right) \qquad \text{(Eq. 7.49)}$$

The integral forms of the equation (see Appendix A) will provide a more accurate calculation than that obtained by the trapezoidal rule. The reader should try to compare these.

Fraction of Available Dose Remaining in the Body

The total area under the plasma concentration: time curve is proportional to the amount of drug absorbed. If the elimination of the drug takes place from the central or blood compartment and the mechanism of elimination does not change with the drug concentration, the amount of drug remaining in the body is proportional to the area remaining to be negotiated on the plasma concentration profile (Fig. 7.7):[1]

$$X/FX_0 = \frac{AUC_t^{\infty}}{AUC_0^{\infty}} \qquad \text{(Eq. 7.50)}$$

where AUC_t^{∞} = area (shaded, Fig. 7.7) from time t to infinity
AUC_0^{∞} = area from time zero to time infinity, e.g., Equation 7.49

This equation is also applicable to intravenous administration and other routes of administration, as well as to those circumstances in which the distribution is not instantaneous—a model which will be discussed later.

Continuous Infusion Data

Constant intravenous infusion involves zero order input since the rate of input remains constant. The change of the amount of drug in the body, assuming instantaneous distribution, is given by:

$$dX/dt = k_0 - KX \qquad \text{(Eq. 7.51)}$$

where k_0 = the infusion rate (e.g., mg/hr)
K = overall elimination rate constant

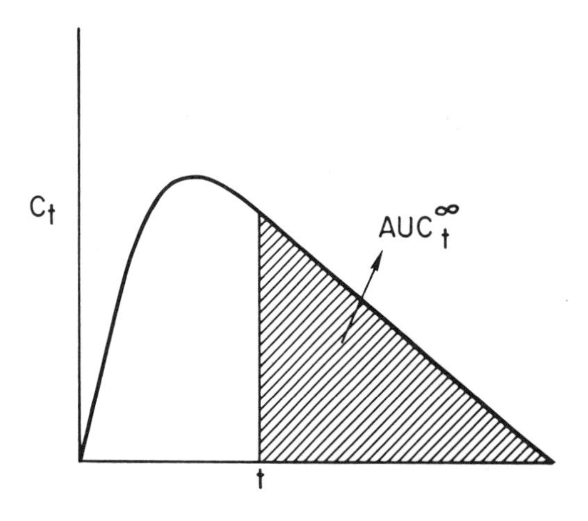

Figure 7.7. Determination of the fraction of the available dose remaining in the body as a function of time from the area under the curve.

An integrated form of Equation 7.51 (Appendix A) is:

$$X = \frac{k_0}{K}(1 - e^{-Kt}) \qquad \text{(Eq. 7.52)}$$

or

$$C = \frac{k_0}{VK}(1 - e^{-Kt}) \qquad \text{(Eq. 7.52)}$$

Thus the drug concentration increases until such time as e^{-Kt} becomes negligible and a steady state is reached. Note that in theory this will take infinite time but in practice a steady state can be assumed after about 6 or 7 disposition half-lives. At steady state:

$$X_{ss} = k_0/K \qquad \text{(Eq. 7.53)}$$

or

$$C_{ss} = k_0/VK \qquad \text{(Eq. 7.54)}$$

Note that X_{ss} is simply the ratio of the input rate and the output rate constant. Since the half-life of a drug can be assumed constant, the absolute value of X_{ss} or C_{ss} is determined only by the rate of infusion. For example, if the rate of infusion is increased by a factor of two, the steady state levels will also increase by a factor of two.

The rate with which the steady state is achieved is, however, independent of the rate of drug infusion. A fraction of the steady state is given by:

$$C/C_{ss} = f_{ss} = 1 - e^{-Kt} \qquad \text{(Eq. 7.55)}$$

$$= 1 - e^{-(0.693/t_{0.5})t} \qquad \text{(Eq. 7.56)}$$

$$= 1 - e^{-(0.693\, nt_{0.5})}$$

where $nt_{0.5}$ = number of half-lives

f_{ss}	$nt_{0.5}$
0.05	0.07
0.10	0.15
0.25	0.42
0.50	1.00
0.60	1.32
0.75	2.00
0.80	2.32
0.90	3.32
0.95	4.32
0.99	6.65
0.999	9.97
0.9999	13.29
1.00	∞

Example 7.12:
Calculate the maximum amount accumulated and the time required for 50 percent and 90 percent of maximum accumulation following intravenous infusion of 100 mg/hr of:

(a) Sodium penicillin (half-life = 30 min)
(b) Tetracycline (half-life = 10 hr)

For penicillin:

$$X_{ss} = k_0/K = (100\ \text{mg/hr})/(0.693/0.5)$$
$$= 72.15\ \text{mg}$$

The time for 50 percent of maximum accumulation is 1 half-life of 30 minutes and for 90 percent it is 3.32 half-lives or about 1.66 hours. Similarly, for tetracycline X_{ss} = 1443 mg and the times for 50 percent and 90 percent of maximum accumulation are 10 hours and 33.2 hours.

Multiple Dosing of Drugs

More frequently, drugs are administered on a multiple dose basis. The purpose of multiple dosing is to achieve and maintain a desired plasma level of drugs for a period long enough to treat an ailment. The plasma levels obtained from each succeeding dose are always higher than those obtained from the previous dose since some drug remains in the plasma when repeat doses are administered. However, since the rate of elimination is proportional to the amount of drug in the body, it increases with the increasing amounts of drug until it approaches the rate of input (such as 100 mg/24 hr). At this point the amount reaches a plateau in the body and no further increase in the amount of drug in the body can be possible unless the rate of administration is changed. Therefore, the pharmacokinetics of multiple dosing is quite similar to that previously described by the constant input model. The main difference is the fluctuation in the plasma concentration due to multiple dosing (Fig. 7.8).

The equations described earlier can be used to calculate the "peaks" and "valleys" in the plasma concentration since all plasma concentrations are additive provided the rate constants do not change.

If the drug is administered intravenously or by another route giving instantaneous input:

$$(X_1)_{\text{max}} = X_0 \qquad \text{(Eq. 7.57)}$$

$$(X_1)_{\text{min}} = X_0 e^{-K\tau} \qquad \text{(Eq. 7.58)}$$

where τ = dosing interval.

$$(X_2)_{\text{max}} = (X_1)_{\text{min}} + X_0 \qquad \text{(Eq. 7.59)}$$

$$= X_0(1 + e^{-K\tau}) \qquad \text{(Eq. 7.60)}$$

$$(X_2)_{\text{min}} = X_0(1 + e^{-K\tau})e^{-K\tau} \qquad \text{(Eq. 7.61)}$$

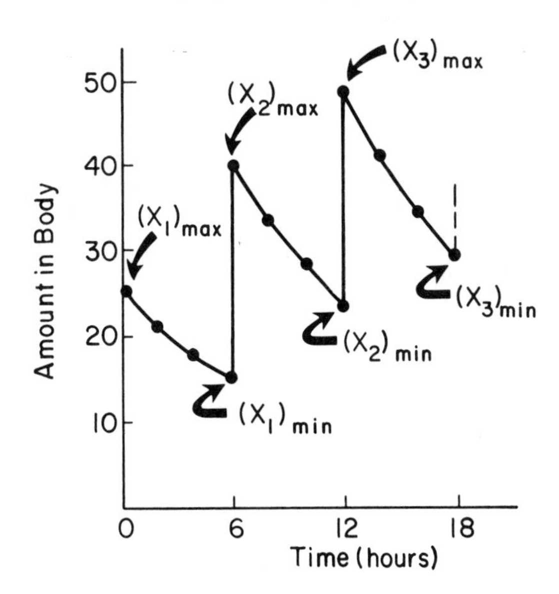

Figure 7.8. Multiple intravenous input profiles of the amount of drug in the body, assuming a single compartment characteristic or instantaneous distribution.

Thus

$$(X_n)_{max} = X_0(1 + e^{-K\tau} + e^{-2K\tau} + \ldots + e^{-(n-1)K\tau}) \quad \text{(Eq. 7.62)}$$

$$(X_n)_{min} = (X_n)_{max}\, e^{-K\tau} \quad \text{(Eq. 7.63)}$$

Since Equation 7.62 represents a geometric series, it can be solved to give:

$$(X_n)_{max} = X_0\left(\frac{1 - e^{-nK\tau}}{1 - e^{-K\tau}}\right) \quad \text{(Eq. 7.64)}$$

Since $C = XV$, the plasma concentration at any time during multiple dose therapy can be described by:

$$C_n = \frac{X_0(1 - e^{-nK\tau})}{V(1 - e^{-K\tau})} e^{-Kt} \quad \text{(Eq. 7.65)}$$

if time $t = \tau$ and C_n becomes equal to $(C_n)_{min}$.

Following multiple dosing, as the number of doses, n, approaches infinity, a plateau is reached at which the plasma concentration is given by:

$$C_\infty = \frac{X_0\,(1)}{V(1 - e^{-K\tau})} e^{-Kt} \quad \text{(Eq. 7.66)}$$

Notice that the exponential term becomes equal to zero in the numerator. Again the maximum and minimum concentrations at the plateau are given by the selected value of t. For maximum concentration, $t = 0$ and for minimum concentration, $t = \tau$ (Fig. 7.8). The equation above can be used to calculate the steady state concentration of a drug if its half-life is known along with the volume of distribution. One interesting approach involves the use of the area under the curve to calculate the steady state plasma concentration. Remember that the area under the curve following a single dose administration (up until time equals infinity) is equal to the area under the curve during the dosing interval at steady state. This can be easily understood since the total amount of drug eliminated from the body during a dosing interval period is equal to the dose administered at each time. In a single dose situation, the area under the curve up until time equals infinity also represents the total dose elimination from the body (Fig. 7.9).

The area under the curve following single instantaneous input is given by:

$$\int_0^{\infty} C \cdot dt = \frac{X_0}{V} e^{-Kt}\, dt \qquad \text{(Eq. 7.67)}$$

$$= X_0/KV \qquad \text{(Eq. 7.68)}$$

The area under the curve at the steady state will be equal to that given by Equation 7.68:

$$\int_0^{\infty} C_{\infty} \cdot dt = X_0/KV \qquad \text{(Eq. 7.69)}$$

Since the area under the curve is a product of concentration and time, an average concentration can be calculated by dividing the area by the dosing interval:

$$\bar{C} = \int_0^{\infty} C_{\infty} \cdot dt/\tau \qquad \text{(Eq. 7.70)}$$

$$= X_0/KV\tau \qquad \text{(Eq. 7.71)}$$

Example 7.13:
After oral administration of a 0.1 g dose the following plasma data were obtained:

Time (hr)	*Plasma Concentration (μg/ml)*
1	5
2	8
3	15
5	15
8	5
24	0

Estimate the total AUC using the trapezoidal rule and estimate the plateau level if a 0.1 g dose is administered every 4 or every 12 hours. Also calculate the average plateau level if a 1 g dose is administered every 4 hours.

Using Equation 3.2, the AUC = 120.5 μg·hr/ml:

$$\bar{C} = \text{AUC}/\tau = 120.5/4 = 30.12\ \mu\text{g/ml}$$

$$= 120.5/12 = 10.04\ \mu\text{g/ml}$$

$$= \text{AUC} \times 10/4 = 301.2\ \mu\text{g/ml for a 1 g dose}$$

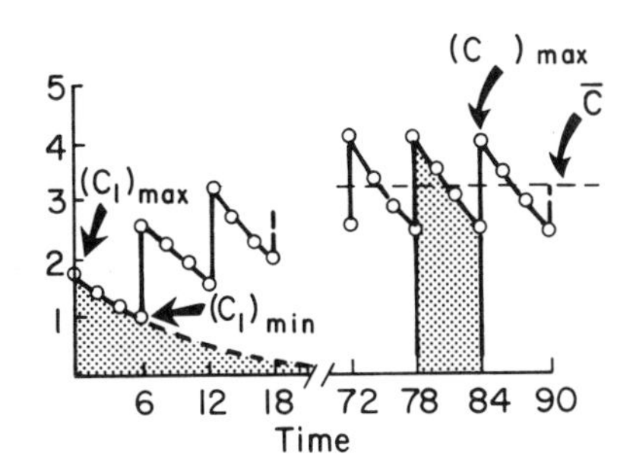

Figure 7.9. Plasma concentration profile following intravenous administration of equal doses of a drug. The respective shaded areas represent the areas under the curves either following a single dose or during the dosing interval at the steady state. Both of these areas are equal in dimension.

Notice that the AUC for a 1 g dose is ten times the AUC for a 0.1 g dose, since linear pharmacokinetics is assumed and the area is therefore proportional to the dose.

The rate with which the plateau level is reached is dependent only on the elimination half-life of the drug, as demonstrated earlier by continuous infusion examples. It will therefore take 1 half-life to reach 50 percent of the plateau level and 3.32 half-lives to reach 90 percent of the maximum level. However, the amount of drug present in the body at the steady state is related to the dose administered as well as to the elimination rate. The maximum accumulation of drug in the body can be given in terms of the administered dose as:

$$\overline{X}/X_0 = 1/K\tau = 1.44\, t_{0.5}/\tau \qquad \text{(Eq. 7.72)}$$

Thus if the dosing interval is equal to the half-life of the drug the maximum accumulation will be 1.44 times the administered dose. This allows calculation of the dose needed at each dosing interval if the desired steady state level is known. As the frequency of dosing increases the accumulation level also increases. For example, if the dosing interval is one-fourth the half-life, the maximum accumulation is 5.76 times the administered dose. Therefore, a desired plateau level can be achieved either by selecting a dosing interval and adjusting the dose or by changing the dosing interval to give a fixed dose.

The fact that the rate with which a plateau level is reached is dependent only on the elimination half-life can be very inconvenient with drugs with long biologic half-lives. For example, assuming a half-life of 60 hours for phenobarbital, it will take about 100 hours of continuous dosing to reach 90 percent of the plateau level. One means of avoiding this delay is to give a large initial dose, called a **PRIMING DOSE**, followed by periodic maintenance doses to keep the concentration stable. A priming dose is simply the dose that provides an effective concentration in the plasma instantaneously. The **MAINTENANCE DOSE** is equal to the amount of drug lost from the body during the dosing interval:

$$X_0 = X_0^* (1 - e^{-K\tau}) \qquad \text{(Eq. 7.73)}$$

where X_0^* = the priming dose. This concept is expressed in Figure 7.10.

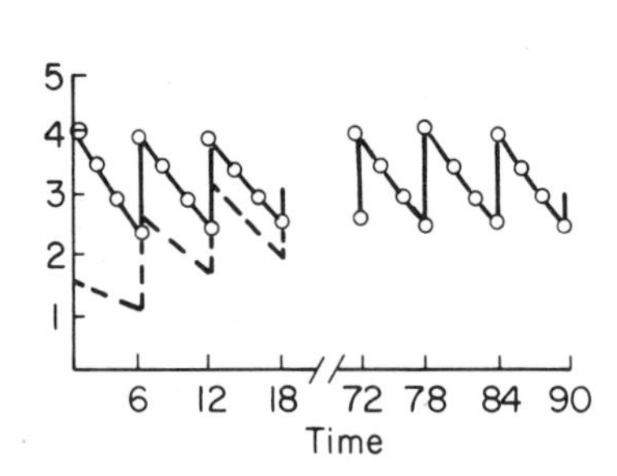

Figure 7.10. The plasma concentration profiles following continuous administration of a maintenance dose (- - - -) and a bolus or priming dose followed by maintenance dose (o).

Example 7.14:

On a once-a-day administration basis, how long will it take to reach 90 percent of the plateau level for digitoxin? What will the priming dose be if the maintenance dose is suggested as 0.2 mg? The half-life is about seven days.

It will take 3.32 half-lives, or more than three weeks, to reach a plateau level for digitoxin. However, the priming dose can be used to reach this level instantaneously:

$$X_0^* = X_0/(1 - e^{-K\tau})$$
$$= 0.2/(1 - e^{-0.099 \times 1}) = 2.12 \text{ mg}$$

Thus if a 2.12 mg priming dose is given, followed by a 0.2 mg maintenance dose, an instantaneous response to digitoxin can be achieved and maintained.

Example 7.15:

An epileptic patient was initially given 300 mg of sodium phenobarbital in tablet form. The disease syndrome was not controlled after the medication and another 300 mg dose was subsequently administered six hours after the first dosing. The patient started to show significant improvement within a couple of hours. However, when a 600 mg maintanance dose was administered the patient showed clear toxic symptoms. What is the desirable maintenance dose for this patient?

The half-life of phenobarbital varies between 50 and 150 hours. Assuming a half-life of about 60 hours (actual half-life to be determined in the patient or ascertained from other functions, as will be shown in later chapters), the total accumulation following a 600 mg maintenance dose will be:

$$\overline{X} = X_0/K\tau$$
$$= 600 \text{ mg}/(0.693/60 \text{ hr})\ (24 \text{ hr}) = 2164.5 \text{ mg}$$

Certainly this level of accumulation is toxic. A maintenance dose which will not give toxic response can be calculated from the loading dose of 600 mg, which gave a satisfactory response. Note that one 300 mg dose followed by another 300 mg dose can be, for practical purposes, considered a 600 mg dose since the half-life of elimination is quite long. The maintenance dose is simply the amount of drug lost from the body during the dosing interval (Eq. 7.73), which amounts to 167.0 mg. The reader should perform this calculation.

For many drugs the range of effective plasma concentration is quite small. One such example is that of phenytoin, whereby the effective range of serum concentration lies between 10 and 20 mg/liter for both the anticovulsant and the antiarrhythmic actions of the drug. Such toxic symptoms as nystagmus almost invariably appear at concentrations between 20 and 30 mg/liter; ataxia appears between 30 and 40 mg/liter; and mental changes appear above 40 mg/liter. The dosage regimen of such drugs should be designed so that high concentrations immediately after the dosing do not reach the toxic level. However, some fluctuation is always present and the percentage of fluctuation is given by:

$$\% \text{ fluctuation} = \frac{100\,\{(X_\infty)_{max} - (X_\infty)_{min}\}}{(X_\infty)_{max}} \qquad \text{(Eq. 7.74)}$$

$$= 100\,(1 - e^{-K\tau}) \qquad \text{(Eq. 7.75)}$$

From the equation above it is quite obvious that the fluctuation in the plasma concentration is only a function of the drug half-life (0.693/K) or the dosing interval. Since the half-life of a drug can be considered constant, an adjustment in the dosing interval is necessary to achieve a desired fluctuation. If a drug is administered more frequently (smaller τ), the fluctuation decreases and reaches a limit of 0 percent fluctuation when the dosing interval is zero or when there is constant intravenous infusion.

The choice of continuous intravenous infusion over multiple dosing is based on the frequency of drug administration required to achieve a desired plasma level and the degree of its fluctuation during the dosing intervals. For example, if the half-life of a drug is only three hours and a fluctuation of less than 20 percent is desired, this fluctuation can be achieved only by continuous intravenous infusion since it requires dosing at less than one-hour intervals. Such dosing is certainly not very desirable, especially if the drug is to be administered intravenously.

The pharmacokinetics of oral multiple dosing is quite similar to that described for the intravenous system. The only difference is that instead of the sharp peaks and valleys in the plasma concentration which follow intravenous administration, a gradual increase in the plasma concentration is noted due to the slow absorption process. If absorption is very fast, however, the plasma concentration profiles are similar to those obtained with intravenous administration. The plasma concentration during the dosing can be calculated by an approach similar to that used for intravenous administration:

$$C_n = \frac{k_a F X_0}{V(k_a - K)} \left(\frac{1 - e^{-nK\tau}}{1 - e^{-K\tau}} e^{-Kt} - \frac{1 - e^{-nk_a\tau}}{1 - e^{-k_a\tau}} e^{-k_a t} \right) \qquad \text{(Eq. 7.76)}$$

This equation is analogous to Equation 7.65, derived for intravenous administration. After oral dosing on a continuous basis, a plateau level will also be reached which can be obtained from Equation 7.76, making n equal to infinity:

$$C_{\infty} = \frac{k_a F X_0}{V(k_a - K)} \left(\frac{1}{1 - e^{-K\tau}} e^{-Kt} - \frac{1}{1 - e^{-k_a\tau}} e^{-k_a t} \right) \quad \text{(Eq. 7.77)}$$

Again, notice the similarity between Equation 7.66 and the equation above. The minimum plasma concentration is described by the equations above when the time $t = \tau$, but the maximum or peak concentration is not given by $t = 0$, since the maximum concentration does not occur immediately after the administration of an oral dose. The time at which the peak occurs can be obtained by differentiating Equation 7.76 and Equation 7.77 with respect to time and making the differential equal to zero:

$$t_p' = \frac{2.303}{k_a - K} \log \left[k_a \frac{1 - e^{-K\tau}}{1 - e^{-k_a\tau}} \right] \quad \text{(Eq. 7.78)}$$

where t_p' is the time at which the plasma peak occurs at the steady state during multiple oral dosing. A comparison of this equation with Equation 7.37 shows that the maximum plasma concentration is reached earlier at the plateau than after a single dose (Eq. 7.37).

The peak plasma concentration can be obtained from the absorption/elimination half-lives, which can be obtained from single dose studies, and then, assuming that these values do not change, t_p' can be obtained by using Equation 7.78, which can then be substituted in Equation 7.77 to obtain the value of peak plasma concentration.

If a therapy is begun with multiple oral dosing, considerable time may be required to reach the plateau concentration, depending on the half-life of the drug. It may therefore be necessary to use a loading dose or a priming dose to achieve an effective level, which can be maintained by using supplemental doses:

$$X_0^* = \frac{X_0}{(1 - e^{-K\tau})(1 - e^{-K_a\tau})} \quad \text{(Eq. 7.79)}$$

where X_0^* is the priming dose and X_0 is the maintenance dose required to maintain the effective concentration obtained by the priming dose. Any difference between respective priming doses in oral and intravenous administration is due to the lag time in the absorption rates and a ratio of the loading doses:

$$\frac{\text{Priming oral dose}}{\text{Priming IV dose}} = 1/(1 - e^{-k_a\tau}) \quad \text{(Eq. 7.80)}$$

As the absorption rate or the dosing interval becomes larger the value of the ratio approaches unity. For example, if the dose is administered at each half-life or $t = \tau$, the above ratio is equal to 2 if $k_a = K$; 1.33 if $k_a = 2K$; and 1.001 if $k_a = 10K$.

An interesting observation is the parameter of average plasma concentration following multiple oral dosing. Since by definition the average concentration is the area divided by the dosing interval, and since the total area under the curve following multiple dosing at the steady state is equal to the total area following a single dose administration (whether oral or intravenous), the average concentration can be easily determined by using the following relationship:

$$\bar{C} = FX_0/VK\tau \qquad \text{(Eq. 7.81)}$$

where F is the fraction of dose absorbed upon oral administration. According to this equation a dose of 150 mg q6h will give the same average level as obtained from 300 mg q12h or 600 mg qd.

Equation 7.81 is also applicable where direct determination of V, K, or F is not possible. If a single dose is given orally and the plasma concentration is monitored for a sufficient length of time, $\bar{C}$ can be obtained by dividing the total area under the curve by τ.

Example 7.16:
Guanethidine has a half-life of about 10 days and it is used to treat moderate to severe hypertension. In outpatient therapy, the dosage regimen begins with a 10 mg dose per day for the first week. An increase of 10 to 20 mg per day is made each week if the medication response is not satisfactory (e.g., 20, 30, 50, and 70 mg per day in the second, third, fourth, and fifth weeks, respectively).

Since the half-life of guanethidine is much longer than the dosing interval of one day, equations describing intravenous infusion can be used to calculate oral administration values. For example, at the end of the first week the total accumulation is given by:

$$X = \frac{k_0}{K}(1 - e^{-Kt})$$

where k_0 is the daily administration rate, e.g., 10 mg/day:

$$X = \frac{10 \text{ mg/day}}{0.693/10 \text{ day}^{-1}}(1 - e^{-0.693 \times 7/10})$$

$$= 55.46 \text{ mg}$$

If in the second week the dose is changed to 20 mg/day for the seven days, the total accumulation at the end will be 110.93 mg plus the remaining amount from the first week's residual, which will be equal to 34.08 mg. Thus the total amount at the end of the second week will be 145.01 mg. Similarly, at the end of the third week the total amount will be 255.65 mg and at the end of the fifth week it will be 655.75 mg, which amount can be maintained in the body by replacing the amount lost per day. This can be calculated either by using an exponential equation or by using the K value, since the dosing interval is much smaller than the half-life of the drug. Since $K = 0.0693$, about 6.93 percent of the drug is removed from the body every day, and thus if at the end of the fifth week an effective dose is achieved it can be maintained by giving 45.44 mg of the drug every day.

DELAYED DISTRIBUTION EQUILIBRIUM MODELS

The preceding discussion pertained to pharmacokinetic models in which an instantaneous distribution equilibrium was reached following drug administration. However, an instantaneous distribution is not realistically possible, since the body is composed of a heterogeneous group of tissues, each of which has a different affinity for the drug molecules and a different rate of equilibration. Tissues which are highly perfused, such as the liver and kidneys, equilibrate with the drug quickly, whereas bones, fat, and cartilage will equilibrate very slowly, depending on either the drug solubility in these phases or the specific drug–tissue interactions. In theory, therefore, a true pharmacokinetic model should have a rate constant for each tissue undergoing equilibration—such a model is therefore impossible. Fortunately, an empirical approach in which various tissues with similar distribution equilibration properties are grouped together leads to simpler models whereby two or three sets of rate constants can describe the disposition of the drug.

One of the most common models encountered for most drugs is referred to as a two compartment model, in which the body tissues are classified into two broad categories: those which equilibrate with the drug instantaneously and those which require some length of time for such equilibration (Fig. 7.11). The group of tissues which equilibrate instantaneously is supposed to reside in the central compartment, which is also the sampled compartment (though such sampling is not always necessary). The tissue or peripheral compartment contains slowly equilibrating tissues. Note that this classification is arbitrary and no attempt should be made to identify various organs of the body according to the compartments. It is possible to have part of an organ in the central compartment and the rest in the tissue compartment. The determining factor for the classifictiation is only the rate of equilibration.

After a drug has been administered intravenously, the plasma concentration declines sharply, due mainly to the distribution effect and also to some elimination from the body. The distribution process, however, is generally faster than the elimination process and is thus about 90 percent completed within 3.3 half-lives of distribution. When the contribution of distribution is negligible in the overall disappearance of the drug from the plasma, a pseudodistribution equilibrium is said to have been achieved. The term PSEUDO refers to the apparent nature of the process. In the postdistributive phase the plasma concentration is always described by a single exponent:

$$C = \text{Distribution Exponent} + \text{Elimination Exponent} \quad \text{(Eq. 7.82)}$$
$$= Ae^{-\alpha t} + Be^{-\beta t}$$

where α and β are the distribution and elimination rate constants and A and B are empirical constants. The distribution rate constant is generally larger than the elimination rate constant and thus the decrease in the first exponent occurs at a faster rate with change in time than is observed for the elimination

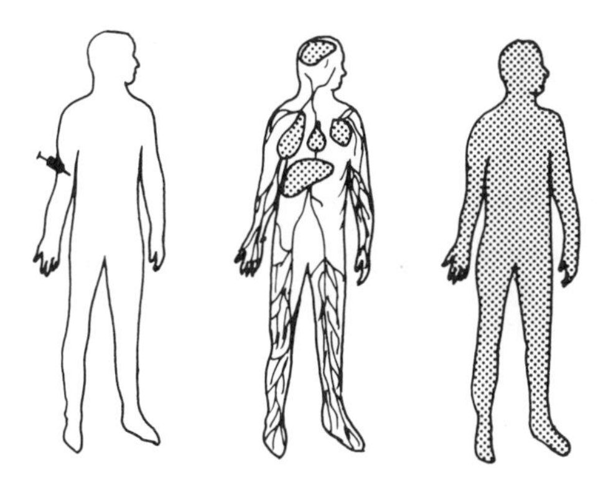

Figure 7.11. Two compartment visualization.

exponent. At pseudodistribution equilibrium the contribution of the distribution exponent is almost negligible and the plasma concentration is given by (Fig. 7.12):

$$C = Be^{-\beta t} \tag{Eq. 7.83}$$

Note the similarity between this equation and the equation describing instantaneous equilibration, or a single compartment model (Eq. 7.2). The rate

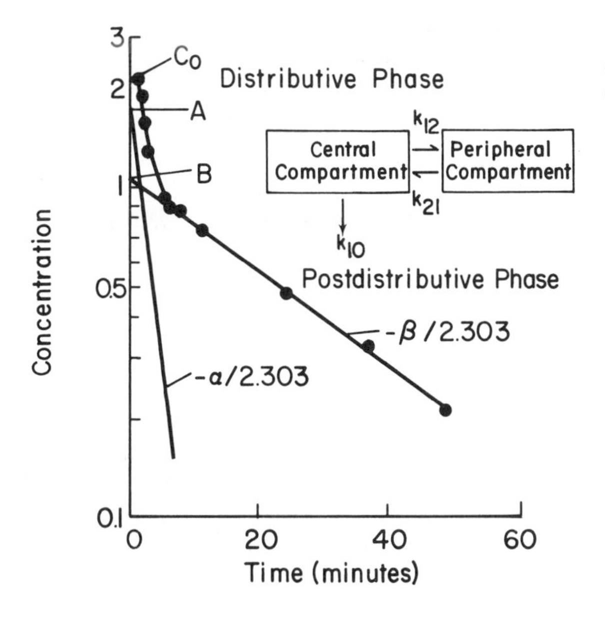

Figure 7.12. Plasma concentration profile of a drug exhibiting delayed equilibrium with body tissues (two compartment model).

constants α and β are, however, not simple rate constants. These are hybrid constants, or constants composed of several constants as described in Fig. 7.12. According to this scheme:

$$dX_c/dt = k_{21}X_p - k_{12}X_c - k_{10}X_c \quad \text{(Eq. 7.84)}$$

where X_c and X_p are the amounts of drug in the central and peripheral compartments. The rate constants describe the movement of drug between compartments (k_{12} and k_{21}) and from the body to the outside (k_{10}). The equation above can be solved for X_c:

$$X_c = \frac{X_0(\alpha - k_{21})}{(\alpha - \beta)}e^{-\alpha t} + \frac{X_0(k_{21} - \beta)}{(\alpha - \beta)}e^{-\beta t} \quad \text{(Eq. 7.85)}$$

where X_0 = dose administered intravenously and:

$$X_c = CV_0 \quad \text{(Eq. 7.86)}$$

where V_0 = volume of distribution at time equals zero. Since at time equals zero all the drug is present in the central compartment, V_0 is also the volume of the central compartment. The hybrid rate constants α and β are the functions of intercompartmental and elimination rate constants:

$$\alpha\beta = k_{21}k_{10} \quad \text{(Eq. 7.87)}$$

$$\alpha + \beta = k_{12} + k_{21} + k_{10} \quad \text{(Eq. 7.88)}$$

Note that the sum of the distribution and elimination rate constants is equal to the sum of all of the component constants. All of these constants can be calculated by plotting the plasma concentration on a semilogarithmic scale (Fig. 7.12). After a pseudodistribution equilibrium has been reached the plasma concentration is essentially described by a single exponent (Eq. 7.83) and, as with the calculation of a single compartmental model, B and β can be obtained from a straight line drawn through the terminal points and extended to the intercept, where:

$$B = (k_{21} - \beta)/(\alpha - \beta)\cdot C_0 \quad \text{(Eq. 7.89)}$$

This extrapolated line is also referred to as the β phase profile, which can be subtracted from the plasma concentration (C) to obtain:

$$C - Be^{-\beta t} = Ae^{-\alpha t} \quad \text{(Eq. 7.90)}$$

where

$$A = (\alpha - k_{21})/(\alpha - \beta)\cdot C_0 \quad \text{(Eq. 7.91)}$$

and A is calculated from the extrapolated line obtained by the difference. The rate constant α is obtained from the slope of the line or from the half-life of the line, which in this case would be termed the distribution half-life. The half-life obtained from the terminal part of the plot is referred to as the disposition half-life or the β phase half-life. All of the component rate constants can now be calculated from the values obtained from a graphical presentation of the plasma concentration profile.

The volume of the central compartment is obtained from:

$$V_0 = X_0/(A + B) \qquad \text{(Eq. 7.92)}$$

The sum $(A + B)$ is equal to the extrapolated initial plasma concentration at time equals zero.

The volume of distribution after pseudodistribution equilibrium can be calculated easily, since at this time the total clearance from the body is equal to the clearance from each of the body compartments:

$$\beta V_d = k_{12} V_0 \qquad \text{(Eq. 7.93)}$$

where V_d is the pseudodistribution equilibrium volume of distribution. Since the total body clearance is also given by:

$$\text{Total Body Clearance} = \text{Dose/Total Area under the Curve} \qquad \text{(Eq. 7.94)}$$

$$V_d = X_0/\beta\,\text{AUC} \qquad \text{(Eq. 7.95)}$$

where AUC is the total area under the curve and can be obtained from the pharmacokinetic parameters:

$$\text{AUC} = A/\alpha + B/\beta \qquad \text{(Eq. 7.96)}$$

The equation above represents the area from time zero to infinity.

The pharmacokinetic parameters can also be calculated by using urinary excretion rate data in place of plasma concentration profiles:

$$dX_u/dt = k_e X_c \qquad \text{(Eq. 7.97)}$$

where X_u is the amount of unchanged drug in the urine and k_e is the urinary excretion rate constant.

$$dX_u/dt = k_e A e^{-\alpha t} + k_e A e^{-\beta t} \qquad \text{(Eq. 7.98)}$$

Therefore, a plot of the logarithm of the urinary excretion rate against time will yield a plot identical in shape to the plasma concentration profile except that the intercepts on the lines obtained after "peeling off" or stripping of the curve now contain an additional term, k_e.

The renal clearance of the drug can be obtained by:

$$Q_r = k_e V_d = X_u^{\infty} \Big/ \int_0^{\infty} C \cdot dt \qquad \text{(Eq. 7.99)}$$

Example 7.17:

A 108 mg dose of griseofulvin dissolved in 4 ml of polyethylene glycol 400 was injected intravenously and the following profile was obtained:

Time (hr)	*Plasma Concentration (μg/ml)*
0.25	1.15
0.50	1.09
1.0	1.01
2.0	0.87
3.0	0.76
4.0	0.68
5.0	0.63
8.0	0.47
10.0	0.43
12.0	0.39
24.0	0.215

A plasma concentration: time plot on a semilogarithmic scale yields a biexponential curve from which the following pharmacokinetic constants are calculated:

$$A = 0.48\ \mu g/ml;\ B = 0.7\ \mu g/ml;\ \alpha = 0.3465\ hr^{-1};\ \beta = 0.048\ hr^{-1}$$

and the plasma concentration profile can be represented as:

$$C(\mu g/ml) = 0.48e^{-0.3465t} + 0.7e^{=0.048t}$$

$$\text{The total AUC} = A/\alpha + B/\beta = 15.57\ \mu g \cdot hr/ml$$

$$\text{and } k_{21} = 0.1694\ hr^{-1};\ k_{10} = 0.0982\ hr^{-1};$$

$$\text{and } k_{12} = 0.1269\ hr^{-1}$$

The disposition half-life $= 0.693/\beta = 14.4$ hours
$V_d = X_0/\beta\ \text{AUC} = 140.9$ liters
$V_0 = X_0/(A + B) = 91.52$ liters
The total body clearance $= k_{10}V_c = 8.99$ liters/hr
The pharmacokinetic model can be schematically represented as:

IV → CENTRAL $V_0 = 91.25$ liters

$k_{12} = 0.1269\ hr^{-1}$ →

← $k_{21} = 0.1694\ hr^{-1}$

TISSUE $V_T = 49.38$ liters

↓ $k_{10} = 0.0982\ hr^{-1}$

Note that the term V_T is the volume of the tissue compartment at pseudodistribution equilibrium and is calculated simply by the difference between the V_d and the V_0.

The computations presented above were applied to a two compartment model. However, in some instances higher order models involving three or four compartments are needed to accurately describe the disposition profile of a drug in the body. The purpose of multicompartmental modeling is to account for the disposition properties of drugs in order to design a rationale for dosage regimens. For example, it is possible that a patient might show higher than normal fat body mass, and if the distribution of a drug is dependent on the fat body content the volume of distribution will change. The dosage regimen should be based on the volume of distribution calculations to provide consistent plasma concentrations. Multicompartmental pharmacokinetics also helps establish the rationale for various toxicologic and pharmacologic actions of drugs. For example, the pharmacologic response of lysergic acid diethylamide (LSD) correlates better with its tissue (slow equilibration compartment) concentration than with the central or fast-equilibrating compartment. The accumulation of drugs as a result of slow equilibration has also been suggested as a possible reason for their prolonged effect and frequently their toxicity. In the case of thiopental this is responsible for its short duration of action. The drugs which have high volumes of distribution generally remain in the body for longer periods of time and can prove more toxic.

NONLINEAR PHARMACOKINETIC MODELS

In the preceding discussion the pharmacokinetic models were considered LINEAR, i.e., the rate constants of such models do not change with the concentration of the drug in the body. However, NONLINEAR traits in these rate constants are always possible, due for instance to the saturation of both the plasma protein binding sites and the enzymes responsible for the biotransformation of drugs. Generally, all active transport processes are subject to saturation and thus nonlinearity in the pharmacokinetics can be estimated from the Michaelis-Menton type of equations.

If the amount of drug in the body is such that it produces nonlinear pharmacokinetics, significant differences in the plasma levels will be observed. The rate of plasma concentration change is given by:

$$-dC/dt = \frac{V_m C}{K_m + C} \quad \text{(Eq. 7.100)}$$

where C is the plasma concentration, K_m is the Michaelis-Menton rate constant (units: concentration), and V_m is the maximum velocity of the reaction (units: concentration/time). For example, if a drug undergoes a saturable biotransformation at low concentration, when C is much smaller than K_m, the rate of reaction is essentially a first order process since the denominator becomes equal only to K_m and the units for the rate are reciprocal time. However, at higher concentrations when the enzyme responsible for the biotransformation can be saturated, the rate of reaction is equal to V_m (a zero order reaction with units of concentration/time) and is then independent of the plasma concentration. The half-life during this zero order process is:

$$t_{0.5} = \frac{C_0}{2V_m} \qquad \text{(Eq. 7.101)}$$

where C_0 is the initial concentration. It should be noted that the rates of decline in the plasma concentration are slower than could be anticipated in first order kinetics. One of the classic examples of nonlinear pharmacokinetics is that of aspirin (Fig. 7.13), which slows a slower elimination rate at higher dosage levels.[86] In many clinical uses of aspirin, doses above 1.5 g are used. Thus a nonlinear pharmacokinetics can be operative, resulting in prolonged effect of the drug as well as its potentiated side effects. Other examples of nonlinear pharmacokinetics include the use of ethanol and phenytoin.[87,88]

The pharmacokinetic parameters of a drug undergoing an exclusive nonlinear disposition can be calculated by using Equation 7.100 when converted to:

$$\frac{1}{\Delta C/\Delta t} = \frac{K_m}{V_m C} + \frac{1}{V_m} \qquad \text{(Eq. 7.102)}$$

where a plot of the left side of the equation against the reciprocal of C (the midpoint concentration) will yield a straight line with an intercept of $1/V_m$ and a slope of K_m/V_m. Another, more reliable transformation is:

$$\frac{C}{\Delta C/\Delta t} = \frac{K_m}{V_m} + \frac{C}{V_m} \qquad \text{(Eq. 7.103)}$$

However, in many instances a saturable process takes place in conjunction with a first order process. For example, in the elimination of salicylates from the body, only the conversion products salicyluric acid and salicyl phenolic

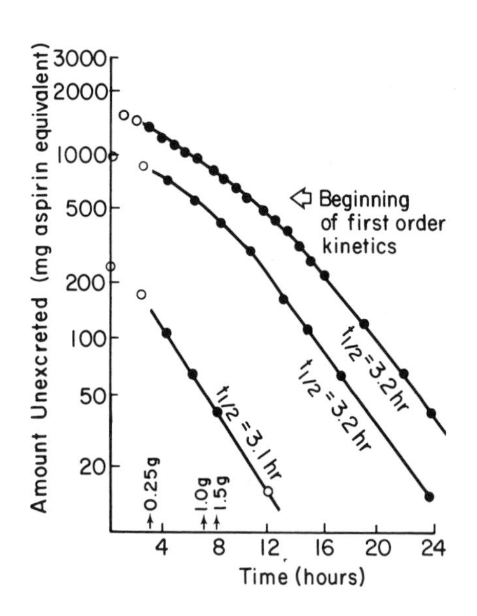

Figure 7.13. Dose-dependent disposition of salicylate in man. The three dose levels studied were 0.25 g, 1.0 g, and 1.5 g. Vertical arrows indicate the time necessary for 50 percent elimination of the administered drug. Note that the term 50 percent elimination is used instead of the half-life, which is generally restricted to first order or dose-independent processes. (From Levy: J Pharm Sci 54:959, 1965)

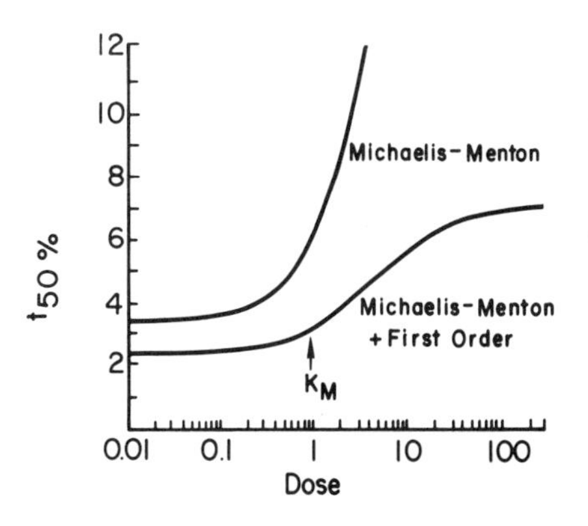

Figure 7.14. Comparison of dose dependency in a Michaelis-Menton type reaction alone and in combination with a first order process. The time for a 50 percent decline in concentration is used in place of half-life to distinguish it from a first order process alone.

glucuronide are saturable, whereas gentisic acid and salicyl acyl glucuronide are not. The following situation therefore arises:

$$-dC/dt = K'C + \frac{V_m C}{K_m + C} \qquad \text{(Eq. 7.104)}$$

where the overall rate of drug disappearance is the sum of a mixed order process and a first order process. The half-life of elimination increases as the dose is increased due to slower rates of elimination (the zero order component at saturation). However, the rate of elimination keeps increasing with the concentration through a first order process. At some point in level of dosage the rate of the first order process becomes so large that the contribution of the zero order component is almost negligible, making the half-life of elimination again independent of the dose (Fig. 7.14). If there were not a parallel first order process the half-life would keep increasing with the dose.

Nonlinear pharmacokinetics are also called DOSE-DEPENDENT pharmacokinetics. One means of minimizing the nonlinear phenomenon is to administer a drug in divided doses, as is demonstrated with para-aminobenzoic acid, which undergoes significantly less biotransformation when a single dose is given than when several divided doses are given. Dose-dependent processes are also attributed to the degree of drug absorption when a drug is given in divided doses. Drugs which are absorbed by an active process will generally show higher bioavailability when given in divided doses than when given in a single large dose (in which case the absorption sites may be saturated, allowing absorption of a smaller fraction of the drug).

The dose dependency of a drug's disposition can also be due to a change in the rates of its excretion. For example, if a drug is partly reabsorbed from the renal tubules by a saturable or capacity-limited process, the elimination rates will be slower at high drug concentrations. The binding and storage of drugs to various body tissues can also induce dose dependency in the disposition because of possible saturation of body storage sites, thus increasing the fraction of unbound drug in the body. In such instances an increased rate of elimination may be observed if the free unbound form of the drug undergoes elimination.

Another mechanism of dose dependency is the phenomenon of product inhibition, whereby the products of biotransformation may inhibit further product formation. This was discussed earlier in Chapter 6.

Dose dependency in drug disposition can also be due to the indirect effects of drugs on blood circulation, urine pH, and other physiologic and biochemical processes if these processes affect either the transport of drug molecules or their transformation into other forms. Drugs which reduce the hepatic blood flow can, for example, affect the first pass phenomenon or the overall yield of a particular biotransformation reaction.

The phenomenon of nonlinear pharmacokinetics is of great importance in multiple dose therapy, in which more significant changes in the plateau levels are wrought by the accumulation of drug in the body than can be anticipated in single dose studies. This accumulation can result in toxic responses, especially when the therapeutic index of the drug is low.

PHYSIOLOGIC PHARMACOKINETIC MODELS

The previous discussion is based primarily on the empirical approach to a mathematical description of drug disposition. The disadvantages of this approach include lack of relevance to physiology of the body, except for some parameters, and inability to predict the disposition behavior in disease states unless specific mechanisms are involved. An added disadvantage is the difficulty in extrapolating the data between species and even within a species.

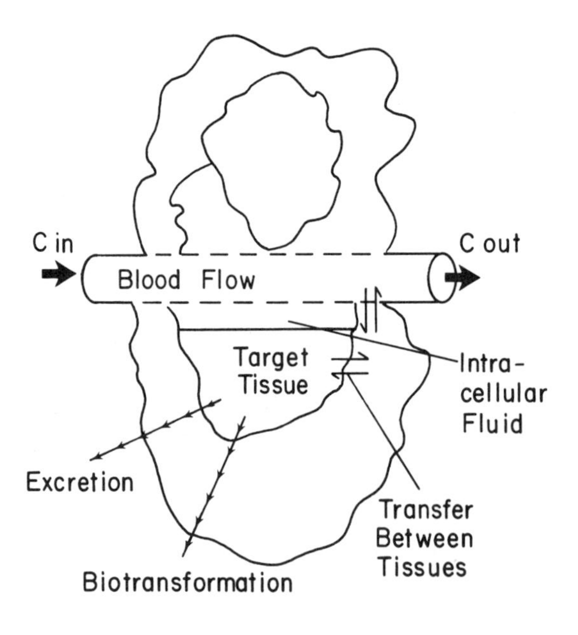

Figure 7.15. Assumptions in a physiologic pharmacokinetic model.

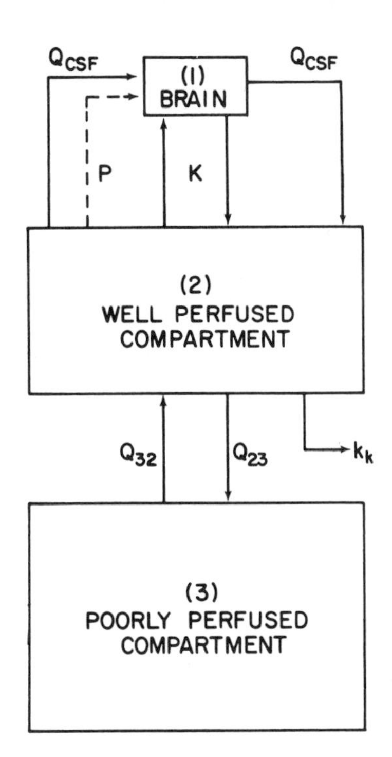

Figure 7.16. Compartmental physiologic model for chloride. K and P are transport parameters and the subscripts refer to the three compartments. Q_{CSF} is the cerebrospinal flow rate. (From Gabelnick et al.: J Appl Physiol 28:636, 1970)

In order to overcome these potential drawbacks some recent studies have attempted to define the disposition patterns of drugs in terms of well-known physiologic principles.[89] It is suggested that the distribution of drugs is dependent on the perfusion of the body tissues, and as long as these fluid perfusion rates are known the equilibration rates of drugs can be reasonably predicted. In pathologic conditions these estimates can be altered to account for differences in the organ blood flow rates,[90,91] and the data between species can be extrapolated based on the anatomic and physiologic differences in the perfusion rates.

The basic approach in physiologic pharmacokinetic modeling involves developing a large data base from animal studies and "scaling up" to apply these data to humans, as with thiopental,[92] methotrexate,[93] lidocaine,[90] and digoxin.[94]

The basic model requires a mass balance equation which will account for:

1. The rate of drug entry into a tissue, which is a function of rate of blood flow, input concentration, permeability, and the output concentration.
2. The rate of drug diffusion from other tissues.
3. The rate of biotransformation and excretion in the tissues.[95,96]

These phenomena are depicted in Figure 7.15.

The rate of accumulation is thus computed as:

(Rate of accumulation) = (Rate of blood flow into tissue × C) +
(Rate of drug diffusion from other tissues) −
(Rate of blood flow out of tissues × C) −
(Rate of drug biotransformation) −
(Rate of drug excretion in the tissue) −
(Rate of drug diffusion to adjacent tissues)

It is apparent from this equation that some knowledge of the physicochemical properties of the drug is needed to establish an initial model which can be refined by comparing with the actual data obtained. It is also imperative that some knowledge of the target tissue be available.

A simpler physiologic model is that of chloride distribution in the brain (Fig. 7.16).[97] The equation describing the amount in brain is given by:

$$V_1(dC_1/dt) = K(C_2 - C_1) - Q_{CSF}\,C_1 + P \qquad \text{(Eq. 7.105)}$$

Since most of the parameters can be determined empirically or are available in the literature, a simulation can easily be made, such as the one in Figure 7.17, which shows good agreement between theoretical and experimental results.

Another model recently developed is shown in Figure 7.18 for digoxin disposition in the rat.[94] The assumptions involved here are that each tissue acts as a well-stirred compartment, drug distribution is plasma-flow dependent, and all processes are linear:*

$$V_p(dC_p/dt) = (Q_hC_h/R_h) + (Q_mC_m/R_m) + (Q_sC_s/R_s) + (Q_kC_k/R_k) + (Q_lC_l/R_l) - Q_pC_p \qquad \text{(Eq. 7.106)}$$

$$V_h(dC_h/dt) = Q_h[C_p - (C_h/R_h)] \qquad \text{(Eq. 7.107)}$$

$$V_m(dC_m/dt) = Q_m[C_p - (C_m/R_m)] \qquad \text{(Eq. 7.108)}$$

$$V_s(dC_s/dt) = Q_s[C_p - (C_s/R_s)] \qquad \text{(Eq. 7.109)}$$

$$V_k(dC_k/dt) = Q_k[C_p - (C_k/R_k)] - (K_kC_k/R_k) \qquad \text{(Eq. 7.110)}$$

$$V_l(dC_l/dt) = (Q_l - Q_g)C_p + (Q_gC_g/R_g) - (Q_lC_l/R_l) - (K_lC_l/R_l) - (K_bC_l/R_l) \qquad \text{(Eq. 7.111)}$$

$$V_g(dC_g/dt) = Q_g[C_p - (C_g/R_g)] + k_aV_cC_c - k_sV_gC_g \qquad \text{(Eq. 7.112)}$$

$$V_c(dC_c/dt) = k_aV_gC_g + (K_bC_l/R_l) - k_aV_cC_c - K_gC_c \qquad \text{(Eq. 7.113)}$$

where K_k, K_l, K_b, and K_g represent renal, metabolic, biliary, and gastrointestinal clearances, respectively; and k_a and k_s represent first order rate constants for gastrointestinal absorption and secretion, respectively. The terms V_i, C_i, Q_i, and R_i represent tissue volumes, drug concentrations, plasma flow

* Equations 7.106 through 7.113 from Harrison LI, Gibaldi M: Physiologically based pharmacokinetic model for digoxin distribution and elimination in the rat. J Pharm Sci 66:1138, 1977.

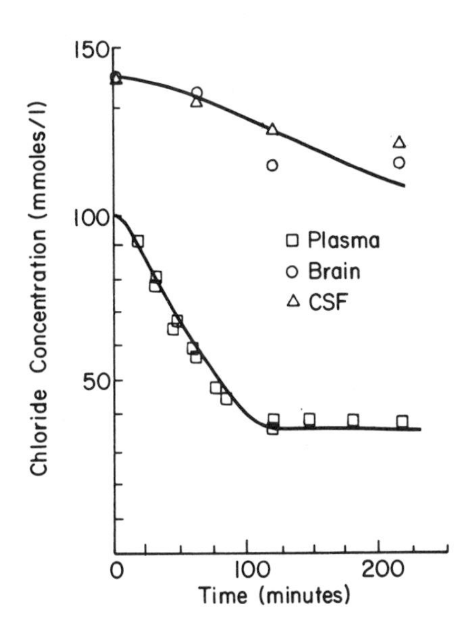

Figure 7.17. Comparison of model simulation (solid line) with experimental data from the cat during and following replacement hemodialysis. (From Gabelnick et al.: J Appl Physiol 28:636, 1970)

rates, and tissue-to-plasma partition coefficients, respectively. The subscripts of these terms are: p = plasma; h = heart; m = skeletal muscle; s = skin, fat; k = kidney; l = liver; g = gastrointestinal tissue; and c = gastrointestinal contents.

The equations described above can be solved to give theoretical estimates, which can be compared with experimental values obtained either by using

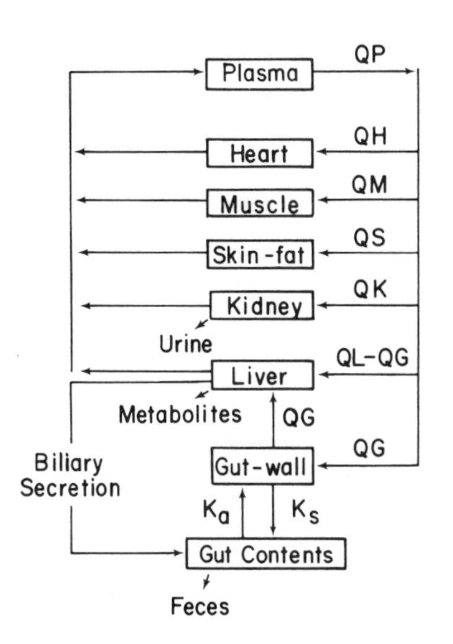

Figure 7.18. Pharmacokinetic model for digoxin disposition in the rat. (From Harrison, Gibaldi: : J Pharm Sci 66:1138, 1977)

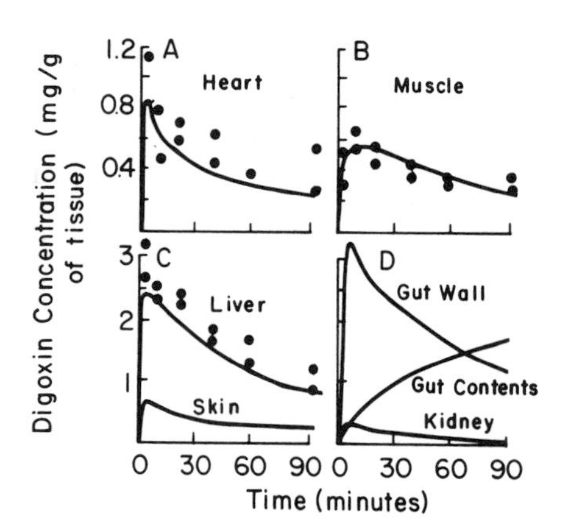

Figure 7.19. Predicted (———) and observed (●) tissue digoxin concentration after 1 mg/kg intravenous injection to control rats. (From Harrison LI, Gibaldi M: J Pharm Sci 66:1138, 1977

a radiolabeled drug or by specific extraction of various body tissues. Good correlations were thus reported for digoxin (Fig. 7.19).

The scaling up of the disposition kinetics involves calculation of the volumes and weights of various body tissues and comparing them with those of the species studied. Some relationships between body weight and various parameters are listed in Table 7.4 that may serve as a good starting point for

Table 7.4. EQUATIONS RELATING QUANTITATIVE PROPERTIES WITH BODY WEIGHTS AMONG MAMMALS

PARAMETER	EQUATION ($A\ B^{exp}$)	
	A	*exp*
Intake of water (ml/hr)	0.10	0.88
Urine output (ml/hr)	0.0064	0.82
Urea clearance (ml/hr)	1.59	0.72
Inulin clearance (ml/hr)	1.74	0.77
Creatinine clearance (ml/hr)	4.2	0.69
Diodrast clearance (ml/hr)	2.14	0.89
Hippurate clearance (ml/hr)	5.4	0.80
Ventilation rate (ml/hr)	120	0.74
Tidal volume (ml)	0.0062	1.01
Kidney wt (g)	0.0212	0.85
Brain wt (g)	0.081	0.70
Heart wt (g)	0.0066	0.98
Lungs wt (g)	0.0124	0.99
Liver wt (g)	0.082	0.87
Thyroid wt (g)	0.00022	0.80
Adrenals wt (g)	0.0011	0.92
Pituitary wt (g)	0.00013	0.76
Stomach + intestine wt (g)	0.112	0.94
Blood wt (g)	0.055	0.99

After Adolph: Science 109:579, 1949.

the extrapolation of data between species. For example, in a cat weighing 5 kg the average weight of the kidneys is 29.54 g, compared to a 70 kg man who will have kidneys weighing 278.34 g.

A recent study simulated the plasma concentration of cephalosporin antibiotics using parameters for the blood flow to various organs and the extracellular volumes (Table 7.5) and showed excellent agreement with experimental results.[99] This study also emphasizes the need to consider such factors as the protein binding of drugs as well as the necessity of obtaining accurate estimates of fluid flow rates to various parts of the body.

Table 7.5. PLASMA FLOWS AND ORGAN VOLUMES FOR PERFUSION MODEL CALCULATIONS

TISSUE	*TOTAL VOLUME (liters)*	*EXTRACELLULAR VOLUME (liters)*	*PLASMA FLOW (liters/hr)*
Bone	3.60	—	6.48
Skin	3.00	0.45	7.60
RET	0.60	0.09	57.60
Muscle	33.00	4.95	43.20
Liver	3.90	0.59	54.00
Blood	5.40	3.24	168.88

From Greene et al.: J Pharm Sci 67:191, 1978.

PHARMACOLOGIC PHARMACOKINETIC MODELS

In a way similar to the physiologic approach to pharmacokinetic modeling, pharmacologic response can also be treated mathematically. One well-known example of the quantitation of pharmacologic response is the familiar log dose-response curve (Fig. 7.20), in which an almost linear relationship exists between the logarithm of dose and the response within the range of 20 and 80 percent. Such response curves are possible for those drugs which exhibit reversible action and produce no development of tolerance at the receptor site, and with which the role of metabolites in eliciting the pharmacologic response is already known. These drugs are also known to show direct effects, the requirement being that the free plasma concentration is in equilibrium with the free concentration at the site of action, which, in turn, is proportional to the response.

In instantaneous distribution (single compartment) models the log dose-response behavior of drugs can be grouped with their disposition kinetics to obtain relationships between the pharmacologic response and time. An excellent account of this is given by Gibaldi and Perrier:[100]

$$E = m \log C + e \qquad \text{(Eq. 7.114)}$$

where E is the intensity of the effect, C is the plasma concentration, and m

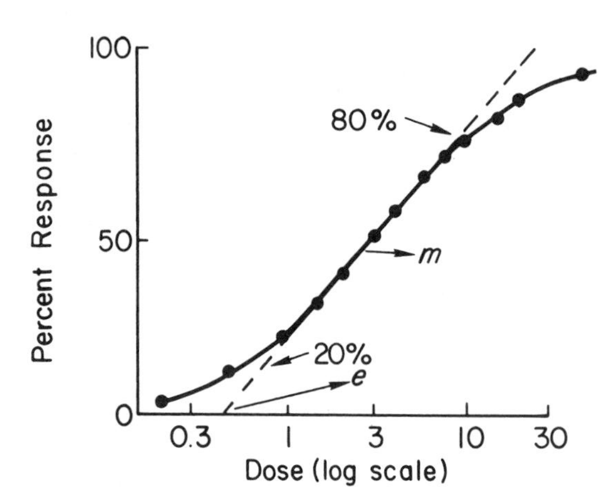

Figure 7.20. A typical log dose-response curve showing linear range between approximately 20 and 80 percent of the response. See text for details of the slopes and intercepts of the extrapolated line.

and e are slopes and intercepts as shown in Figure 7.20. Rearrangement of Equation 7.114 leads to:

$$\log C = \frac{E - e}{m} \qquad \text{(Eq. 7.115)}$$

$$= \log C_0 - \frac{Kt}{2.303} \qquad \text{(Eq. 7.116)}$$

Substituting the intensity of action term in place of the concentration leads to:

$$\frac{E - e}{m} = \frac{E_0 - e}{m} - \frac{Kt}{2.303} \qquad \text{(Eq. 7.117)}$$

where E_0 is the initial intensity immediately after the injection of the drug.[101] The equation above also simplifies to:

$$E = E_0 - \frac{Kmt}{2.303} \qquad \text{(Eq. 7.118)}$$

Thus a plot of intensity against time will yield a straight line with a slope of $Km/2.303$. An excellent example in which this concept is applicable is the average degree of muscular paralysis following intravenous administration of succinylcholine chloride (Fig. 7.21).[102] Similar observations have been made for the loss of the stimulant action of amphetamine after intravenous administration.[103] It should be noted that the half-life of a drug can also be calculated (from K) if accurate estimates of m are made from the plot. This method therefore provides an additional means of calculating a pharmacokinetic parameter if other methods are not applicable, in the absence of a specific or sensitive analytical technique.

The duration of a pharmacologic response can also be predicted if the amount of drug in the body is related to the intensity of response:

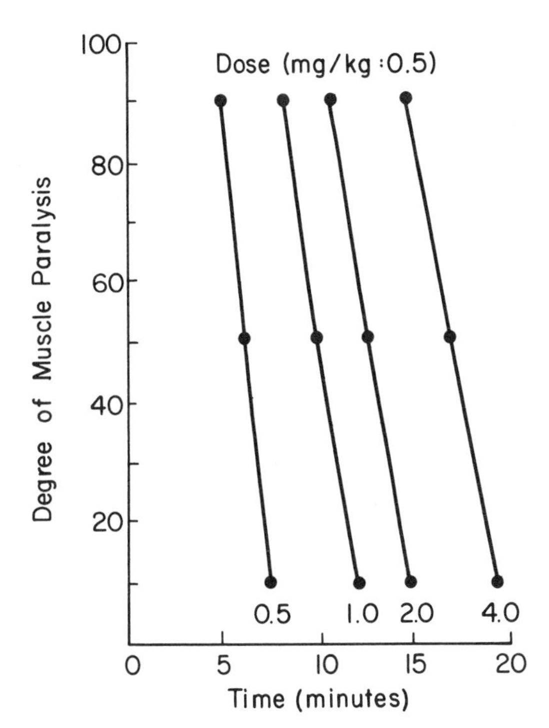

Figure 7.21. Average degree of muscular paralysis as a function of dose and time following intravenous administration of succinylcholine chloride in humans. (From Levy: J Pharm Sci 56:1687, 1967)

$$t_d = \frac{2.303}{K} \log X_0 - \frac{2.303}{K} \log X^*_{min} \qquad \text{(Eq. 7.119)}$$

where, X^*_{min} is the minimum effective level which can be easily calculated by measuring the duration of action at different levels of X_0 (intravenous dose) and by plotting the data as shown for pentobarbital in Figure 7.22.[104] The minimum effective level is obtained from the intercept of the x-axis (dose) at which the duration of action is zero. The slope of the line is equal to 2.303/K, allowing direct measurement of the half-life of the drug.

It is not necessary to follow a pharmacologic end point such as duration of sleep or maximal muscular contraction. Instead, an arbitrary end point can be set, such as a definite percent rise in blood pressure or heart rate, etc.

The duration of action and intensity of a pharmacologic response are generally higher if a second dose is administered immediately after the cessation of the effect of the first dose. This is due to the fact that an amount equal to X^*_{min} still remains in the body and the successive dose will elevate the levels much higher than those obtained from the first dose (Fig. 7.23). If, however, the successive dose is equal to $X_0 - X^*_{min}$ then an identical effect will be obtained.

The pharmacologic responses to many drugs cannot be measured in terms of the intensity or the duration of action (such as weight loss, retention or excretion of electrolytes, or diuresis). These responses can, however, be

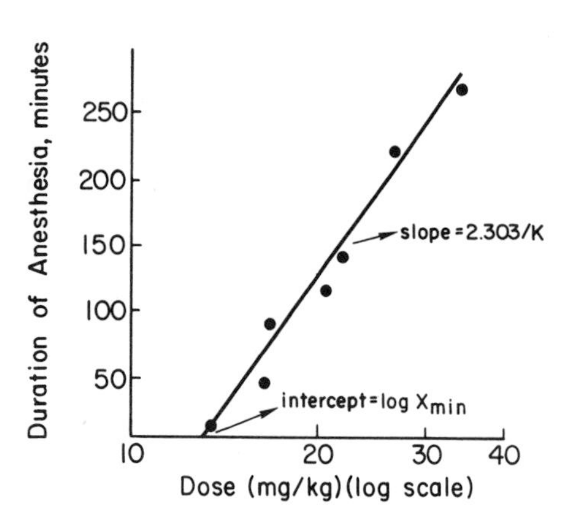

Figure 7.22. Relationship between dose and duration of anesthesia following intravenous administration of pentobarbital. (From Levy: Clin Pharmacol Ther 7:362, 1966)

related to the dose in terms of total pharmacologic activity (TPA), which is defined as:

$$TPA = \int_0^\infty E \cdot dt \qquad \text{(Eq. 7.120)}$$

The relative pharmacologic activity is thus simply TPA/dose, which takes into account any nonlinear relationships that might exist between the dose and the total pharmacologic activity. It is interesting to note that the TPA can be increased if a drug is administered in divided doses rather than in a single dose. For example, a 1 g dose of chlorothiazide given twice a day produces greater diuretic response than a 2 g dose given daily.[105]

The discussion presented above was applied to single compartment or instantaneous distribution models. It can, however, be extended to delayed distribution or multicompartment models, but with additional source of complexity: the site of action can lie in the central compartment or in the tissue

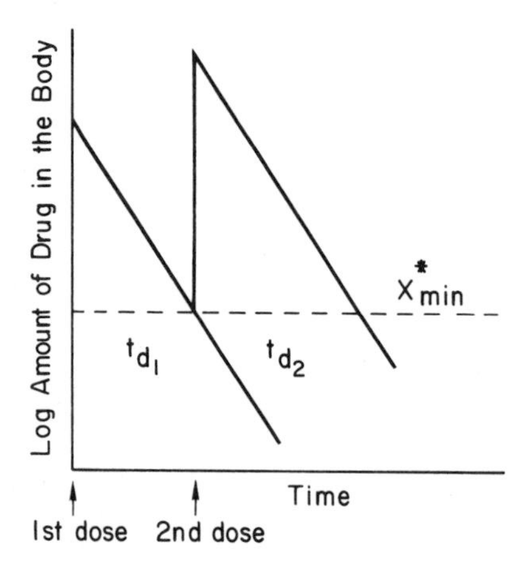

Figure 7.23. Increase in duration of action following second dose in a single compartment model. See text for details.

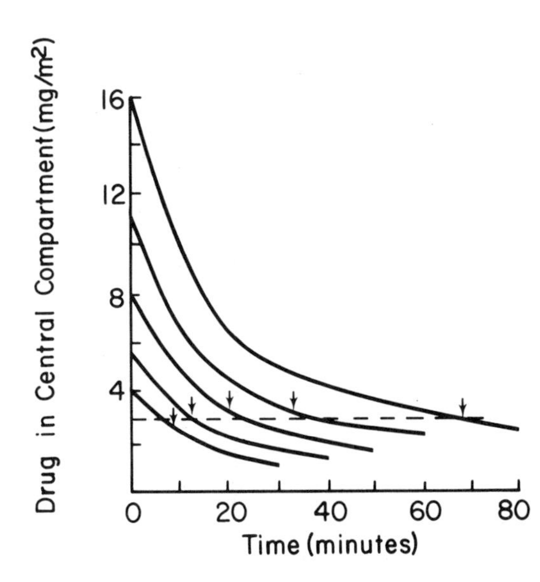

Figure 7.24. Plasma concentration profiles for the central compartment following intravenous administration of various doses of *d*-tubocurarine. The arrows indicate the duration of neuromuscular block effects and the dotted line indicates the average plasma concentration corresponding with the effect. (From Gibaldi et al.: Anesthesiology 36:213, 1972)

compartment and thus the response can be correlated only with the levels in these compartments. For example, an excellent linear relationship exists between the "tissue" concentration of LSD and the performance score.[106]

The duration of action in multicompartment systems may not be linear to the dose, as is observed for single compartment models. This is mainly due to the nature of the plasma concentration decline in relation to the dose (Fig. 7.24).[107] Since the plasma concentration declines rapidly in the initial phase in multicompartment models, higher doses will lead to proportionally greater durations of action. The minimum effective levels will be pushed into the elimination phase, rather than into the distribution phase where they are reached sooner. This relationship between the duration of action and the dose is also dependent on the degree of the monitored pharmacologic response. If the degree of response monitored is, say, only 10 percent, a linear relationship can be expected, but at higher levels of response the duration of action invariably increases.

References

1. Niazi S: Volume of distribution as a function of time. J Pharm Sci 65:452, 1976
2. Rowland M, Riegelman S, Harris PA, Sholkoff SD: Absorption kinetics of aspirin in man following oral administration of an aqueous solution. J Pharm Sci 61:379, 1972
3. Wan SH, Pentikainen P, Azarnoff DL: Bioavailability studies on para-aminosalicylic acid and its various salts in man. I: Absorption from solution and suspension. J Pharmacokinet Biopharm 2:1, 1974
4. Zarowny D, Ogilivie R, Tamblyn D, Macleod C, Ruedy J: Pharmacokinetics

of amoxacillin. Clin Pharmacol Ther 16:1045, 1974

5. Cole M, Kenig MD, Hewitt VA: Metabolism of penicillins to penicilloic acids and 6-aminopenicillanic acid in man and its significance in assessing penicillin absorption. Antimicrob Agents Chemother 1973:463, 1973
6. Kirby WMM, Regamey C: Pharmacokinetics of cefazolin compared with four other cephalosporins. J Infect Dis (Suppl) 128:341, 1973
7. Naumann P, Reintjens E: Antibacterial activity and pharmacokinetic behavior of cefazolin as compared with other cephalosporin antibiotics. Infection 2:19, 1974
8. Rosenblatt JE, Kind AC, Brodie JL, Kirby WMM: Mechanisms responsible for the blood level differences of isoxazodyl penicillins. Arch Intern Med 122:345, 1968
9. Peterson RE, Pierce CE, Wyngaarden JB, Bunim JJ, Brodie BB: The physiological disposition and metabolic fate of cortisone in man. J Clin Invest 36:1301, 1957
10. Dittert LW, Griffen WO, LaPiana JC, Shainfeld FJ, Dolusio JT: Pharmacokinetic interpretation of penicillin levels in serum and urine after intravenous administration. Antimicrob Agents Chemother 1969:42, 1969
11. Cutler RE, Forrey AW, Christopher TG, Kimpel BM: Pharmacokinetics of furosemide in normal subjects and functionally anephric patients. Clin Pharmacol Ther 15:588, 1974
12. Smith JT, Hamilton-Miller JMT: Hetacillin: A chemical and biological comparison with ampicillin. Chemotherapy 15:366, 1970
13. Burt LR, Davidson IW: Insulin half-life and utilization in normal pregnancy. Obstet Gynecol 43:161, 1974
14. Kind AC, Tupasi TE, Standifor HC, Kirby WMM: Mechanisms responsible for plasma levels of nafcillin lower than those of oxacillin. Arch Intern Med 125:685, 1970
15. Moore WE, Portmann GA, Stander H, McChesney EW: Biopharmaceutical investigation of nalidixic acid in man. J Pharm Sci 54:36, 1965
16. Kampmann J, Hansen JM, Siersbaek-Nielson K, Laursen H: Effect of some drugs on penicillin half-life in blood. Clin Pharmacol Ther 13:516, 1972
17. Loo JCK, Foltz EL, Wallick H, Kwan KC: Pharmacokinetics of pivampicillin and ampicillin in man. Clin Pharmacol Ther 16:35, 1974
18. Schuppan D, Riegelman S, von Lehman B, Pilbrant A, Becker C: Preliminary pharmacokinetic studies of propylthiouracil in humans. J Pharmacokinet Biopharm 1:307, 1973
19. Prescott LF, Sansur M, Levin W, Conney AH: The comparative metabolism of phenacetin and N-acetyl-p-aminophenol in man, with particular reference to effect on the kidney. Clin Pharmacol Ther 9:605, 1968
20. Ablad B, Ervik M, Hallgren J, Solvell L: Pharmacological effects and serum levels of orally administered alprenolol in man. Eur J Clin Pharmacol Ther 5:44, 1972
21. Clarke JT, Libke RD, Regamey C, Kirby WM: Comparative pharmacokinetics of amikacin and kanamycin. Clin Pharmacol Ther 15:610, 1974
22. Mather LE, Long JG, Thomas J: The intravenous toxicity and clearance of bupivacaine in man. Clin Pharmacol Ther 12:935, 1971
23. Azzollini F, Gazzaniga A, Lodola E, Natangelo R: Elimination of chloramphenicol and thiamphenicol in subjects with cirrhosis of the liver. Int J Clin Pharmacol Biopharm 6:130, 1972
24. Forist AA, DeHaan RM, Metzler CM: Clindamycin bioavailability from clindamycin-2-palmitate and clindamycin-2-hexadecyclocarbonate in man. J Pharmacokinet Biopharm 1:89, 1973
25. Froman J, Gross L, Curatola S: Serum and urine levels following parenteral administration of sodium colistimethate to normal individuals. J Urol 103:210, 1970

26. Adamson RH: Metabolism of anticancer agents in man. Ann NY Acad Sci 179:432, 1971
27. Ho DHW, Frei E: Clinical pharmacology of 1-D-arabino-furanosyl cytosine. Clin Pharmacol Ther 12:944, 1971
28. Walker SR, Evans ME, Richards AJ, Paterson JW: The fate of (^{14}C)disodium cromoglycate in man. J Pharm Pharmacol 24:525, 1972
29. Dume T, Wagner C, Wetzels E: Zur Pharmakokinetik von Ethambutol bei Gesunden und Patienten mit erminaler Niereninsuffizienz. Dtsch Med Wochenschr 96:1430, 1971
30. Lockwood WR, Bower JD: Tobramycin and gentamicin concentrations in the serum of normal and anephric patients. Antimicrob Agents Chemother 3:125, 1973
31. Estes JW, Pelikan EW, Kruger-Thiemer E: A retrospective study of the pharmacokinetics of heparin. Clin Pharmacol Ther 10:329, 1969
32. Peterson RE: Metabolism of adrenocorticosteroids in man. Ann NY Acad Sci 82:846, 1959
33. Champion GD, Paulus HE, Mangan E, Okun R, Pearson CM, Sarkissian E: The effect of aspirin on serum indomethacin. Clin Pharmacol Ther 13:329, 1972
34. Acocella G, Bonollo P, Garimoldi M, et al: Kinetics of rifampicin and isoniazid administered alone and in combination to normal subjects and patients with liver disease. Gut 13:47, 1972
35. Rowland M, Thomson PD, Guichard A, Melmon KL: Disposition kinetics of lidocaine in normal subjects. Ann NY Acad Sci 179:383, 1971
36. Mather LE, Tucker GT, Pflug AE, Lindop MJ, Wilkerson C: Meperidine kinetics in man: Intravenous injection in surgical patients and volunteers. Clin Pharmacol Ther 17:21, 1975
37. Alkalay D, Khemani L, Bartlett MF: Spectrophotofluorometric determination of methyltestosterone in plasma or serum. J Pharm Sci 61:1746, 1972
38. Brunk SF, Delle M, Wilson WR: Morphine metabolism in man: Effect of aspirin. Clin Pharmacol Ther 15:283, 1974
39. Laclercq R, Copinschi G: Patterns of plasma levels of prednisolone after oral administration in man. J Pharmacokinet Biopharm 2:175, 1974
40. Weily HS, Genton E: Pharmacokinetics of procainamide. Arch Intern Med 130:366, 1972
41. Lowenthal DT, Briggs WA, Gibson, TP, Nelson H, Cirskena WJ: Pharmacokinetics of oral propanolol in chronic renal disease. Clin Pharmacol Ther 16:761, 1974
42. Song CS, Bonkowsky HL, Tschudy DP: Salicylamide metabolism in acute intermittent porphyria. Clin Pharmacol Ther 15:431, 1974
43. Rowland M, Riegelman S: Pharmacokinetics of acetylsalicylic acid and salicylic acid after intravenous administration in man. J Pharm Sci 57:1313, 1968
44. Kunin CM, Finland M: Persistence of antibiotics in blood of patients with acute renal failure. III: Penicillin, streptomycin, erythromycin, and kanamycin. J Clin Invest 38:1509, 1959
45. Sandburg AA, Salunwhite WR Jr: Metabolism of 4-C^{14}-testosterone in human subjects. I: Distribution in bile, blood, feces, and urine. J Clin Invest 35:1331, 1956
46. Levy G, O'Reilly RA, Aggeler PM, Keech GM: Pharmacokinetic analysis of the effect of barbituate on the anticoagulant action of warfarin in man. Clin Pharmacol Ther 11:372, 1970
47. Weiss P, Hersey RM, Dujovne CA, Bianchine JR: The metabolism of amiloride hydrochloride in man. Clin Pharmacol Ther 10:401, 1969
48. Kunin CM, Dornbush AC, Finland M: Distribution and excretion of four tetracycline analogues in normal young men. J Clin Invest 38:1487, 1959
49. Wagner JG, Northam JI, Sokoloski WT: Biological half-lives of the antibiotic lincomycin observed in repetitive experiments in the same subjects. Nature 207:201, 1965

50. Kaplan SA, Weinfeld RE, Abruzzo CW, Lewis M: Pharmacokinetic profile of sulfisoxazole following intravenous, intramuscular, and oral administration in man. J Pharm Sci 61:773, 1972
51. Jaffe JM, Colaizzi JL, Poust RI, MacDonald RH: Effect of altered urinary pH on tetracycline and doxycyline excretion in humans. J Pharmacokinet Biophar 1:267, 1973
52. Mitenko PA, Ogilvie RI: Rapidly achieved plasma concentration plateaus, with observations on theophylline kinetics. Clin Pharmacol Ther 13:329, 1972
53. Brook R, Schrogie JJ, Soloman HM: Failure of probenecid to inhibit the rate of metabolism of tolbutamide in man. 9:314, 1968
54. Craig WA, Kunin CM: Trimethoprim-sulfamethoxazole: Pharmacodynamic effects of urinary pH and impaired renal function. Ann Intern Med 78:491, 1973
55. Anggard E, Jonsson LE, Hogmark AL, Gunne LM: Amphetamine metabolism in amphetamine psychoses. Clin Pharmacol Ther 14:870, 1973
56. Karad D, Inaba T, Endrenyi L, Johnson GE, Kalow W: Comparative drug elimination capacity in man—glutethimide, amobarbital, antipyrine, and sulfinpyrazone. Clin Pharmacol Ther 14:552, 1973
57. Schwartz MA, Postma E, Gaut Z: Biological half-life of chlordiazepoxide and its metabolite, demoxepam, in man. J Pharm Sci 60:1500, 1971
58. Mattila MJ, Nieminen E, Titinen H: Serum levels, urinary excretion, and side effects of cycloserine in the presence of isoniazid and p-aminosalicylic acid. Scand J Respir Dis 50:291, 1969
59. Gelber R, Peters JH, Gordon GR, Glazko AJ, Levy R: The polmorphic acetylation of dapsone in man. Clin Pharmacol Ther 12:225, 1971
60. Huffman DH, Benjamin RS, Bachur NR: Danuorubicin metabolism in acute nonlymphocytic leukemia. Clin Pharmacol Ther 13:895, 1972
61. Dolusio JT, Dittert LW: Influence of repetitive dosing of tetracyclines on biologic half-life in serum. Clin Pharmacol Ther 10:690, 1969
62. Alexanderson B: Pharmacokinetics of desmethylimipramine and nortriptyline in man after single and multiple oral doses—a cross-over study. Eur J Clin Pharmacol 5:1, 1972
63. Curry SH, Riddal D, Gordon JS, et al: Disposition of glutethimide in man. Clin Pharmacol Ther 12:849, 1971
64. Riegelman S, Rowland M, Epstein WL: Griseofulvin-phenobarbital interaction in man. JAMA 213:426, 1970
65. Jack DB, Riess W: Pharmacokinetics of idochlorhydroxyquin in man. J Pharm Sci 62:1929, 1973
66. Groth U, Prellwitz W, Janchen E: Estimation of pharmacokinetic parameters of lithium from saliva and urine. Clin Pharmacol Ther 16:490, 1974
67. Hollister LE, Levy G: Kinetics of meprobamate elimination in humans. Chemotherapia 9:20, 1964
68. Inturissi CE, Verebely K: Disposition of methadone in man after a single oral dose. Clin Pharmacol Ther 13:923, 1972
69. MacDonald H, Kelly RG, Allen ES, Noble JF, Kanegis LA: Pharmacokinetic studies on minocycline in man. Clin Pharmacol Ther 14:852, 1973
70. Carruthers SG, Kelly JG, McDevitt DG, Shanks RG: Blood levels of practolol after oral and parenteral administration and their relationship to exercise heart rate. Clin Pharmacol Ther 15:497, 1974
71. Bunger VP, Diller W, Fuhr J, Kruger-Thiemer E: Vergleichende Unterschingen an neuren Sulfanilamiden. Arzneim Forsch 11:247, 1961
72. Lous P: Plasma levels and urinary excretion of three barbituric acids after oral administration to man. Acta Pharmacol Toxicol (Kbh) 10:147, 1954
73. Kalser SC, McLain PL: Atropine metabolism in man. Clin Pharmacol Ther 11:214, 1970
74. Palmer L, Bertilsson L, Collste P, Rawlins M: Quantitative determination of carbamezapine in plasma by mass frag-

mentography. Clin Pharmacol Ther 14:827, 1973
75. Maxwell JD, Carrella M, Parkes JD, et al: Plasma disappearance and cerebral effects of chlorpromazine in cirrhosis. Clin Sci Mol Med 43:143, 1972
76. Taylor JA: Pharmacokinetics and biotransformation of chlorpropamide in man. Clin Pharmacol Ther 13:710, 1972
77. Hillestad L, Hansen T, Melsom H: Diazepam metabolism in normal man. Clin Pharmacol Ther 16:485, 1974
78. Vessel ES, Passananti GT, Greene FE, Page JG: Genetic control of drug levels and the induction of drug-metabolizing enzymes in man: Individual variability in the extent of allopurinol and nortriptyline inhibition of drug metabolism. Ann NY Acad Sci 179:752, 1971
79. Storstein L: Studies on digitalis. II: The influence of impaired renal function on the renal excretion of digitoxin and its cardioactive metabolites. Clin Pharmacol Ther 16:25, 1974
80. Koup JR, Greenblatt DJ, Jusko WJ, Smith TW, Koch-Weser J: Pharmacokinetics of digoxin in normal subjects after intravenous bolus and infusion doses. J Pharmacokinet Biopharm 3:181, 1975
81. Buchanan RA, Kinkel AW, Smith TC: The absorption and excretion of ethosuximide. Int J Clin Pharmacol Biopharm 7:213, 1973
82. Cressman WA, Bianchine JR, Slotnick VB, Johnson PC, Plostnieks J: Plasma level profile of haloperidol in man following intramuscular administration. Eur J Clin Pharmacol 7:99, 1974
83. Alvan G, Lindgren JE, Bogentoft C, Ericsson O: Plasma kinetics of methaqualone in man after single oral doses. Eur J Clin Pharmacol 6:187, 1973
84. Smith RB, Dittert LW, Griffen WO Jr, Dolusio JT: Pharmacokinetics of pentobarbital after intravenous and oral administration. J Pharmacokinet Biopharm 1:5, 1973
85. Hvidberg EF, Andreasen PB, Ranek L: Plasma half-life of phenylbutazone in patients with impaired liver function. Clin Pharmacol Ther 15:171, 1974
86. Levy G: Pharmacokinetics of salicylate elimination in man. J Pharm Sci 54:959, 1965
87. Lundquist F, Wolthers H: The kinetics of alcohol elimination in man. Acta Pharmacol Toxicol (Kbh) 14:265, 1958
88. Gerber N, Wagner JG: Explanation of dose-dependent decline of diphenylhydantoin plasma levels by fitting to the integrated form of Michaelis-Menton equation. Res Commun Chem Pathol Pharmacol 3:455, 1972
89. Bischoff KB, Brown RG: Drug distribution in mammals. Chem Eng Prog Symp No 66, 62:33, 1966
90. Benowitz N, Forsyth RP, Melmon KL, Rowland M: Lidocaine disposition kinetics in monkey and man. I: Prediction by perfusion model. Clin Pharmacol Ther 16:87, 1974
91. Benowitz N, Forsyth RP, Melmon KL, Rowland M: Lidocaine disposition kinetics in monkey and man. II: Effect of hemorrhage and sympathomimetic drug administration. Clin Pharmacol Ther 16:99, 1974
92. Bischoff KB, Dedrick RL: Thiopental pharmacokinetics. J Pharm Sci 57:1346, 1968
93. Bischoff KB, Dedrick RL, Zaharko DS, Longstrength JA: Methotrexate pharmacokinetics. J Pharm Sci 60:1128, 1971
94. Harrison LI, Gibaldi M: Physiologically based pharmacokinetic model for digoxin distribution and elimination in the rat. J Pharm Sci 66:1138, 1977
95. Dedrick RL, Zaharko DS, Lutz RJ: Transport binding of methotrexate in vivo. J Pharm Sci 62:882, 1973
96. Dedrick RL: Animal scale-up. J Pharmacokinet Biopharm 1:435, 1973
97. Gabelnick HL, Dedrick RL, Bourke RS: In vivo mass transfer of chloride during exchange hemodialysis. J Appl Physiol 28:636, 1970
98. Adolph EF: Quantitative relations in the physiological constituents of mammals. Science 109:579, 1949
99. Greene DS, Quitiliani R, Nightingale CH: Physiological perfusion model for cephalosporin antibiotics. I: Model se-

lection based on blood drug concentrations. J Pharm Sci 67:191, 1978
100. Gibaldi M, Perrier D: Pharmacokinetics. New York, Marcel Dekker, 1975, p 189
101. Levy G: Relationship between elimination rate of drugs and rate of decline of their pharmacologic effects. J Pharm Sci 53:342, 1964
102. Levy G: Kinetics of pharmacologic activity of succinylcholine in man. J Pharm Sci 56:1687, 1967
103. Van Rossum JM, Van Koppen ATJ: Kinetics of psychomotor stimulant drug action. Eur J Pharmacol 2:405, 1968
104. Levy G: Kinetics of pharmacologic effects. Clin Pharmacol Ther 7:362, 1966
105. Murphy J, Casey W, Lasagna L: The effect of dosage regimen on the diuretic efficacy of chlorothiazide in human subjects. J Pharmacol Exp Ther 134:286, 1961
106. Wagner JG: Relations between drug concentrations and response. J Mond Pharm 4:14, 1971
107. Gibaldi M, Levy G, Hayton W: Kinetics of the elimination and neuromuscular blocking effect of d-tubocurarine in man. Anesthesiology 36:213, 1972

Questions

1. What is the basic principle of kinetics that is applied to pharmacokinetic studies of drugs?
2. Derive Equation 7.2 from Equation 7.1. What are the required limits of integration?
3. What are the assumptions involved in the development of instantaneous distribution models?
4. How long does it take for the blood volume to undergo one complete circulation through the body?
5. What is the meaning of the term "compartment"?
6. Show that in a first order removal process the half-life is independent of the dose.
7. Cite the half-lives of five drugs.
8. Why must the data be collected for at least seven half-lives in order to calculate the bioavailability in urinary excretion rate studies?
9. What does a change in the half-life indicate in a patient?
10. What are the ranges of half-lives of the four classifications of drugs based on their half-lives (UFD, MD, SD, and VSD)?
11. Why may the UFD drugs be administered by continuous intravenous administration?
12. Under what circumstances will the elimination rate constant represent the fraction of drug removed from the body per unit of time (same units as those of the rate constant)?
13. Discuss the sources of error in Table 7.2.
14. What are the various components of an elimination rate constant? What is meant by the additive nature of rate constants?
15. What is the physiologic meaning of the term "volume of distribution," if any?

16. How can the total body clearance be calculated from the volume of distribution? What are the components of the TBC?
17. If the absorption rate constant is smaller than the elimination rate constant in the oral administration of drugs, what kind of plasma concentration profile will be obtained?
18. If the time for the peak plasma concentration and the elimination half-life are known, can one calculate the absorption rate constant?
19. What is the peak plasma concentration following oral administration if the absorption rate constant is equal to the elimination rate constant?
20. What routes of administration will provide a first order input?
21. Can Equation 7.50 be applied to multicompartment systems?
22. During continuous infusion of drugs, what determines the rate with which a plateau is reached?
23. Solve Equation 7.55 for f_{ss} of 0.3, 0.55, and 0.85.
24. Derive Equation 7.64.
25. Explain the observation in Figure 7.9.
26. Under those conditions whereby the AUC is proportional to the dose, what kind of pharmacokinetics is applicable (linear or nonlinear)?
27. What are the reasons for using a priming dose?
28. Fluctuation of the plasma concentration is dependent on what factors? How can it be minimized?
29. Cite examples of drugs with which you would recommend intravenous infusion over multiple dosing.
30. By using Equations 7.78 and 7.37, show that the peak plasma concentration achieves an earlier plateau than after administration of a single dose. Can you explain this?
31. The difference between an intravenous and an oral priming dose is determined by only one variable parameter. What is this variable?
32. How does one arrive at the average plasma concentration?
33. What is the delayed distribution equilibrium model?
34. How does the blood flow or perfusion determine the compartmental nature of drug disposition?
35. Why is the clearance following pseudodistribution equilibrium constant and given by: $k_{12}V_0$?
36. Show that total body clearance = dose/area.
37. What are the pharmacologic and physiologic applications of multicompartmental pharmacokinetic systems?
38. How does the nonlinear pharmacokinetic model differ from a linear model?
39. Why is the Michaelis-Menton equation used to describe the nonlinear pharmacokinetic behavior of a system?
40. What are the possible saturable processes in the elimination of drugs?
41. Under what conditions does a saturable process behave as a first order process or as a zero order process?
42. In parallel first order and saturable process, why does the half-life (time for 50 percent decay) become constant at higher dose levels?

43. The following data were obtained following intravenous administration of a 50 mg dose of a hydrobromide salt of a drug to a 65 kg male:

Time (hr)	*C (mg/liter)*
1	1.222
2	0.954
3	0.864
4	0.671
6	0.447

a) Using graphical method calculate K, $t_{0.5}$, and C_{12}.
b) What is the volume of distribution and what is its percentage of the body weight?
c) What is the equation describing the plasma concentration above?
d) What is the total body clearance?
e) What is the renal clearance if 80 percent of the dose is excreted unchanged?

44. Digoxin has a half-life of 40.8 hours. What percentage of the total amount in the body is lost daily?
45. The following data of the salicylic acid plasma levels were obtained from a 55 kg male subject following an intravenous injection of 500 mg of salicylic acid, administered in its sodium form.

Time (hr)	*C (μg/ml)*
1	41
2	32
3	29
4	22.5
6	15

a) Determine the biologic half-life of the drug in the above subject by a graphic method on a semilog scale.
b) What is the apparent volume of distribution of the drug in this subject? Calculate also the apparent volume of distribution in terms of percentage of the body weight.
c) Estimate its plasma concentration at 20 hours after administration by the graphic method and by computation.
d) Show the pharmacokinetic model of the drug disposition above, including the elimination rate constant.

46. Evans blue, a dye, distributes only into the plasma and is sometimes used to measure the plasma volume. A 60 kg 30-year-old white female presenting with symptoms of blood loss was given 25 mg IV. The results obtained were:

Time (min)	*C (mg%)*
10	1.90
20	1.80
30	1.73

a) What is the plasma volume of this patient? How does this compare with normal subjects?
b) Her Hct was 30. What is her blood volume? Is it within normal limits? What is a usual female Hct?
c) Is there any reason to prefer collecting three samples at 10-minute intervals rather than a single sample at 1 minute after injection?

47. The following urinary excretion data were collected following an IV dose of 500 mg Vancomycin to a 50 kg black female whose renal function has been found to be within normal limits.

Time at End of Collection Period (hr)	*Urine Volume (ml)*	*Drug Conc. in Urine (μg/ml)*
1	100	520
2	80	578
3	75	548
4	110	333
5	90	363
7	175	306

From these data determine:
a) K; k_e; k_b; percent metabolized; and plasma half-life.
b) The population average for the fraction excreted unchanged is about 95 percent. Is this patient "normal"?
c) If some trauma occurred which reduced renal function to about 40 percent of normal, what would be the new K, $t_{1/2}$, and percent metabolized?

48. After an intravenous injection of 1.5 mg of digoxin dissolved in dilute alcohol solution to a patient with congestive heart disease, the following urinary data were obtained:

Time (day)	*Urine Volume (ml)*	*Urinary Conc. of Digoxin (μg/ml)*
0–1	1000	0.01386
1–2	1200	0.00815
2–3	1400	0.00500
3–5	2520	0.00333

a) By the graphic method, estimate the biologic half-life of the digoxin in this subject.
b) Calculate the fraction of the dose excreted unchanged up until infinite time.
c) Assuming that the unexcreted portion of the digoxin is metabolized in the liver, determine the rate constants for k_e, k_b, and K.
d) If this patient became anuric, what would be the $t_{1/2}$?

49. The following data were obtained following administration of 490 mg of proxyphylline as an oral solution to a 70 kg white male. Volume of distribution coefficient = 0.50 liters/kg.

Time (hr)	C (μg/ml)
0.25	3.2
0.5	4.7
1.0	7.3
1.5	9.5
2.0	8.8
4	6.2
6	4.5
10	2.4

a) What are K, $t_{1/2}$ elimination, and k_a?
b) What fraction of the dose is absorbed?
c) Estimate the area under the curve (AUC).

50. The following data were obtained following administration of several test dosage forms to the same subject in a cross-over design. This subject was a male, 78 kg, 38 years old.

Route:	IV	Oral	Oral	Rectal	Oral
Dose (mg):	300	300	350	400	500
Dosage Form:	Solution	Tablets	Capsule	Suppository	Solution
Time (hr)	C (μg/ml)	C (mg%)	C (ng/ml)	C (μg/ml)	C (mg/liter)
0	30	0	0	0	0
1	27	0.20	3300	3.76	30.0
2	24.5	0.26	6250	7.12	35.0
3	22	0.32	8400	9.58	35.0
5	17.55	0.42	11,000	12.54	28.75
7	14.5	0.48	10,500	11.97	23.75
9	11.5	0.54	10,000	11.40	18.75
13	7.6	0.53	7600	8.66	12.5
15	6.2	0.49	6200	7.07	10.375
18	4.5	0.40	4500	5.13	7.5
22	2.95	0.30	2950	3.36	5.0
25	2.1	0.21	2100	2.39	3.5

a) For each dosage form, compute all the pharmacokinetic parameters appropriate to the dosage form.
b) Calculate the bioavailability of each of the dosage forms normalized to dose.
c) Postulate the differences in bioavailability of the three oral dosage forms.
d) What is the V_d for this drug? Express this as a percentage of body weight.

51. Following IM administration of 100 mg of a new aminoglycoside antibiotic in solution to a 70 kg 35-year-old black male, the following data were collected:

Time (hr)	C (μg/ml)
0.2	1.65
0.4	2.33
0.6	2.55
0.8	2.51
1.0	2.40
1.5	2.00
2.5	1.27
4.0	0.66
5.0	0.39
6.0	0.25
7.0	0.15

a) Compute $t_{1/2}$ for elimination and absorption.
b) Compute the V_d and its percentage of the body weight.
c) If the AUC following IV administration of 100 mg was found to be 30 μg·hr/ml, what is the percent bioavailability?
d) From certain IV studies it has been found that 85 percent of the administered dose is excreted unchanged. What would the elimination half-life be for a patient in total renal failure?

52. Using the data given below, find plateau levels for administration of Gentamicin 1 mg/kg q 8h IV to a 170 lb 60-year-old white man with apparently normal renal function.
Contrast this with the same dose given IM.

Gentamicin Data from the Literature:
$t_{1/2}$ = 2 hr
V_d = 28% BW
Fraction eliminated unchanged = 0.90
k_a = 2.16 hr^{-1}
MEC = 2–8 μg/ml
Fraction absorbed = 1.0

53. After an oral administration of 0.1 g of drug A to a patient, the following plasma data of the intact drug were obtained:

Time (hr)	C (μg/ml)
1	9.5
2	23
3	20.3
5	12.5
8	5
10	2.55
12	1.28

a) Estimate the total area under the curve (AUC) by a trapezoidal method.
b) Estimate the average plateau level if 0.1 g of drug A is administered to the same patient every (1) 4 hours, (2) 12 hours, and (3) day.

c) Estimate the average plateau level if 1 g of the drug A is administered to the same patient every 4 hours.

54. A drug has a V_d of 20 liters in a patient and the $t_{0.5}$ is estimated to be 10 hours. The patient is given a bolus dose of 250 mg IV.
    a) Calculate C_0.
    b) Calculate the constant infusion rate which is required to maintain the C of 8 μg/ml.
    c) Prescribe a dosage regimen that will maintain a C of 8 μg/ml following the administration of the bolus dose.
55. In a single compartment open model the maximum accumulation $(X_\infty)_{max}$ of the drug in the body after intravenous administration of repeated dose X_0 at interval τ is expressed as:

$$(X_\infty)_{max} = \frac{X_0}{1 - e^{-K\tau}}$$

where K is the first order elimination constant.
    a) Estimate $(X_\infty)_{max}$, if $\tau = t_{0.5}$.
    b) Estimate $(X_\infty)_{max}$, if $\tau = 2t_{0.5}$.
    c) Estimate the time required in terms of $t_{0.5}$ to reach 50, 75, and 90 percent of the $(X_\infty)_{max}$.
56. A patient has been on a successful maintenance dose of 0.15 mg per day of digitoxin during the last six months.
    a) If $t_{0.5}$ = 6 days, calculate the average amount of digitoxin present in the body during the maintenance therapy.
    b) Estimate the dose required to redigitalize the patient if the patient skips the drug for 1, 2, 5, or 30 days.
57. A 500 mg IV dose of spectinomycin was administered to a patient. These data were collected:

Time (hr)	*C (μg/ml)*
0.1667	74.4
0.333	54.8
0.50	45.2
1	40.9
2	23.4
4	9.1
6	3.7
8	1.9

    a) Calculate A, B, α, β, and V_0.
    b) Write the equation which summarizes these data.
58. A single dose of 200 mg of warfarin was administered to a 70 kg male in order to study the drug's pharmacokinetics. The following data were collected:

Time (hr)	C (μg/ml)	Time (hr)	C (μg/ml)	Time (hr)	C (μg/ml)
0.25	41.1	8.5	26	90	7.3
0.50	33.9	12.5	21.9	117	4.7
0.75	30.2	24	17.2	145	3.0
1	28	37	15.6	168	1.8
3	25.7	48	12.8	192	1.3
6	23.9	72	9.9		

a) Using these data, compute: A, B, α, β, k_{12}, k_{21}, k_{10}, and V_0.
b) What is the plasma:tissue ratio?
c) What is the AUC?

59. What are the inherent disadvantages of the empirical approach in describing disposition kinetics?
60. What is the perfusion-limited pharmacokinetic model?
61. What is the "scaling-up" of physiologic models? Can this be applied to all organ systems?
62. What are the basic mass balance requirements in a physiologic model?
63. Why is it necessary to assume that each tissue acts as a well-stirred compartment?
64. Calculate the weights of the hearts in a 5 kg cat and a 70 kg man.
65. Based on the data given in Table 7.5, arrange the various organs in order of decreasing rate of equilibration with blood. Assuming equal partitioning (an erroneous assumption?), which organ will contain the highest content of a drug?
66. Why is it necessary to use a pharmacologic response range of 20 to 80 percent for correlation with the plasma concentration?
67. Is the relationship between pharmacologic response and time a first order process or a zero order process? Justify your answer.
68. How would you calculate the disposition half-life of a drug by the pharmacologic response?
69. What parameters are associated with the threshold dose for a given activity?
70. Derive Equation 7.119
71. What is meant by a pharmacologic end point? Is it always necessary to use this in correlating with plasma levels?
72. It is suggested that when a dose is repeated, the second dose provides an increased duration of action when compared to the first one. Why? Would this phenomenon continue with succeeding doses?
73. What is the TPA? How would you account for nonlinearity in the process using the TPA?
74. Explain the observation that divided doses can provide a greater TPA/dose than single doses.
75. Why is the duration of action nonlinear with the dose in multicompartment systems?
76. At what level of pharmacologic response can a linear relationship between the dose and the duration of action be obtained?

CHAPTER 8

The Pharmacokinetic Basis of Variability in Clinical Response

The clinical response to a given dose varies widely between and within individuals. As early as 1913 large variations in the doses of sodium salicylate were seen to produce identical toxic symptoms in several hundred subjects.[1] It is in most instances difficult to establish a quantitation of the variability in human drug response because of the limited possibility of controlled studies and the inherent errors in such estimations. Generally, the response to a drug can be described in terms of a normal distribution curve (Fig. 8.1). However, in many instances skewed distributions are frequently observed, such as those shown for phenylephrine and warfarin, where as much as a six-fold variation in dose needed on a daily basis was recorded.[2,3]

The source of variation in human drug response is a two-fold phenomenon, including both the variation in plasma or tissue concentration (site of action) and intensity of tissue response to a given concentration. However, in most instances it is the plasma or tissue level which is responsible for a variation, and therefore the pharmacokinetic principles discussed earlier can be applied to understand the reasons for this variation and means can be developed to rectify any problems arising from it.

The reasons for variability in plasma concentration to a given dose can be explained in terms of the four processes of pharmacokinetics: absorption, distribution, biotransformation, and excretion.

ABSORPTION

Drug molecules must be free to be absorbed in the gastrointestinal tract. The factors which affect the absorption rates from various sites of administration were discussed earlier and will not be repeated here. The main concern in the variability in drug absorption is with the drug interactions with the contents of the gastrointestinal tract. For example, the absorption of tetracyclines is reduced in the presence of antacids, ferrous sulfate, and dairy products. Multiple drug therapy further complicates the problem of interactions at the absorption site. For example, kaolin and other adsorbents, when given

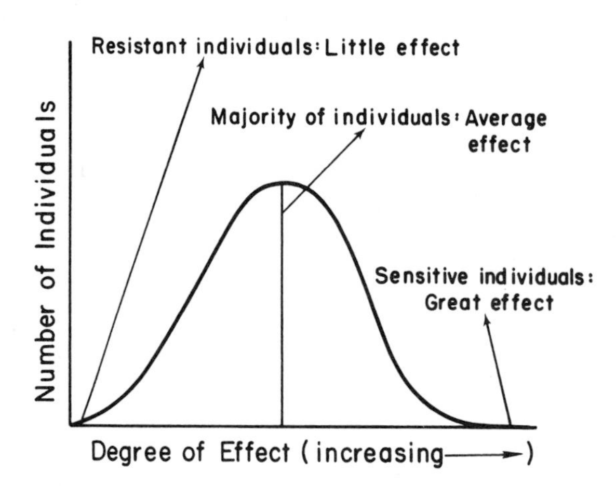

Figure 8.1. Distribution of drug effects in normal population.

with drugs like lincomycin or promazine, show decrease in the absorption of these drugs.[4,5] Phenobarbital reduces the absorption of griseofulvin and heptabarbital decreases the total bioavailability of dicumarol.[6,7] An interesting and frequently used combination is that of the antacid Mylanta with quinidine sulfate, apparently to decrease gastrointestinal irritation—the combination can be expected, however, to result in decreased bioavailability of quinidine.

Variation in the gastric pH has often been cited as one reason for variability in the total bioavailability of drugs. However, chances are that this factor is unimportant since most drugs are absorbed from the small intestine. There are examples, however, in which the gastric acidity plays an important role either in the release of the drug from a dosage form or in dissolving the contents. In one example, which is an exception, a total block of absorption is noted for carbenoxolone when the gastric pH rises above 2.[8] The gastric pH also determines the emptying rate of the stomach and thus induces possible differences in absorption rates.

The presence of food delays the absorption of drugs. For example, cloxacillin shows delayed absorption even though the total bioavailability remains the same (Fig. 8.2).[9] Similar observations have been noted for many penicillins and for acetaminophen.[10–12] Specific interactions between food and drugs have also been noted—for instance, eggs reduce the absorption of iron and fatty meals increase the absorption of griseofulvin.[13,14]

Administration of substances, such as cholestyramine, which will interfere with the absorption of fat also affects the absorption of vitamins, such as K, as well as such drugs as warfarin.[15] Chymotrypsin increases the blood levels of phenethicillin,[16] and para-aminosalicylic acid decreases the absorption of rifampicin.[17]

Gastrointestinal motility is often affected by drugs and in most instances a delay in gastric emptying will decrease the absorption of drugs. Anticholinergic drugs, such as phenothiazines and tricyclic antidepressants, may delay the absorption of other drugs. These interactions are important because

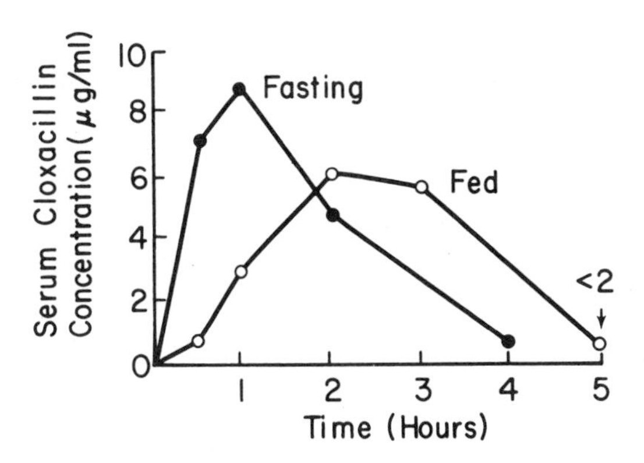

Figure 8.2. Serum cloxacillin concentrations (mean of 10 subjects) following oral administration of 500 mg cloxacillin, fasting or fed. The AUCs are the same in both instances.(After Knudsen, Brown, and Rolinson: Lancet 2:632, 1962)

in many instances several of these drugs are administered concomitantly. Table 8.1 lists the most frequently used drugs in psychiatric care. Most of these drugs are administered concomitantly, sometimes with as many as nine other drugs. It is almost certain that the drug absorption profiles in these patients will not fit into any studied norm. Little work has been reported regarding the absorption efficiencies of drugs during such multiple drug therapies.

Whereas intuitive explanations for most of the interactions in the absorption of drugs are possible, little knowledge exists of inherent differences in the absorptivity of the intestine between individuals and even within an individual. The physiologic properties of intestinal membranes responsible for the absorption of drugs change, often abruptly, in disease states, leading to malabsorption syndromes. One such factor affecting absorption is disease of

Table 8.1. MOST COMMONLY USED DRUGS IN PSYCHIATRIC CARE COMBINATION THERAPY

Diazepam (Valium)	Pyrrobutamine compound (Co-pyrinol)
Amitriptyline hydrochloride (Elavil)	Perphenazine (Trilafon)
Phenobarbital	Desipramine hydrochloride (Norpramin)
Protriptyline hydrochloride (Vivactil)	Lithium carbonate (Lithonate)
Trimethobenzamide hydrochloride (Tigan)	Thiothixene hydrochloride (Navane)
Flurezapam hydrochloride (Dalmane)	Chlorpromazine (Thorazine)
Levothyroxine sodium (Synthroid)	Chlorprothixene (Taractan)
Perphenazine-amitriptyline (Etrafon)	Haloperiodol (Haldol)
Chlorthalidone (Hygroton)	Imipramine pamoate (Tofranil)
Tranylcyclopromine (Parante)	Thioridazine hydrochloride (Mellaril)
Doxepin hydrochloride (Sinequan)	Ethchlorvinyl (Placidyl)
Trifluoperazine hydrochloride (Stelazine)	Fluphenazine hydrochloride (Prolixin)
Benztropine mesylate (Cogentin)	Hydroxyzine pamoate (Vistaril)
Triamterine/hydrochlorothiazide (Dyazide)	Multivitamins

From a survey of local hospitals by the author.

the small intestine and its adjacent secretory organs (pancreas, biliary tract, and gall bladder). Since deficiencies of minerals (calcium and iron) and vitamins (folic acid, B_{12}, D, and K) are common clinical features of malabsorption syndromes complicated by steatorrhea, it is possible that abnormalities of drug absorption might also occur.[18]

In several disease states, such as Parkinsonism, gastric emptying rates are decreased, thus increasing the exposure of the drug to gastric fluid which can increase the decomposition rate of drugs, as with levodopa,[19] with which therapeutic failure can occur. However, administration of levodopa with decarboxylase inhibitor reduces this problem.

In gastrectomy, a significant decrease in the absorption of drugs occurs, as is shown for iron,[20] folic acid,[21] ethionamide,[22] and cephalexin.[23] The absorption of para-aminosalicylic acid and ampicillin is reported unchanged upon gastrectomy.[22,23] However, recent studies in our laboratory have shown a significant decrease in the absorption of ampicillin in patients who have undergone ileojejunal bypass. In some instances the total amount of drug (ampicillin) absorbed was only 5 percent of the administered dose.

Intestinal obstruction significantly decreases the absorption of drugs and may be caused by irritant drugs. Undetected changes in gastric emptying due to pyloric stenosis or a carcinoma of the pyloric antrum may also alter absorption rates and the extent of drug absorption.

Steatorrhea may be due to impaired micelle formation, various disease states, or to the absence or reduction in the supply of bile, which is important in the absorption of fat. Obstructive jaundice prevents the secretion of bile into the gut and leads to steatorrhea. The absorption of digoxin is generally impaired in steatorrhea,[24] but drugs such as penoxymethylpenicillin, ampicillin, and cephalexin are unaffected, suggesting that micellization is important only in the absorption of digoxin, among these compounds.

The absorption of phenoxymethylpenicillin is significantly reduced in acute and chronic pancreatitis and the absorption of cephalexin is reduced in cystic fibrosis.[25,26]

The absorption of drugs is also affected in various coeliac diseases, such as dermatitis herpetiformis, small bowel diverticulosis, and lactose intolerance. The altered pathophysiology in coeliac diseases may change normal drug absorption (Table 8.2). Gastric emptying is more rapid in coeliac diseases,[27] and luminal pH is sometimes increased.[28] For example, the absorption of practolol is delayed in coeliac disease, possibly due to reduced solubility at increased luminal pH.[29] The absorption of propranolol (unchanged) is increased, possibly due to reduced first pass effect.[30] The absorption of cephalexin, sodium fusadate, sulfamethoxazole, and trimethoprim is also increased, due mainly to increased mucosal permeability in coeliac diseases. An interesting example of decreased absorption is that of pivampicillin, whereby enzyme deficiencies of small gut esterases are responsible for the hydrolysis of pivampicillin (which is a prodrug) to its parent compound. Reduced plasma albumin levels are often noted in coeliac diseases which

Table 8.2. ABNORMALITIES IN GLUTEN ENTEROPATHY AND CROHN'S DISEASE THAT MAY ALTER DRUG ABSORPTION

ABNORMALITY	*POSSIBLE EFFECT*	*ABNORMALITY*	*POSSIBLE EFFECT*
Gluten enteropathy		*Crohn's disease*	
Increased rate of gastric emptying	Drugs delivered more rapidly to their site of absorption	Reduced surface area available for absorption	Malabsorption of drugs whose optimum site of absorption is at the site of disease/resection
Increased permeability of gut wall	Increased transport of those drugs absorbed by passive nonionic diffusion	Thickening of the bowel wall	Impaired drug diffusion
Enzyme deficiencies at the brush border	Impaired hydrolysis of esterified drugs (pivampicillin) to their constituents		Malabsorption
Altered intestinal drug metabolism	Increased absorption of unchanged drug	Bowel flora changed to a predominantly anaerobic population	The absorption patterns of drugs which are active against anaerobes (lincomycin, metronidazole) would be important
Steatorrhea	Malabsorption of lipid-soluble drugs (vitamins A, D, and K)	Slower intestinal transit	Unpredictable patterns of absorption of orally administered drugs
Reduced enterohepatic recycling of bile acids	Impaired absorption of drugs which require normal micelle formation for optimum absorption		

From Parsons: Clin Pharmacokinet 2:45, 1977.

can affect the concentration gradient across the gastrointestinal membrane due to changes in protein binding. Variability can also be introduced by differences in the distribution characteristics at lowered albumin concentration.

The disorders of the ileum, such as Crohn's disease, also affect drug absorption (Table 8.2) and delay peak plasma concentration of lincomycin, trimethoprim, and sulfamethoxazole.[26,32]

The effect of age on drug absorption has also been noted. In general, children have a somewhat greater efficiency of absorption that decreases with age. However, since biotransformation and excretion are also rapid in younger children, there may not be a significant difference in the blood levels with doses calculated on body weight or body surface area.

The rates of drug absorption are also dependent on the site of administration, since blood flow rates vary widely between sites. For example, plasma levels of lidocaine are much higher when the drug is injected into an arm than when it is injected into the leg, due to differences in the blood flow rates.[33]

Drugs themselves cause changes in the gastric emptying rates and absorption efficiency and are responsible for specific interactions which can lead to significant changes in absorption profiles (Table 8.3). For example, laxatives may interfere with gastrointestinal motility and may also result in loss of electrolytes, steatorrhea, and decreased intestinal uptake of glucose. Mineral oil decreases absorption of oil-soluble vitamins. Hypocholesterolemic agents such as cholestyramine or clofibrate may decrease the absorption of cholesterol as well as vitamin B_{12}, d-xylose, iron, and sugar (Table 8.3). Surfactants such as dioctyl sodium sulfosuccinate (Colace) and Tween 80 may increase absorption of vitamin A and cholesterol. Antimicrobial agents, such as neomycin, bind fatty acids and bile acids and may cause steatorrhea. They may also affect the absorption of folic acid and vitamin B_{12}. Cytotoxic agents and colchicine may destroy or injure intestinal epithelial lining and decrease absorption of disaccharides, fat, vitamin B_{12}, folic acid, and d-xylose. Anticonvulsants and oral contraceptives interfere with polyglutamate conjugase, resulting in decreased absorption of dietary folic acid. Alcohol reduces absorption of folic acid, vitamin B_{12}, and magnesium.

Table 8.3. CHANGES IN ABSORPTION RESULTING FROM DRUG INTERACTIONS, DRUG-INDUCED ALTERATION OF GASTRIC EMPTYING OR INTESTINAL MOTILITY, AND DRUG-INDUCED MALABSORPTION

DRUG ALTERING ABSORPTION	*DRUG WHOSE ABSORPTION ALTERED*	*EFFECT*	*MECHANISM**
Amitriptyline, nortriptyline	Dicoumarol	Increased amount absorbed	?GE†
Antacids (some compounds)	Chlorpromazine	Decreased amount absorbed	DI‡
	Digoxin		DI
	Iron		DI
	Isoniazid		GE
	Tetracyclines		DI
Atropine	Lignocaine	Decreased rate of absorption	GE
Cholestyramine (also colestipol)	Acetaminophen	Decreased amount absorbed	DI
	Aspirin		
	Cephalexin		
	Chlorothiazide		
	Clindamycin		
	Digitoxin		
	Digoxin		
	Phenobarbital		
	Phenylbutazone		
	Sodium fusidate		
	Tetracyclines		
	Thyroid		
	Trimethoprim		
	Warfarin		
Desipramine	Phenylbutazone	Decreased rate of absorption	GE

Table 8.3. (*CONTINUED*)

DRUG ALTERING ABSORPTION	*DRUG WHOSE ABSORPTION ALTERED*	*EFFECT*	*MECHANISM**
Diphenhydramine	PAS	Decreased rate of absorption	GE
Kaolin-pectin	Digoxin	Decreased amount absorbed	DI
	Lincomycin		
Metoclopramide	Acetaminophen	Increased rate of absorption	GE
	Aspirin	Decreased rate of absorption	GE
	Digoxin (slowly dissolving tablets)	Increased amount absorbed	GE
	Levodopa	Increased rate of absorption	GE
	Lithium (slow release tablets)	Increased rate of absorption	GE
	Pivampicillin	Increased rate of absorption	GE
	Tetracycline	Increased rate of absorption	GE
Narcotic and strong analgesics (heroin, morphine, pethidine, pentazocine)	Acetaminophen	Decreased rate of absorption	GE
	Mexiletine	Decreased rate and ? amount absorbed	GE
Neomycin	Digoxin	Decreased amount absorbed	?MS§
	Penicillin V	Decreased amount absorbed	?MS
Phenobarbital	Griseofulvin	Decreased amount absorbed	DI
Propantheline	Acetaminophen	Decreased rate of absorption	GE
	Digoxin (slowly dissolving tablets)	Increased rate of absorption	GE
	Lithium (slow release tablets)	Decreased rate of absorption	GE
	Pivampicillin	Decreased rate of absorption	GE
	Riboflavin	Decreased rate but increased amount absorbed	GE
	Sulfamethoxazole	Decreased rate of absorption	GE
	Tetracycline	Decreased rate of absorption	GE

* See references 34–37 for details of mechanisms.
† GE: Change in gastric emptying or intestinal motility
‡ DI: Interaction in intestinal lumen (various mechanisms)
§ MS: Malabsorption syndrome induction
From Parsons: Clin Pharmacokinet 2:45, 1977.

A significant variation in the absorption profiles can be expected for drugs which undergo a first pass effect during the absorption process. Since the extent of the first pass effect is dependent on the enzyme levels in the liver and on blood flow rates, variation can be especially expected in the disease states where both of these parameters can vary.

A relatively unnoticed factor in the variability of absorption is the dissolution of poorly water-soluble drugs in the gastrointestinal fluids. At low dosage levels (calculated on the basis of body weight) the drugs will dissolve faster and more completely than at higher dosage levels, since the fluid content in the gastrointestinal tract does not increase in proportion to the body weight. This was discussed earlier (Fig. 3.6).

DISTRIBUTION

The distribution of drugs initiates the drug action and in many instances determines the intensity of action as well. The factors which affect the distribution of drugs include body composition, plasma protein and red blood cell binding, variability in the apparent volume of distribution, and the hemodynamic states of individuals.

Body Composition

Body composition varies with age, sex, and weight. For example, elderly patients are on the average smaller than younger patients and, obviously, with standard dosing this in itself would lead to their having higher plasma and tissue concentrations of a drug. With age the total body water and lean body mass decrease, body fat increasing from 18 to 36 percent of the total body weight in men and and from 23 to 48 percent in women as they progress from 18 to 55 years of age (Fig. 8.3).[38,39] Note that the percentage of body fat increases and then approaches a plateau as weight increases. On the other hand, the percentage of interstitial fluid declines with body weight. Similar changes are noted for fat and interstitial fluid content as a function of age. The blood volume is about 5 liters in adults (about 3 liters of plasma and 2 liters of blood cells). However, there is significant variation depending on age, weight, sex, etc. The blood volume per unit of body weight is higher for males than for females, due mainly to the higher fat content of females where the circulation is poor. The normal hematocrit is 47 ± 5 percent for males and 42 ± 5 percent for females, but it changes in disease states, thus altering the distribution patterns of drugs.

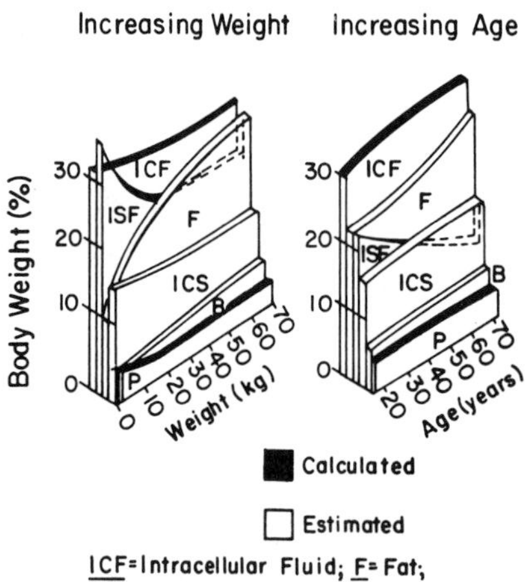

Figure 8.3. Body composition with increasing age and weight. (From Altman and Dittmer, eds.: Blood and Other Body Fluids, 1961. Courtesy of the Federation of American Societies for Experimental Biology)

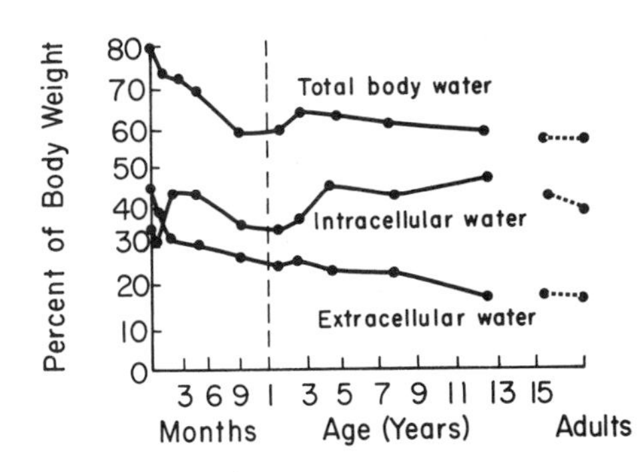

Figure 8.4. Developmental changes in various body waters. (After Friis-Hansen: Pediatr 28:169, 1961)

The total body water and its distribution are entirely different in the newborn and the fully mature individual. In newborns, total body water constitutes 70 to 75 percent of the body weight, whereas the corresponding value for adults is 50 to 55 percent. This is due mainly to incomplete calcification of bones in infants. The extracellular water content is also significantly greater in the neonatal period, some 40 percent as compared with about 20 percent in adults when calculated with thiosulfate or inulin (Fig. 8.4).[40] Since distribution throughout the extracellular water is important in the access of drug molecules to their sites of action, the intensity of drug action can often be correlated with this parameter. Changes in these fluid volumes are most likely to affect the clearance of the drug from the body by changes in the volume of distribution:

$$\text{Clearance} = \text{volume of distribution} \times (0.693/t_{0.5}) \qquad \text{(Eq. 8.1)}$$

If, however, the clearance remains constant, the elimination half-life must increase to compensate for increase in the volume of distribution. An interesting phenomenon is the variation in the drug disposition half-life as a function of excretion mechanisms, as shown in Table 8.4. It is apparent that the drug half-lives vary with age for several reasons, one of which is that the volume of distribution changes as a function of age and also of body weight.[41]

Table 8.4. CALCULATION OF HALF-LIFE FOR A PROPOSED DRUG*

		ECW		*INULIN*		*PAH*	
	WEIGHT (kg)	*% of Weight*	*Total Vol (ml)*	*Clearance ($ml \cdot min^{-1}$)*	*$t_{1/2}$ (min)*	*Clearance ($ml \cdot min^{-1}$)*	*$t_{1/2}$ (min)*
Infant	4.5	32	1,440	about 10	100	about 25	40
Adult	70.0	18	12,600	130	67	650	13

* Same clearance as inulin (glomerular filtration) or paraaminohippuric acid (tubular secretion) in an infant (1½ months old) and an adult. The drug is supposed to distribute in the extracellular water.
From Rane and Wilson: Clin Pharmacokinet 1:2, 1976.

The half-life of lipid-soluble drugs can be expected to increase with increasing age, based on their distribution patterns in the elderly, due mainly to a decrease in lean tissue mass. On the other hand, the half-lives can also be decreased (as with thyroxine) when the volume of distribution decreases.

Binding to Blood Components

The most important distributional effect lies in the protein binding of drugs. Since the unbound fraction of drug is generally the active form of the drug, even slight changes in the percentage binding will seriously affect the pharmacologic response of drugs. For example, in hypoproteinemic disorders (such as the nephrotic syndrome, liver disease, coeliac disease, and starvation), a rise in the unbound fraction of drug occurs. Concomitant with a possible decrease in the elimination rates, this variation can lead to serious toxic symptoms. Patients with nephrotic syndrome eliminate the dye Evans blue more rapidly than do normal individuals.[42] An almost five-fold increase in the elimination rate of warfarin has been noted for warfarin in nephrotic patients, due to binding variability.[43] Hypoalbuminemia (less than 2.5 g/100 ml) is responsible for an almost 200 percent increase in the frequency of side effects from prednisolone therapy, since its active metabolite prednisone is highly plasma protein bound.[44]

Although in most instances little interindividual difference in the protein binding exists, as with desipramine and phenytoin, a large variation has been noted for such drugs as chlorpromazine and nortriptyline.[45,46] In the case of notriptyline this is also due to genetic differences.

The binding of drugs decreases in renal failure, due partly to hypoalbuminemia, as is shown for sulfonamides,[47] digitoxin,[48] and phenytoin.[49] It has also been suggested that the binding capacity of albumin molecules also decreases, for reasons still debated in the literature. Some examples of reduced plasma protein binding in patients with renal diseases are with administration of aspirin, salicylic acid, phenylbutazone, sulfadiazine, and thiopental.[50] Differences in the binding have also been noted for furosemide and diazoxide.[51,52] It is interesting to note that all reports of abnormal drug binding in plasma from uremic subjects have concerned acidic drugs. Weakly basic drugs seem to be little affected, as in the case of desipramine, which is extensively bound to plasma proteins (Table 8.5).

It is likely that the factors affecting the plasma protein binding in nephrotic patients also affect the binding of drugs to tissues. One study has alluded to this phenomenon in connection with the serious reduction of the uptake of digoxin by the myocardium in patients with poor renal function.[53]

Apart from renal malfunction, liver diseases also affect the plasma protein binding of drugs (Table 8.6). The plasma from patients with alcoholic liver disease shows substantial reduction in the binding of fluorescein, dapsone, quinidine, and triamterene.[54] The main reason for this reduction is hypoalbuminemia, which occurs in patients with liver disease. Such other factors

Table 8.5. EXAMPLES OF DRUGS WHOSE BINDING CHANGES IN RENAL MALFUNCTION

ACIDIC DRUGS		*NEUTRAL DRUGS*
Congo red	Fluorescein	Digitoxin
Phenol red	Salicylates	Digoxin
Methyl red	Benzylpenicillin	
Methyl orange	Dicloxacillin	*BASIC DRUGS*
Sulfonamides	Phenobarbital	Triamterene
Phenytoin	Thiopental	Morphine (slight decrease)
Thyroxin	Diazoxide	
Tryptophan	Phenylbutazone	
Clofibrate		

as the high levels of bilirubin in such patients can also account for lower binding due to the competitive relationship of the drug binding process and bilirubin.[55] One study involved estimation of the binding of phenytoin in normal and hyperbilirubinemic infants. An approximately 50 percent higher unbound fraction was present when bilirubin content was high. A good correlation was observed between the bilirubin concentration and the unbound fraction of phenytoin.[56]

Table 8.6. DECREASED PLASMA PROTEIN BINDING OF DRUGS IN PATIENTS WITH LIVER DISEASE

Sulfonamides	Dapsone
Digitoxin	Phenytoin
Tolbutamide	(+)-Propranolol
Quinidine	Diazepam
Fluorescein	Morphine
Triamterene	

The free fatty acid concentration in plasma usually varies between 200 and 500 μEq/liter, but in disease states levels as high as 2500 μEq/liter have been recorded, as with diabetes mellitus, infections, hyperthyroidism, pheochromocytoma, and myocardial infarction. Since these fatty acids are extensively bound to albumin and have higher affinity constants than many drugs, the binding of drugs may be decreased in hyperlipidemic states, as is shown for warfarin and phenytoin.[57]

The effect of variation in plasma protein binding of drugs on the pharmacologic response varies widely between drugs. For example, it is possible to obtain similar steady state levels of drugs in patients with impaired plasma protein binding and in patients with normal profiles. However, the peak plasma concentration can significantly vary in accordance with the half-life and the volume of distribution.[58] The elevated peaks can cause significant side effects, and in some instances elevated peaks in impaired binding may lead to increased effectiveness of drugs, as is demonstrated for phenindione and diazoxide.[59,60] However, various studies have concluded that impaired protein binding will have no effect on drug disposition, since identical steady state levels are obtained, and that it should not be an important factor in the dosage adjustment of drugs. This position is misleading since impaired protein binding is invariably accompanied by other complications which may require dosage adjustment. Since it is possible to obtain different peak plasma concentrations yet similar steady state levels, changes in plasma protein binding should be given full consideration in determining the dosage regimen in disease states.

The effect of body stress has also been studied in relation to drug distribution, since it alters the hydration and thus the volumes of distribution of drugs. It is possible that during stress drugs might enter body spaces which are otherwise inaccessible and initiate unpredictible drug actions. For example, during stress the volume of central compartment for antipyrine decreases but the tissue distribution volume increases.

Hemodynamic Changes

The differences in the distribution of drugs can also be attributed to variations among individuals in the blood flow rates to various organs. The effect of age is very clear on the cardiac output, where an approximately 1 percent decrease is recorded every year from age 19 to 86.[61] Since the vital organs receive preference for circulation distribution, abnormal distribution of drugs can be expected in the elderly. The hemodynamic changes, accompanied by possible changes in the permeability of various tissues and barriers (such as blood–brain barriers) can also contribute to abnormal distribution profiles in the elderly.

Drug Interactions

Plasma protein binding interactions are of great significance in the distribution of drugs. Earlier in this chapter, interactions with such endogenous substances as bilirubin and free fatty acids were discussed. Various drugs share common binding sites and are prone to interact, displacing each other from the binding sites. For example, sulfinpyrazone displaces sulfamethoxypyridazine from albumin in rats, causing a decrease in the plasma levels and an increase in the tissue levels of sulfamethoxypyridazine.[62] Sulfamethoxypyridazine, sulfaethylthiodazole, and acetylsalicylic acid reduce penicillin binding and cause a decrease in total serum penicillin, as well as a rise in the free or unbound fraction.[63] A change in the free or unbound fraction can result in significant pharmacologic response variability, especially if the drug is extensively bound. For example, if a drug is 99 percent bound in plasma and interacts with another drug (which displaces the first drug), a 1 percent change to 98 percent binding will result in a 100 percent increase (from 1 percent unbound to 2 percent unbound) in pharmacologic response if the unbound fraction is proportional to the response. Typical interactions are listed in Table 8.7 along with possible pharmacologic or clinical responses. Generally, sulfonamides, salicylates, and phenylbutazone derivatives are very strongly bound drugs which displace many moderately bound drugs and endogenous compounds, such as bilirubin, which particular case can result in a toxic response (kernicterus) due to the penetration of bilirubin into the brain.

Table 8.7. BINDING DRUG INTERACTIONS AND RESULTANT EFFECT

THIS DRUG	*DISPLACES*	*RESULTING IN*
Sulfonamides/phenylbutazone/ salicylates	Tolbutamide	Hypoglycemia
Oxyphenylbutazone/ phenylbutazone/sulfinpyrazone	Warfarin	Hemorrhage
Sulfonamides/salicylates	Methotrexate	Cytopenia/blood dyscrasias
Quinacrine	Pamaquin	Gastrointestinal distress/ anemias
Pyrimethamine	Quinine	Cinchonism/neutropenia
Phenylbutazone	Sulfaethidole	Shortened action
Ethacrynic acid	Oral antidiabetics	Hypoglycemia
Coumarin anticoagulants	Sulfonamides	Enhanced activity

However, not all drug interactions result in decreased binding. Pempidine, which is not normally bound, becomes extensively bound in the presence of chlorothiazide,[64] and tetracyclines increase the binding of both promazine and chlorpromazine to albumin.[65]

Binding interactions also take place at tissue levels. For example, the binding of pamaquin to tissues is decreased in patients previously treated with mepacrine, resulting in increased plasma levels of pamaquin and its toxicity.[66]

BIOTRANSFORMATION

The most important factor in the variability of drug response among individuals is due to differences in the biotransformation of drugs. The reasons for these differences are: quantity of enzymes available, enzyme inhibition, enzyme stimulation, genetic effects, and physiologic and disease state factors.

Enzyme Content

Cytochrome P-450 is an important component of various biotransformation reactions. Significant differences in its levels are reported among individuals, sometimes with as much as a five- to six-fold difference.[67–69] The variations in the half-lives of drugs such as antipyrine (6.0 to 15.1 hours), phenylbutazone (1.2 to 7.3 days), and dicoumarol (7.0 to 74.0 hours) can be correlated with the cytochrome levels.[70]

Table 8.8. OBSERVED AND PREDICTED FIRST PASS METABOLISM OF NORTRIPTYLINE IN SIX SUBJECTS

			% FP (PREDICTED)	
SUBJECT	*AUC (μg/liter × hr)*	*% FP† (OBSERVED)*	*Blood Flow Model*	*Plasma Flow Model*
5	895	54	38	53
6	860	50	39	54
7	925	41	37	53
8	670	53	45	61
9	885	49	38	54
10	1500	50	27	41
Mean ±	—	49.50 ±	37.33 ±	52.67 ±
SEM		1.88	2.38	2.64
T, predicted: observed		—	4.02	0.98
			$p < 0.0025$	(n.s.)

† FP: First pass metabolism.
From Niazi: J Pharm Sci 65:1535, 1976.

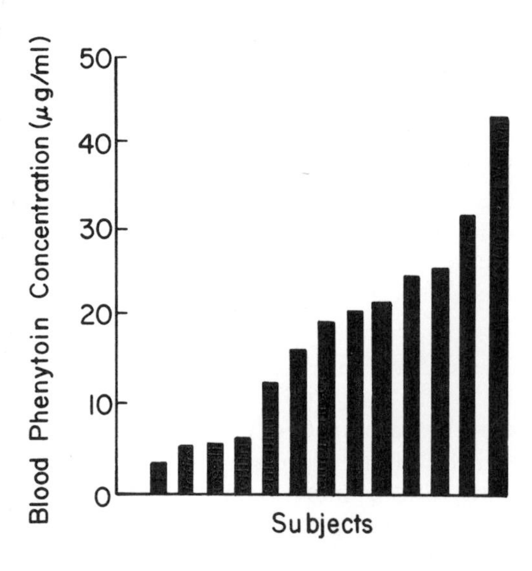

Figure 8.5. Steady state plasma concentrations of phenytoin in 13 subjects, following oral administration of 400 mg phenytoin daily for at least 14 days. (From Loeser: Neurology (Minneap) 11:424, 1961)

The activity of microsomal enzymes can often be correlated to the steady state levels of various drugs (Fig. 8.5), such as phenytoin and desipramine.[71,72] The levels of microsomal enzymes are often determined by the previous exposure of the body to various drugs. Examples of biotransformation inhibition (Table 6.11) and stimulation (Table 6.12) have already been discussed. Variability in the drug levels in the blood can also be attributed to differences in the extent of the first pass effect (following oral and intraperitoneal administration) due to the level of enzymes as well as the blood flow rates. The blood flow rates to the liver and to other organs in which biotransformation takes place changes significantly in the disease states, but unfortunately few data have been reported regarding these parameters which are so vital to the understanding of the disposition process. In many instances, however, the blood flow rates do not affect the first pass process significantly and therefore rates of drug removal in the liver can be predicted which correlate well with actual rates (Tables 8.8, 8.9).[73,74]

Table 8.9. PREDICTED AND OBSERVED FIRST PASS METABOLISM OF IMIPRAMINE IN FOUR SUBJECTS

SUBJECT	*SEX*	*AGE*	*DOSE (mg)*	*BODY WEIGHT (kg)*	$\int_0^\infty C_0\,dt$ *(mg·min/liter)*	*FIRST PASS PREDICTED (%)*	*FIRST PASS OBSERVED (%)*
A.B.	f	28	35	52	11.97	66	71
G.A.	f	28	40	55	44.24	37	23
U.F.	m	25	50	68	15.13	68	52
P.L.B.	m	59	50	71	19.84	62	65
Mean ± *SD*						58.25 ± 14.38	52.75 ± 21.36

From Niazi: J Pharm Sci 65:1063, 1976.

Age

Age has an important effect on the biotransformation of drugs. Although no direct evidence has been reported, it is apparent from indirect studies that elderly patients have a reduced ability to biotransform a number of drugs, such as antipyrine,[75] aminopyrine,[76] and acetaminophen,[77] whose biologic half-lives increase significantly in the elderly due to this depressed biotransformation. The drug which has been most extensively studied to elucidate this phenomenon is antipyrine, which shows rapid absorption and distribution, low protein binding, and extensive biotransformation in the liver which is independent of the hepatic blood flow rates.[78] The liver function can often be correlated with the plasma half-life of its main metabolite, 4-hydroxyantipyrine.[79]

Neonates show poor development of a microsomal enzyme system and are thus unable to conjugate chloramphenicol,[80] sulfonamides,[81,82] acetaminophen,[83] and para-aminobenzoate.[84] The well-known cyanosis or gray syndrome in young babies is attributed to chloramphenicol biotransformation depression.

Genetic Effects

Genetic factors can significantly affect the disposition of drugs. Two types of variations are possible: continuous and discontinuous. Continuous variation leads to unimodal curves, as shown in Fig. 8.1, and polymodal curves are found for discontinuous variation—this is best exemplified by isoniazid, which undergoes conjugation with acetyl coenzyme A in the liver to form acetylated isoniazid. The bimodal distribution shows that after administration of a 9.8 mg/kg dose, one group of individuals will show plasma concentrations of about 1 μg/ml and the other group will show concentrations of 4.5 μg/ml.[85] The populations are thus termed either slow inactivators or fast inactivators. Slow inactivators have smaller quantities of enzyme N-acetyl transferase, which is identical in its potency to fast acetylators. The difference in the enzyme content can be attributed to both slow production and degradation.

Polymorphic acetylation affects biotransformation as is shown for sulfamethiazine, for which a direct correlation with the biotransformation of isoniazid has been established. The specificity of correlation is so good that sulfamethazine is used to phenotype individuals as slow or fast acetylators. Following a 45 mg/kg dose, subjects who excrete more than 64 percent of the dose as the acetyl metabolite are typed as rapid acetylators and those who excrete less are slow acetylators.[87] Other drugs which are presumed to undergo acetylation by the same enzyme are sulfapyridine, hydralazine, procainamide, and dapsone, all of which show polymorphism in their biotransformations.[88–90]

Incidences of genetic effect on drug biotransformation have also been noted in the dealkylation of drugs. For example, marked methaemoglobinemia following the administration of drugs such as phenacetin or phenytoin can be attributed to dealkylation differences.[91,92]

The biotransformation of drugs appears to be genetically influenced by polygenic control. A number of studies have shown that identical twins show similar handling of drugs, but fraternal twins do not, as is shown with phenylbutazone (Fig. 8.6),[93] dicoumarol,[94] antipyrine,[95] ethanol,[96] halothane,[97] and nortriptyline.[98]

Disease States

The effect of liver diseases on drug biotransformation is not as pronounced as may be expected. For example, in the case of liver cirrhosis the half-lives

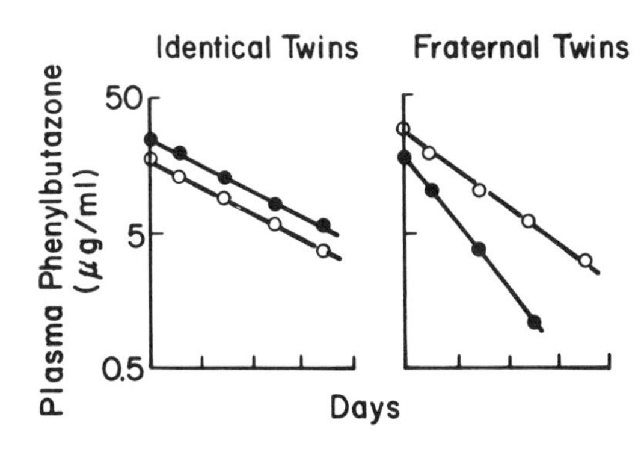

Figure 8.6. Phenylbutazone plasma levels in identical and fraternal twins. (From Vesell and Page: Sci NY 159:1479, 1968)

of phenobarbital, antipyrine, dicoumarol, salicylate, and phenylbutazole remain the same.[97,98] However, chloramphenicol, rifamycin, and tolbutamide show depressed biotransformation in the disease states of liver.[99–101] The decreased biotransformation can lead to such complications of increased toxicity as erythropoietic depression due to chloramphenicol treatment in liver disease.[102] Generally, the effect of liver disease on biotransformation occurs when the disease condition is severe,[103] where decreased levels of cytochrome P-450, aminopyrine demethylase, pseudocholinesterase, and cholinesterase activity have been reported.[103,104] In many instances the effect of liver disease is indirectly noticed in terms of hypoalbuminemia, as with the decreased biotransformation of amylobarbital.[105]

Although a large number of functional tests exist to monitor renal performance, few correlations have been made regarding liver function. For example, in the case of hepatic necrosis due to acetaminophen poisoning, the half-life of acetaminophen is related to the extent of damage, which is also reflected in the disposition characteristics of barbiturates, phenytoin, and antipyrine.[106,107]

The enzyme induction in liver diseases seems to be unaffected since response to such drugs as phenobarbital remains within the normal range.[108]

The effect of renal diseases on drug biotransformation is unclear. There is some indication of decreased pseudocholinesterase activity in uremic patients,[104] acetylation of sulfafurazole,[109] and decreased reduction of cortisol.[110]

Since the effect of hepatic diseases on drug disposition is complicated through protein binding interactions, enzyme level variability, and intrinsic activity of enzymes, it is important to observe the total effect in terms of the half-lives of drugs in hepatic diseases (Table 8.10), as is recently reported in a review article which should be consulted for a detailed list of references.[111] This list includes accounts of patients who are free of renal diseases and thus represent the overall effect of liver disease on disposition half-lives. Note that each study represents a rather small number of patients and can therefore be considered only as trend-setting rather than as clearly establishing results. The diagnoses can also differ depending on the interpretation.

Table 8.10. TERMINAL HALF-LIVES OF DRUGS STUDIED IN PATIENTS WITH HEPATIC DISEASES

DRUG	*DOSE AND ROUTE*	*CONTROL* $t_{1/2}$	*DIAGNOSIS*	*NO. OF PATIENTS*	*DISEASED* $t_{1/2}$
Analgesics					
Acetaminophen	1.5 g P.O. (S)	2.0 hr	Mostly cirrhosis	7	↑ 3.3 hr (1.5–7.0 hr)
Meperidine	0.8 mg/kg I.V. (S)	3.37 ± 0.82 hr	Acute viral hepatitis	14	↑ 6.99 ± 2.74 hr
	0.8 mg/kg	3.21 ± 0.80 hr	Cirrhosis	10	↑ 7.04 ± 0.92 hr
Antibiotics					
Aminosalicylic acid	I.V.	—	Chronic liver disease	12	↔ 20–36 min
Ampicillin	600 mg I.V. (S)	1.31 ± 0.15 hr	Alcoholic cirrhosis	9	↑ 1.90 ± 0.56 hr
Carbenicillin	2.0 g I.V. (S)	1.0 ± 0.25 hr	Severe hepatic dysfunction	2[a]	↔ 1.25 hr
Chloramphenicol	I.V.	—	Chronic liver disease	12	↔ 131–213 min
	500 mg I.V. (S)	2.9 ± 0.9 hr	Advanced cirrhosis	11	↑ 5.2 hr
	10 mg/kg I.V. (S)	2.29 hr (1.72–2.82 hr)	Cirrhosis	8	↑ 4.05 hr (2.19–6.42 hr)
Clindamycin	300 mg I.V. (S)	3.42 ± 0.45 hr	Alcoholic cirrhosis	7	↑ 4.46 ± 0.93 hr
	300 mg I.M. (S)	3.0 hr (2.4–4.2 hr)	Acute/chronic hepatitis, cirrhosis	10	↑ 6.4 hr (2.6–14.2 hr)
	300 mg I.V. (M)	1.8 hr	Acute hepatitis	7	↔ 2.6 hr
	300 mg I.V. (M)	1.8 hr	Chronic hepatitis	6	↔ 2.1 hr
	300 mg I.V. (M)	1.8 hr	Cirrhosis	9	↔ 2.5 hr
Isoniazid	600 mg P.O. (S)	3.24 ± 0.14 hr	Chronic liver disease	13	↑ 6.74 ± 0.33 hr
	20 mg/kg I.V.	[b]	Acute/chronic liver disease	68	↑ [b]
Nafcillin	7.5 mg/kg I.V. (S)	1.0 hr	Cirrhosis	1	↑ 1.4 hr
Rifampin	600 mg P.O. (S)	2.80 ± 0.22 hr	Chronic liver disease	13	↑ 5.42 ± 0.55 hr
Sulfamethoxazole	160 mg P.O. (S)	12.1 hr (7.9–17.4 hr)	Minor hepatic injury	4	↔ 11.9 hr (10.3–14.3 hr)
	160 mg P.O. (S)	12.1 hr	Severe decompensated liver damage	7	↔ 12.3 hr (9.3–15.1 hr)
Trimethoprim	800 mg P.O. (S)	11.4 hr (8.4–14.8 hr)	Minor hepatic injury	4	↔ 11.0 hr (8.4–13.5 hr)
	800 mg P.O. (S)	11.4 hr	Severe decompensated liver damage	7	↑ 14.3 hr (6.6–23.7 hr)
Anticonvulsants					
Phenobarbital	0.85 mg/kg P.O. (S)	86 ± 3 hr	Cirrhosis	6	↑ 130 ± 15 hr
	2.55 mg/kg P.O. (S)	88 hr	Cirrhosis	2	↑ 130 ± 3 hr
	0.85 mg/kg P.O. (S)	86 ± 3 hr	Acute viral hepatitis	8	↔ 104 hr[c] (60–127 hr)
Phenytoin	250 mg I.V. (S)	12.6 ± 5.5 hr	Acute viral hepatitis	5	↔ 13.2 ± 6.5 hr

Antiinflammatory Agents					
Colchicine	2 mg I.V. (S)	19.3 ± 7.5 min	Halothane jaundice	1	↓ 9.8 min
	2 mg I.V. (S)	19.3 ± 7.5 min	Chronic hepatitis	1	↓ 8.9 min
	2 mg I.V. (S)	19.3 ± 7.5 min	Cirrhosis	1	↓ 9.1 min
Cortisone	200 mg I.V. (S)	28 min (23–35 min)	Cirrhosis	3	↔ 30 min
Hydrocrotisone	—	98 min	Cirrhosis	1	↑ 320 min
Phenylbutazone	300–600 mg P.O. (M)	81.8 hr (51–104 hr)	Cirrhosis	6	↔ 94.5 hr (48–149 hr)
	400–600 mg P.O. (S)	78 hr	Liver disease, unpretreated[d]	34	↑ 100 hr
	400–600 mg P.O. (S)	57 hr	Liver disease, pretreated[d]	61	↔ 56 hr
Prednisolone	10 mg P.O. (S)	—	Chronic hepatitis, biliary cirrhosis	3	↔[e]
	20 mg I.V. (S)	175 min	Active chronic hepatocellular disease	1	↑ 250 min
	20 mg I.V. (S)	175 min	Inactive chronic hepatocellular disease	2	↑ 207.5 min
	10 mg I.V. (S)	3.4 hr	Chronic active liver disease	16	↑ 4.0 hr
Prednisone	10 mg P.O. (S)	—	Chronic hepatitis, biliary cirrhosis	8	↑[f]
	20 mg P.O. (S)	—	Active chronic hepatocellular disease	9	↑[g]
	20 mg P.O. (S)	—	Inactive chronic hepatocellular disease	7	↔[g]
	20 mg P.O. (S)	—	Acute hepatitis	6	↑[g]
	10 mg I.V. (S)	—	Chronic active liver disease	16	↔[h]
Cardiac Agents					
Digitoxin	0.6 mg I.V. (S)	8.1 days	Chronic active hepatitis	6	↓ 4.4 days
Digoxin	1.0 mg I.V. (S)	—	Cirrhosis	3	↔[i]
	0.375–0.75 mg I.V. (M)	—	Acute hepatitis	15	↔[j]
Lidocaine	400 mg P.O. (S)	1.4 hr	Mostly cirrhosis	10	↑ 6.7 hr (2.3–19.0 hr)
	50 mg I.V. (S)	107.8 min	Chronic alcoholic hepatic disease	8	↑ 296 min
	1 mg/kg I.V. (S)	90 min[k]	Viral hepatitis	6	↔ 160 min[c]
Phenytoin	250 mg I.V. (S)	12.6 ± 5.5 hr	Acute viral hepatitis	5	↔ 13.2 ± 6.5 hr

(Continued)

Table 8.10. (*CONTINUED*)

DRUG	DOSE AND ROUTE	CONTROL $t_{1/2}$	DIAGNOSIS	NO. OF PATIENTS	DISEASED $t_{1/2}$
Propranolol	40 mg I.V. (S)	2.9 ± 0.6 hr	Mild chronic liver disease[l]	11	↑ 9.8 ± 5.1 hr
	40 mg I.V. (S)	2.9 ± 0.6 hr	Severe chronic liver disease[m]	9	↑ 22.7 ± 9 hr
	80 mg P.O. (S)	3.3 hr	Portacaval anastomosis	1	↑ 18 hr
Sedative-Hypnotic Agents					
Amobarbital	3.23 mg/kg I.V. (S)	21.1 ± 1.2 hr	Chronic liver disease[n]	5	↑ 39.4 ± 6.6 hr
	3.23 mg/kg I.V. (S)	21.1 ± 1.2 hr	Chronic liver disease[o]	5	↓ 17.7 ± 1.84 hr
Diazepam	10 mg I.V. (S)	38.0 ± 20.2 hr	Hepatitis	2	↑ 90.0 ± 63.6 hr
	10 mg P.O. (S)	32.7 ± 8.9 hr[p]	Acute viral hepatitis	8	↑ 74.5 ± 27.5 hr
	10 mg P.O. (S)	32.7 hr[p]	Postviral hepatitis, LFT recovery	5	↑ 53.8 ± 6.1 hr
	0.1 mg/kg I.V. (S) or 10 mg P.O.	32.7 hr[p]	Chronic active hepatitis	4	↑ 59.7 ± 23.0 hr
	10 mg I.V. (S)	38.0 ± 20.2 hr	Cirrhosis	6	↑ 100.5 ± 54.3 hr
	0.1 mg/kg I.V. (S)	46.6 ± 14.2 hr[f]	Alcoholic cirrhosis	9	↑ 105.6 ± 15.2 hr
Hexobarbital	6.45 mg/kg I.V. (S)	261 ± 69 min	Acute hepatitis	13	↑ 490 ± 185 min
	6.13 mg/kg I.V. (S)	261 min	Postacute hepatitis, LFT recovery	6	↑ 385 ± 128 min
Meprobamate	8 mg/kg I.V. (S)	12.6 ± 2.5 hr	Chronic liver disease	9	↑ 24.3 ± 4.4 hr
Oxazepam	45 mg P.O. (S)	5.1 ± 0.5 hr[p]	Acute viral hepatitis	7	↔ 5.3 ± 0.3 hr
	45 mg P.O. (S)	5.6 ± 0.3 hr[p]	Cirrhosis	6	↔ 5.8 ± 0.5 hr
Pentobarbital	4 mg/kg I.V. (S)	15–32 hr	Chronic hepatitis, cirrhosis	12	↔ 17–36 hr
	100 mg I.V. (S)	21.0 hr	Decompensated cirrhosis	10	↑ 38.5 hr
Phenobarbital	0.85 mg/kg P.O. (S)	86 ± 3 hr	Acute viral hepatitis	8	↔ 104 hr[c] (60–127 hr)
	0.85 mg/kg P.O. (S)	86 ± 3 hr	Cirrhosis	6	↑ 130 ± 15 hr
	2.55 mg/kg P.O. (S)	88 hr	Cirrhosis	2	↑ 130 ± 3 hr
Miscellaneous Agents					
Adriamycin	60 mg/m² I.V. (S)	16.7 ± 8.22 hr	Malignancies with liver disease	8	↑ [q]
Chlorpromazine	25 mg I.V. (S)	31 hr	Cirrhosis	24	↔ 24 hr
Disulfiram	500 mg P.O. (S)	r	Active alcoholic hepatitis	4	↔ r

Theophylline	2.8 mg/kg I.V. (S)	9.19 ± 1.5 hr	Cirrhosis	4	↑ 30.0 ± 17.8 hr
	5.6 mg/kg I.V. (S)	6.68 hr (2.77–12.4 hr)	Stable cirrhosis	6	↑ 30.8 hr (11.5 ± 55.8 hr)
Tolbutamide	250–500 mg I.V. (S)	6.1 ± 1.7 hr[k]	Acute viral hepatitis	5	↓ 4.4 ± 0.9 hr
	12 mg/kg I.V. (S)	7.21 ± 2.15 hr	Mostly acute hepatitis	9	↓ 3.47 ± 0.96 hr
	20 mg/kg I.V. (S)	4.4 ± 0.7 hr	Cirrhosis	10	↑ [q]
	500 mg P.O. (S)	—	Laennec's cirrhosis	10	↔ 6.0 ± 1.9 hr
Warfarin	15 mg P.O. (S)	23 hr[k]	Acute viral hepatitis	5	↔ 25 hr

ABBREVIATIONS USED: I.M., intramuscular; I.V., intravenous; LFT, liver function tests; (M), multiple doses; P.O., oral; (S), single dose; $t_{1/2}$, terminal half-life.

[a] Only two patients in this study had normal renal function with hepatic dysfunction.

[b] Because of genetically determined bimodal acetylation rate of INH metabolism, no average $t_{1/2}$ would be useful. In a distribution of half-lives curve ($t_{1/2}$ vs. number of subjects), the effect of liver disease was that "there was no bimodal curve and no antimode, and the whole group was shifted slightly to a longer $t_{1/2}$."

[c] Not statistically significantly different from controls.

[d] Treatment with barbiturates, antihistamines, or corticosteroids prior to the study obliterated the effect of liver disease on phenylbutazone half-life.

[e] No difference in peak serum values or disappearance rates of prednisolone in patients or controls.

[f] Since prednisone is active only when converted to prednisolone, the listed arrows denote not $t_{1/2}$, but an increase (↑) or decrease (↓) in conversion time to produce peak prednisolone serum levels.

[g] In this report, lower serum levels of prednisolone after prednisone dose than after prednisolone dose were interpreted to mean delayed prednisone conversion to prednisolone. For consistency within this table, the delayed conversion is displayed as an increased prednisone $t_{1/2}$.

[h] No difference in peak serum values or disappearance rates of prednisone in patients or controls.

[i] Serum half-life not reported; fecal and urinary half-lives similar in patients and controls.

[j] Serum half-life not reported; multiple daily serum digoxin concentration determinations suggested no difference in drug elimination between patients and controls.

[k] Patients served as own controls after recovery.

[l] Patients with serum albumin greater than 3.0 g/100 ml.

[m] Patients with serum albumin less than 3.0 g/100 ml.

[n] Patients with serum albumin levels less than 3.5 g/100 ml.

[o] Patients with serum albumin levels greater than 3.5 g/100 ml.

[p] Age-matched controls.

[q] Exact serum half-lives not reported.

[r] Serum half-lives not reported: radioactivity in stool, urine, and breath samples showed different metabolic routes, but about the same rate in patients and controls.

From Closson: Am J Hosp Pharm 34:520, 1977.

EXCRETION

A significant variation in the degree of response obtained from different individuals given the same dose is due to the possible role of the excretion process in maintaining a certain plasma concentration. Whereas all routes of

Table 8.11. EXAMPLES OF DRUGS WHOSE DOSAGE REGIMEN SHOULD BE CHANGED IN VARIOUS DEGREES OF RENAL IMPARIMENT

MILD IMPAIRMENT	*MODERATE IMPAIRMENT*	*SEVERE IMPAIRMENT*
Acetohexamide	Acetazolamide	Acetaminophen*
Cefazolin	Acetasalicylic acid	Acetazolamide*
Chlorpropamide	Acetohexamide*	Amphotericin B
Clofibrate	Allopurinol	Azathioprin
Colistimethate	Aminosalicylic acid*	Cephalexin
Gentamicin	Amoxacillin	Cephalothin
Kanamycin	Ampicillin	Colchicine
Methadone	Carbenicillin	Digitoxin
Streptomycin	Chlordiazepoxide	Diphenhydramine
Tetracycline	Chlorpropamide*	Ethacrynic acid*
Vancomycin	Cyclophosphamide	Glutethimide
	Digoxin	Hydralazine
	Ethambutol	Lincomycin
	Flucytosine	Methicillin
	Gentamicin	Neostigmine
	Gold sodium thiomalate*	Nitrofurantoin*
	Guanethidine	Penicillin G
	Insulin	Phenformin*
	Lithium carbonate*	Phenobarbital
	Meprobamate	Quinine
	Mercurials*	Spironolactone*
	Methenamine mandalate*	Sulfamethoxazole-trimethoprim*
	Methotrexate	Thiazides*
	Methyldopa	Triamterene*
	Minocycline	Tolbutamide
	Neomycin	
	Ouabain	
	Penicillamine	
	Pentamidine	
	Phenazopyridine*	
	Phenylbutazone*	
	Primidone	
	Phenothiazines	
	Probenecid*	
	Procainamide	
	Propylthiouracil	
	Sulfamethoxazole-trimethoprim	
	Sulfisoxazole	
	Trimethadione	

* Indicates drugs which should be avoided in the given status of renal impairment.

drug excretion are important, only biliary and renal excretion contribute to an extent sufficient to make a difference in the drug response.

The most important factor in the variability of drug response due to kidney malfunction is the change in the glomerular filtration rate, which is determined by the hydrostatic pressure gradient generated between the glomerular capillaries and the early proximal tubule and the oncotic pressure in the glomerular capillaries. Fortunately, changes in the renal blood flow rates do not have a significant effect on the glomerular filtration rate until a significant change occurs.[112] For example, a 10 to 20 percent decrease or increase in the renal blood flow does not affect the glomerular filtration rate. However, severe renal ischemic manifestations lead to reduction in both the renal blood flow and the glomerular filtration rate. The glomerular filtration rate of drugs is dependent on the protein binding since only the free drug is filtered through the membranes. The age of the individual also determines the level of glomerular filtration. For example, neonates have a lower renal blood flow and glomerular filtration rate (after correction for body weight) than adults, and in old age the filtration rate also decreases significantly.[113]

The tubular secretion of drugs is unaffected by the degree of protein binding, but the renal blood flow affects the degree of active tubular secretion. Newborns and elderly people have a lesser capacity to transport substances across the proximal tubules.[113] The tubular reabsorption of some compounds is affected during saline infusion or when osmotic diuretics are used.[112] For example, proximal tubular absorption of fluoride, bromide, and lithium is inhibited under the conditions above. However, the state of volume depletion and/or reduced sodium intake may lead to enhanced proximal reabsorption, requiring dosage adjustments.

The action of some drugs is dependent on their delivery to the kidneys, as when antibiotics are used in the treatment of urinary tract infections. Thus the efficiency of urinary antiseptics such as methenamine mandelate or hippurate, nalidixic acid, and sulfonamides is not recommended in patients with creatinine clearance of less than 20 ml/min. Few antibiotics are present at sufficient concentrations in the urine at glomerular filtration rates below 25 ml/min. These few include the groups of aminoglycosides, cephalosporins, and penicillins. Cephalosporins are ineffective when glomerular filtration falls below 12 ml/min.[114] Chloramphenicol becomes ineffective at filtration rates below 40 ml/min and nitrofurantoin at rates below 50 ml/min.

The status of renal function is described in terms of creatinine clearance:

Status of Renal Function	*Creatinine Clearance (ml/min)*
Mild impairment	80–50
Moderate impairment	50–10
Severe impairment	<10
Essentially no function	<2

Changes in dosage regimen can be recommended on the basis on this classification. Table 8.11 reports the drugs cleared from the kidneys at varied levels, thus indicating a need requiring dosage adjustment at different levels of renal function. This table also lists the drugs which should be avoided in patients with modified renal function. It should be noted that these are general recommendations which do not take into account the possibility of other complications as a result of either renal impairment or other disease states that might affect the disposition profiles of drugs.

The relationship between creatinine clearance and the dosage adjustment is based on the fact that the half-life of the drug is related to its glomerular filtration rate as reflected in the creatinine clearance. Drugs which are eliminated primarily through the kidneys will show significant changes in their half-lives even if the impairment is mild. The drugs which are less than 20 percent cleared through the kidneys, or which are extensively metabolized, require little change in their dosage regimens. For example, the drugs listed in Table 8.12 require no dosage adjustment due to renal malfunction of any degree, as can be concluded from the renal clearance of the drugs. However, it is well documented that renal impairment is invariably associated with other disease states or alterations in the body physiology which can affect the disposition profiles. Few examples lend themselves to the clear explanation shown for tetracycline in patients with kidney diseases:[115]

Creatinine Clearance (ml/min)	*Tetracycline Half-life (hours)*
100	6
50	9
40	11
30	13
20	14
10	16
5	22
2	42
anuric	4 days

In most instances such direct relationships are not possible, due to modifications of pharmacokinetic parameters not directly related to the decrease of glomerular filtration rate but which occur in renal insufficiency. For ex ample, edema as a result of renal insufficiency considerably disrupts the behavior of drugs, such as the volumes of distribution shown for example in the compounds sulfonamides and thiocyanate.

The effect of renal disease on plasma protein binding of drug was alluded to earlier in the chapter. Any change in the binding parameters will significantly affect the clinical response to a drug which is extensively protein bound. Chronic uremia also causes anatomic changes in the gastrointestinal tract which can affect the absorption of drugs. For example, the serum concentrations of such drugs as chlorpropamide,[116] doxycycline,[117] and penicillin are lower in patients with deficient kidneys than in normal subjects who receive the same dose of the drug. The reasons for this phenomenon are still unclear.

Table 8.12. DRUGS GENERALLY ADMINISTERED WITHOUT REGARD TO RENAL MALFUNCTION

Amitriptyline	Methaqualone
Amobarbital	Methylprednisolone
Atropine	Metolazone
Chloramphenicol	Minoxidil
Chloroquine	Morphine
Chlorpheniramine maleate	Nafcillin
Clindamycin	Nalidixic acid
Codeine	Naloxone
Cortisone	Nortriptyline
Dexamethasone	Oxacillin
Diazoxide	Pentazocine
Dicloxacillin	Phenytoin
Erythromycin	Prednisolone
Ethchlorvinyl	Prednisone
Furosemide	Propranolol
Heparin	Pyrimethamine
Hydrocortisone	Quinidine
Imipramine	Reserpine
Indomethacin	Rifampin
Isoniazid	Theophylline
Levodopa	Trihexyphenidyl
Lidocaine	Tubocurarine
	Warfarin

Whereas a dosage adjustment is reasonable when renal insufficiency exists, such calculations are extremely difficult to make for drugs whose site of action lies in the tissue or peripheral compartments, where not only the volumes of distribution change but also the rate constants for transfer of drug between compartments in renal impairment. One such example is that of digoxin, whereby the volume of distribution decreases and half-life increases along with variation in the intercompartmental transfer rate constants, resulting in only a slight change in the tissue levels compared to a significant change in the total body levels of the drug in renal impairment.[118]

One indirect effect of renal malfunction is variation in the biotransformation of drugs. An increased rate of biotransformation partially compensates for the lower rate of drug clearance from the kidneys shown for such drugs as atropine and doxycycline.[119,120] A rather unusual example is that of phenytoin, which is administered without regard to kidney function (Table 8.12). Its principle metabolite, 5-phenyl-5-parahydroxy-phenylhydantoin, is excreted at a higher concentration in uremic states, resulting in decreased plasma levels of the parent compound, phenytoin. However, the activity of phenytoin does not change significantly because of an increase in the free fraction of the drug in the blood. This increased free drug concentration is not related to change in protein levels or to electrolyte disturbances in uremic states.[121–123]

In many instances the biotransformation of drugs is decreased in renal diseases, as is shown in the vitamin D conversion to its active form.[124] The breakdown of insulin generally occurs in the kidneys and in renal diseases this also is decreased.[125,126] The acetylation of sulfisoxazole, para-aminosalicylic acid, and isoniazid is similarly decreased.[127–129]

The decreased renal elimination of drugs is often compensated by increased clearance through nonrenal routes of elimination (that is, if one is possible). For example, the fecal excretion of digoxin and glibenclamide increases significantly in renal insufficiency, possibly due to the increased fraction of biliary excretion.[130,131]

It should be noted that in renal insufficiency significant variation in the biotransformation profile may occur. Often the presence of otherwise unrecognized metabolites is recorded and invariably the proportion of each metabolite of the drug changes. This phenomenon is further complicated by the various drug interactions resulting from multiple drug therapy, pathologic disorders other than renal insufficiency (which are most often the major cause of drug treatment), and physiologic changes, all of which contribute to the overall variability in the observed clinical response through change in the active drug content in the plasma. However, in many instances the sensitivity to a given drug also changes in renal insufficiency, independent of the effects on disposition profiles. For example, uremic patients show higher sensitivity to most narcotics, which is not always due to variation in the protein binding. The change in receptor sensitivity is also substantiated since the renal insufficiency cannot induce changes in the disposition within a short time after administration of the drug. Examples include the use of thiopental (sometimes related to protein binding), phenobarbital, promethazine, chlorpromazine, and antibiotic colistin.[132,133]

The nervous system generally becomes hypersensitive to drugs in the uremic state due to changes in the permeability of cerebral membranes.[134] Reduced renal function has also been reported to alter the concentration and exchange of electrolytes, enzymes, glucose, and lactate in the brain.[135]

The action of some drugs is similar to the physiologic variability in renal malfunction, such as increased blood urea in tetracycline or corticosteroid administration or water and salt retention by phenylbutazone and other pyrazolone derivatives. These actions of drugs are therefore magnified in the state of renal function and can lead to toxic symptoms. Another example of aggravated response is the effect of xanthine derivatives and digitalis on the gastrointestinal lesions produced in uremia. These drugs increase both the inflammation and the nausea.[136] The action of nephrotoxic drugs is naturally intensified, as with some antibiotics, phenylbutazone, gold salts, mercurial diuretics, and some anesthetics.[137–141]

The role of biliary excretion in the disposition of drugs was discussed earlier. It is understandable that disease states leading to the blocking of the biliary duct will cause accumulation of drugs in the body, especially if a significant fraction of the drug is cleared by this route. One such example is the ovulation-inducing agent clomiphene citrate (Clomid), which is almost entirely eliminated from the body by this route. The drugs whose duration of action is prolonged due to biliary recycling will show a shortened duration of action in biliary obstruction and other diseases of the gall bladder.

SUMMARY

A number of review articles have summarized the effects of renal failure on drug pharmacokinetics and drug action, as well as the interrelationship between renal hemodynamics, drug kinetics, and drug action.[142,143]

It is apparent from the preceding discussion of absorption, distribution, biotransformation, and excretion that there are two major categories of factors which influence clinical responses: the effect of the drug on the organism and the effect of the organism on the drug action. In addition to disease states and normal physiologic variables, the circadian rhythm or diurnal variation also affects the action of drugs.[144] The term CIRCADIAN refers to a 24-hour cycling of events. That there are changes in the excretion patterns of several endogenous substances within the time of a day is a well established fact. Recent data suggest similar patterns for various drugs. For example, the fraction of benzypenicillin excreted in the urine is lower (about 67 percent) when the patient is in the ambulatory state than when bed-ridden (about 82 percent).[145] The half-life of the sulfasymazine is about three times higher during the daytime than during the night.[146] This difference is attributed to non-ionic back diffusion in the kidney, lower urinary pH during the night, and non-ionic diffusion into and out of body cells. An interesting diurnal effect is noted for tolbutamide, whereby significantly higher insulin levels are seen in the morning and the hypoglycemic response is more prolonged in the evening, suggesting diurnal variation in pancreatic beta cell sensitivity, insulin sensitivity, and compensatory responses to hypoglycemia.[147]

There are also temporal variations in the half-lives of drugs. The plasma half-lives of phenacetin and acetaminophen are approximately 15 percent shorter at 2:00 PM than at 6:00 AM.[148] For antipyrine, which has a relatively longer half-life, there are significant variations which result in the decrease and increase of the half-life.[149] Diurnal disposition of ethanol has also been studied.[150]

Most of the discussion above relates to factors which are not always negotiable in the course of drug therapy. However, one of the most important sources of variability in clinical response is entirely controllable. It is patient compliance. According to Dr. Gibaldi, this is the single most important factor contributing to individual differences in plasma drug concentrations in outpatients.[151] It is not uncommon to see a patient compliance of less than 50 percent. This is especially problematic with psychiatric, pediatric, and geriatric patients. On the other hand, some patients feel compelled to take the medication more often than recommended. It is possible to provide patient education that will lead to a higher degree of compliance. The need for this cannot be overemphasized.

References

1. Hanzlik PJ: A study of the toxicity of the salicylates based on clinical statistics. JAMA 60:957, 1913
2. Bertler A, Smith SE: Genetic influences in drug responses of the eye and the heart. Clin Sci 40:403, 1971
3. Nichol ES, Keyes JN, Borg JF, et al: Long-term anticoagulant therapy in coronary atherosclerosis. Am Heart J 55:142, 1958
4. Wagner JG: Biopharmaceutics: Absorption aspects. J Pharm Sci 50:539, 1961
5. Sorby DL, Liu G: Effect of adsorbents on drug absorption. II: Effect of an antidiarrhea mixture in promazine absorption. J Pharm Sci 55:504, 1966
6. Riegelman S, Rowland M, Epstein WL: Griseofulvin-phenobarbital interaction in man. JAMA 213:426, 1970
7. Aggeler PM, O'Reilly RA: Effect of heptabarbital on the response to bishydroxycoumarin in man. J Lab Clin Med 74:229, 1969
8. Downer HD, Galloway RW, Horwich L, Parke DV: The absorption and excretion of carbenoxolene in man. J Pharm Pharmacol 22:479, 1970
9. Knudsen ET, Brown DM, Rolinson GN: A new orally effective penicillinase-stable penicillin—BRL 1621. Lancet 2:632, 1962
10. McCarthy CG, Finland M: Absorption and excretion of four penicillins: penicillin G, penicillin V, phenethicillin and phenylmercaptomethylpenicillin. N Engl J Med 263:315, 1960
11. Sutherland R, Croydon EAP, Rolinson GN: Flucoxacillin, a new isoxazolyl penicillin, compared with oxacillin, cloxacillin, and dicloxacillin. Br Med J 4:455, 1970
12. McGilveray IJ, Mattock GL: Some factors affecting the absorption of paracetamol. J Pharm Pharmacol 24:615, 1972
13. Anon.: Iron in flour. Lancet 2:495, 1968
14. Crounse RG: Human pharmacology of griseofulvin: the effect of fat intake on gastrointestinal absorption. J Invest Dermatol 37:529, 1961
15. Robinson DS, Benjamin DM, McCormack JJ: Interaction of warfarin and nonsystemic gastrointestinal drugs. Clin Pharmacol Ther 12:491, 1971
16. Avakian S, Kabacoff BL: Enhancement of blood antibiotic levels through the combined oral administration of phenethicillin and chymotrypsin. Clin Pharmacol Ther 5:716, 1964
17. Boman G, Hanngren A, Malmborg AS, Borga O, Sjoqvist F: Drug interaction: decreased serum concentrations of rifampicin when given with PAS. Lancet 1:800, 1971
18. Parsons RL: Drug absorption in gastrointestinal diseases with particular reference to malabsorption syndromes. Clin Pharmacokinet 2:45, 1977
19. Rivera-Calimlim L, Dujovne CA, Morgan JP, Lasagna L, Bianchine JR: L-dopa treatment failure: explanation and correction. Br Med J 4:93, 1970
20. Turnberg LA: The absorption of iron after partial gastrectomy. Q J Med 25:107, 1966
21. Elsborg L: Malabsorption of folic acid following partial gastrectomy. Scand J Gastroenterol 9:271, 1974
22. Matilla MJ, Friman A, Larmi TKI, Koskinen R: Absorption of ethionamide, isoniazid and aminosalicylic acid from the post-secretion gastrointestinal tract. Ann Med Exp Bio Fenn 47:209, 1969
23. Lode H, Frish D, Naumann P: Oral antibiotic therapy in patients with partial gastrectomy. In Daikos AA (ed): Progress in Chemotherapy, Vol 1. Athens, Hellenic Society of Chemotherapy, 1974, p 543
24. Heizer WD, Smith TW, Goldfinger SE: Absorption of digoxin in patients with malabsorption syndromes. N Engl J Med 258:257, 1971
25. Lupinsky I, Berthoud S: Absorption of penicillin V in relation to digestive dis-

orders. Schweiz Rundschau Med (Praxis) 62:959, 1973
26. Parsons RL, Paddock GM: Absorption of two antibacterial drugs, cephalexin and co-trimoxazole in malabsorption syndromes. J Antimicrob Chemother 1(Suppl):59, 1975
27. Moberg S, Carlberger G: Gastric emptying in healthy subjects and in patients with various malabsorption states. Scand J Gastroenterol 9:17, 1974
28. Bunn A, Cooke WT: Intraluminal pH of duodenum and jejenum in fasting subjects with normal and abnormal gastric or pancreatic function. Scand J Gastroenterol 6:313, 1971
29. Parsons RL, Kaye CM: Plasma propanolol and practolol in adult coeliac disease. Br J Clin Pharmacol 1:348P, 1974
30. Parsons RL, Kaye CM, Raymond K, Trounce JR, Turner P: Absorption of propanolol and practolol in coeliac disease. Gut 17:139, 1976
31. Parsons RL, Hossack GA, Paddock GM: The absorption of antibiotics in adult patients with coeliac disease. J Antimicrob Chemother 1:39, 1975
32. Parsons RL, Paddock GM, Hossack GA, Hailey DM: Antibiotic absorption in Crohn's disease. In Williams JD, Geddes AM (eds): Chemotherapy, Vol 4. Pharmacology of Antibiotics. New York, Plenum, 1976, p 219
33. Prescott LF: Drug therapy—physiological considerations. Paper presented at a symposium on optimizing drug activity. 31st International Congress of Pharmaceutical Sciences, Washington D.C., September 1971
34. Prescott LF: Gastrointestinal absorption of drugs. Med Clin North Am 58:907, 1974
35. Prescott LF: Gastric emptying and drug absorption. Br J Clin Pharmacol 1:189, 1974
36. Prescott LF: Clinically important drug interactions. In Avery GS (ed): Drug Treatment, Sydney, Adis, 1976, p 193
37. Nimmo WS: Drugs, diseases and altered gastric emptying. Clin Pharmacokinet 1:189, 1976
38. Novak LP: Aging, total body potassium, fat free mass and cell mass in males and females between ages 18 and 55 years. J Gerontol 27:438, 1972
39. Altman PL, Dittmer DS (eds): Blood and Other Body Fluids. Bethesda Md, Federation of American Societies for Experimental Biology, 1961 pp 61, 497, 500
40. Friis-Hansen B: Body water components in children: Changes during growth and related changes in body composition. Pediatrics 28:169, 1961
41. Rane A, Wilson JT: Clinical pharmacokinetics in infants and children. Clin Pharmacokinet 1:2, 1976
42. Wyers PJH, van Munster PJJ: The disappearance of Evans blue dye from the blood in normal and nephrotic subjects. J Lab Clin Med 58:375, 1961
43. Lewis RJ, Spivack M, Spaet TH: Warfarin resistance. Am J Med 42:620, 1967
44. Lewis GP, Jusko WJ, Burke CW, Graves L: Prednisone side-effects and serum-protein levels. Lancet 2:778, 1971
45. Curry SH: Plasma protein binding of chlorpromazine. J Pharm Pharmacol 22:193, 1971
46. Aladanderson B, Borga O: Interindividual differences in plasma protein binding of nortriptyline in man—a twin study. Eur J Clin Pharmacol 4:196, 1972
47. Anton AH, Corey WJ: Plasma protein binding of sulfonamides in anephric patients. Fed Proc 30:629, 1971
48. Shoeman DW, Azarnof DL: The alteration of plasma proteins in uraemia as reflected by their activity to bind digitoxin and diphenylhydantoin. Pharmacology 7:169, 1972
49. Solomon HM, Schrogie JJ: The effect of phenyramidol on the metabolism of diphenylhydantoin. Clin Pharmacol Ther 8:554, 1967
50. Andreasen F: Protein binding of drugs in plasma from patients with acute renal failure. Acta Pharmacol Toxicol (Kbh) 32:417, 1973
51. Andreasen F, Jakobsen P: Determination of furosemide in blood plasma and its binding to proteins in normal plasma

and in plasma from patients with acute renal failure. Acta Pharmacol Toxicol (Kbh) 35:49, 1974

52. O'Malley K, Velasco M, Pruitt A: Decreased plasma protein binding of diazoxide in uremia. Clin Pharmacol Ther 18:53, 1975
53. Jusko WJ, Weintraub M: Myocardial distribution of digoxin and renal function. Clin Pharmacol Ther 16:449, 1974
54. Affrime M, Reidenberg MM: The protein binding of some drugs in plasma from patients with alcoholic liver disease. Eur J Clin Pharmacol 8:267, 1975
55. Odell GB: Dissociation of bilirubin from albumin and its clinical implications. J Pediatr 55:268, 1959
56. Rane A, Lunde PKM, Jalling B: Plasma protein binding of diphenylhydantoin in normal and hyperbilirubinemic infants. J Pediatr 78:877, 1971
57. Gugler R, Shoeman DW, Azarnoff D: Effect of in vivo elevation of free fatty acids on protein binding of drugs. Pharmacology 12:160, 1974
58. Levy G: Effect of plasma protein binding of drugs on duration and intensity of pharmacologic activity. J Pharm Sci 65:1264, 1976
59. Varma DR, Gupta RK, Gupta S: Prothrombin response to phenindione during hypalbuminaemia. Br J Clin Pharmacol 2:467, 1975
60. Pearson RM, Breckenridge AM: Renal function, protein binding and pharmacologic response to diazoxide. Br J Clin Pharmacol 3:169, 1976
61. Bender AD: Effect of increasing age on the distribution of peripheral blood flow in man. J Am Geriatr Soc 13:192, 1965
62. Anton AH: A drug-induced change in the distribution and renal excretion of sulfonamides. J Pharmacol Exp Ther 134:291, 1961
63. Kunin CM; Clinical pharmacology of the new penicillins. II: Effect of drugs which interfere with binding to serum proteins. Clin Pharmacol Ther 7:180, 1966
64. Breckenridge A, Rosen A: The binding of chlorothiazide to plasma proteins. Biochem Biophys Acta 229:610, 1971
65. Franz JW, Jannchen E, Krieglstein J: Der Einfluss verschiedener Pharmaka aug das Bindungsvermogen einer Albuminlosung fur Promazin und Chlorpromazin. Naunyn-Schmiedebergs Arch Expt Path Pharmak 264:462, 1969
66. Zubrod CG, Kennedy TJ, Shannon JA: Studies on the chemotherapy of the human malarias. VIII: The physiological disposition of pamaquine. J Clin Invest 27 (s):114, 1948
67. Nelson EB, Raj PP, Belfi KJ, Masters BSS: Oxidative drug metabolism in human liver microsomes. J Pharmacol Exp Ther 178:580, 1971
68. Remmer H, Schoene B, Fleischmann RA, Olderhausen HF: Drug metabolizing enzymes determined in needle biopsy material of human liver. In: Abstracts Proceedings 5th International Congress of Pharmacology. Basel, Karger, 1972, p 191
69. Thorgeirsson SS, Davies DS: Kinetic studies of the N-demethylation of ethylmorphine by a cytochrome P-450 dependent enzyme system in human liver microsomes. Biochem J 112:30P, 1971
70. Vesell ES, Passananti GT, Greene FE, Page JG: Genetic control of drug levels and of the induction of drug-metabolizing enzymes in man: Individual variability in the extent of allopurinol and nortriptyline inhibition of drug metabolism. Ann NY Acad Sci 179:752, 1971
71. Loeser EW: Studies on the metabolism of diphenylhydantoin (Dilantin). Neurology (Minneap) 11:424, 1961
72. Hammer W, Sjoqvist F: Plasma levels of monomethylated tricyclic antidepressants during treatment with imipramine-like compounds. Life Sci 6:1895, 1967
73. Niazi S: Comparison of observed and predicted first-pass metabolism of imipramine in humans. J Pharm Sci 65:1063, 1976
74. Niazi S: Comparison of observed and predicted first-pass metabolism of nortriptyline in humans. J Pharm Sci 65:1535, 1976
75. Briant RH, Liddle DE, Dorrington R, Williams FM: Plasma half-life of two an-

algesic drugs in young and elderly adults. NZ Med J 82:136, 1975

76. Jori A, DiSalle E, Quadri A: Rate of aminopyrine disappearance from plasma in young and aged humans. Pharmacology 8:273, 1972
77. Triggs EJ, Nation RL: Pharmacokinetics in the aged: A review. J Pharmacokinet Biopharm 3:387, 1975
78. Nies AS, Shand DG, Wilkinson GR: Altered hepatic blood flow and drug disposition. Clin Pharmacokinet 1:135, 1976
79. Vestal RE, Norris AH, Tobin JD, et al: Antipyrine metabolism in man: Influence of age, alcohol, caffeine, and smoking. Clin Pharmacol Ther 18:425, 1975
80. Weiss CF, Glazko AJ, Weston JK: Chloramphenicol in the newborn infant: A physiological explanation of its toxicity when given in excessive dose. N Engl J Med 262:787, 1960
81. Fichter EG, Curtis JA: Sulfonamide administration in newborn and premature infants. Am J Dis Child 90:596, 1955
82. Kauer B, Spring P, Dettli L: Zur Pharmakokinetik der sulfonamide in ersten Lebensjahr. Pharmacologia Clin 1:47, 1968
83. Vest M, Streiff RR: Studies on glucuronide formation in newborn infants and older children. Measurement of p-aminophenol glucuronide levels in the serum after oral dose of acetanilid. Am J Dis Child 98:688, 1959
84. Vest MR, Rossier R: Detoxification in the newborn: The ability of the newborn infant to form conjugates with glucuronic acid, glycine, acetate, and glutathione. Ann NY Acad Sci 111:183, 1963
85. Evans DAP, Manley KA, McKusick VA: Genetic control of isoniazid metabolism in man. Br Med J 2:485, 1960
86. Evans DAP, White TA: Human acetylation polymorphism. J Lab Clin Med 63:387, 1964
87. Varley H: Practical Clinical Biochemistry, 3rd ed. London, Heinemann, 1962, p 634
88. Das KM, Eastwood MA, McManus JPA: Adverse reactions during salicylazosulfapyridine therapy and the relation with drug metabolism and acetylator phenotype. N Engl J Med 289:491, 1973
89. Perry HM, Tan EM, Carmedy S: Relationship of acetyltransferase activity to antinuclear antibodies and toxic symptoms in hypertensive patients treated with hydralazine. J Lab Clin Med 76:114, 1970
90. Karlsson E, Molin L, Norlander B: Acetylation of procaine amide in man studied with a new gas chromatographic method. Br J Clin Pharmacol 1:467, 1974
91. Shahidi NT: Acetophenetidin sensitivity. Am J Dis Child 113:81, 1967
92. Kutt H, Winters W, Kokenge R, McDowell F: Diphenylhydantoin metabolism, blood levels, and toxicity. Arch Neurol 11:642, 1964
93. Vesell ES, Page JG: Genetic control of drug levels in man: phenylbutazone. Sci NY 159:1479, 1968
94. Vesell ES, Page JG: Genetic control of dicoumarol levels in man. J Clin Invest 47:2657, 1968
95. Vessell ES, Page JG: Genetic control of drug levels in man: antipyrine. Sci NY 161:72, 1968
96. Vesell ES, Page JG, Passananti GT: Genetic and environmental factors affecting ethanol metabolism in man. Clin Pharmacol Ther 12:192, 1971
97. Sessions JT, Minkel HP, Bullard JC, Ingelfinger FJ: The effect of barbituates in patients with liver disease. J Clin Invest 333:1116, 1954
98. Brodie BB, Burns JJ, Weiner M: Metabolism of drugs in subjects with Laennec's cirrhosis. Med Exp 1:290, 1959
99. Kunin CM, Glazko AJ, Finland M: Persistance of antibiotics in blood of patients with acute renal failure. II: Chloramphenicol and its metabolic products in the blood of patients with severe renal disease or hepatic cirrhosis. J Clin Invest 38:1498, 1959
100. Acocella G, Baroni GC, Muschio R: Clinical evaluation of the therapeutic activity of rifamycin SV in the treatment of biliary tract infections. G Mal Infett Parassit 14:552, 1962

101. Ueda H, Sakurai T, Ota M, et al: Disappearance rate of tolbutamide in normal subjects in diabetes mellitus, liver cirrhosis and renal disease. Diabetes 12:414, 1963
102. Suhrland LG, Weisberger AS: Chloramphenicol toxicity in liver and renal disease. Arch Intern Med 112:747, 1963
103. choene B, Fleischmann RA, Remmer H, Olderhausen HFV: Determination of drug metabolizing enzymes in needle biopsies of human liver. Eur J Clin Pharmacol 4:65, 1972
104. Reidenberg MM, James M, Dring LG: The rate of procaine hydrolysis in serum of normal subjects and diseased patients. Clin Pharmacol Ther 13:279, 1972
105. Mawer GE, Miller NE, Turnberg LA: Metabolism of amylobarbitone in patients with chronic liver disease. Br J Pharmacol 44:549, 1972
106. Prescott LF, Wright N, Roscoe P, Brown SS: Plasma-paracetamol half-life and hepatic necrosis in patients with paracetamol overdosage. Lancet 1:519, 1971
107. Prescott LF: The modifying affects of physiological variables and disease upon pharmacokinetics and/or drug response. Liver Disease Abstr Proc 5th International Congress of Pharmacology. Basel, Karger, 1972, p 73
108. Levi AJ, Sherlock S, Walker D: Phenylbutazone and isoniazid metabolism in patients with liver disease in relation to previous drug therapy. Lancet 1:1275, 1968
109. Reidenberg MM, Kostenbauder H, Adams W: Rate of drug metabolism in obese volunteers before and during starvation and in azotemic patients. Metabolism 13:209, 1969
110. Englert E, Brown H, Willardson DG, Wallach S, Simons EL: Metabolism of free and conjugated 17-hydroxycorticosteroids in subjects with uremia. J Clin Endocrinol Metab 18:36, 1958
111. Closson RG: Terminal half-lives of drugs studied in patients with hepatic diseases. Am J Hosp Pharm 34:520, 1977
112. Knox FG, Cuche J, Ott CE, Diaz-Buxo JA, Marchand G: Regulation of glomerular filtration and proximal tubule reabsorption. Circ Res 36:1, 1975
113. Cafruny EJ: Renal tubular handling of drugs. Am J Med 62:490, 1977
114. Bennett WM, Craven R: Urinary tract infections in patients with severe renal disease. Treatment with ampicillin and trimethorprim-sulfamethoxazole. JAMA 236:946, 1976
115. Kunin CM, Finland M: Restrictions imposed on antibiotic therapy by renal failure. Arch Int Med 104:1030, 1959
116. Petitpierre B, Fabre J: Effect de l'insuffisance renale sur l'action hypoglycemiante des sulfonylurees. Cinetique de la chlorpropamide en cas de nephropathie. Schweiz Med Wochenschr 102:570, 1972
117. Fabre J, Pitton JA, Vireux C, et al: Absorption, distribution, et excretion de la doxycycline chez l'homme. Schweiz Med Wochenschr 97:915, 1967
118. Wagner JG: Loading and maintenance doses of digoxin in patients with normal renal function and those with severely impaired renal function. J Clin Pharmacol 14:329, 1974
119. Kalser SC, Kelvington EJ, Randolph MM, Santomenna DM: Drug metabolism in hypothermia I: Biliary excretion of C^{14} atropine metabolites in the intact and nephrectomized rat. J Pharmacol Exp Ther 147:252, 1965
120. Giromini M, Wasem R, Merier G, Fabre J: Influence de l'anurie et des hemodialyses sur le comportment des antibiotiques. Praxis 58:1181, 1969
121. Mellk HM, Letteri JM, Durante P: Diphenylhydantoin metabolism in chronic uremia. Ann Intern Med 72:801, 1970
122. Odar-Cederlof I, Lunde P, Sjoqvist F: Abnormal pharmacokinetics of phenytoin in patients with uremia. Lancet 2:831, 1970
123. Odar-Cederlof I, Borga O: Kinetics of diphenylhydantoin in uremic patients: Consequences of decreased plasma protein binding. Eur J Clin Pharmacol 7:31, 1974
124. Avioli LV, Birge S, Lee SW, Slatopolsky E: The metabolic fate of vitamin D3 -^{3}H

in chronic renal failure. J Clin Invest 47:2239, 1968
125. O'Brien JP, Sharp AR: The influence of renal disease on the insulin (I^{131}) disappearance curve in man. Metabolism 16:76, 1967
126. Rabkin R, Simon NM, Steiner S, Colwell JA: Effect of renal disease on renal uptake and excretion of insulin in man. N Engl J Med 282:182, 1970
127. Reidenberg MM, Kostenbauder H, Adams W: Rate of drug metabolism in obese volunteers before and during starvation and in azotemic patients. Metabolism 18:209, 1969
128. Ogg CS, Toseland PA, Cameron JS: Pulmonary tuberculosis in patients on intermittent hemodialysis. Br Med J 2:283, 1968
129. Fabre J, Berthoud S: Les variations individuelles de las response aux medicaments dans les conditions pathologiques. Brux Med 55:519, 1975
130. Bloom PM, Nelp WP: Relationship of the excretion of tritiated digoxin to renal function. Am J Med Sci 251:133, 1966
131. Schmidt FH, Hrstka VE, Heesen D, Schultz O, Schultz E: Plasma-spiegel und Ausscheidung von Glibenclamid bei niereninsuffizienten und leberkranken Patienten im akuten Versuch 9. Travemunde, Kongress der Deutschen Diabetes-Gesellschaft, 1974
132. Fabre J, de Freundenreich J, Duckert A, et al: Influence of renal insufficiency on the excretion of chloroquine, phenobarbital, phenothiazines and methacycline. Helv Med Acta 33:307, 1967
133. Richet G, Ardaillou R, Sultan Y: Accidents neuropsychiques chex des uremiques chroniques traites par le methane-sulfonate de colistine. Bull Mem Soc Med Hop Paris 113:1199, 1962
134. Fishman RA: Permeability changes in experimental uremic encephalopathy. Arch Intern Med 126:835, 1970
135. Rashkin NH, Fishman RA: Neurologic disorders in renal failure. N Engl J Med 294:143, 1976
136. Lawrason FD, Alpert E, Mohr FL, McMahon FG: Ulcerative obstructive lesions of the small intestine. JAMA 191:641, 1965
137. Kleinknecht D, Fillastre JP: La tolerance renale de antibiotiques. Paris, Masson, 1973
138. Scheitlin W, Jeanneret I: Ueber akute Nierenschadigung unter Phenylbutazontherapie. Schweiz Med Wochenschr 87:881, 1957
139. Katz A, Little H: Gold nephropathy. Arch Pathol 96:133, 1973
140. Freeman RB, Maher JF, Schreiner GE, Mostofi FK: Renal tubular necrosis due to nephrotoxicity of organic mercurial diuretics. Ann Int Med 57:34, 1962
141. Gauert WB, Buschman D, Parmley RT: Renal function associated with methoxyflurane anesthesia. South Med J 62:1487, 1969
142. Fabre J, Balant L: Renal failure, drug pharmacokinetics and drug action. Clin Pharmacokinetics 1:99, 1976
143. Duchin KL, Schrier RW: Interrelationship between renal hemodynamics, drug kinetics and drug action. Clin Pharmacokinet 3:58, 1978
144. Moore MC: Circadin rhythms of drug effectiveness and toxicity. Clin Pharmacol Exp Ther 14:925, 1973
145. Levy G: Effect of bed rest on distribution and elimination of drugs. J Pharm Sci 56:928, 1967
146. Dettli L, Pring P: Diurnal variations in the elimination rate of a sulfonamide in man. Helv Med Acta 33:291, 1966
147. Johnson BF, Chura C: Diurnal variation in the effect of tolbutamide. Am J Med Sci 268 (2):93, 1974
148. Shively CA, Vesell ES: Temporal variations in acetaminophen and phenacetin half-life in man. Clin Pharmacol Ther 18:413, 1975
149. Vesell ES, Shively CA: Temporal variation in antipyrine half-life in man. Clin Pharmacol Exp Ther 22:843, 1977
150. Zeiner AR: The time of day effects in ethanol metabolism. Alcohol Tech Rep 3:7, 1974
151. Gibaldi M: Biopharmaceutics and Clinical Pharmacokinetics. Philadelphia, Lea and Febiger, 1977, p 144

Questions

1. What are the reasons for difficulty in quantitation of the variability in clinical drug responses?
2. Is it possible to have 6-fold variation in the response all contained under a normal distribution curve?
3. There are two levels of variability in clinical response. What are these? To which can pharmacokinetic principles be applied?
4. Cite examples of drug interactions leading to decreased drug absorption.
5. Why would Mylanta decrease the absorption of quinidine?
6. Why isn't the variation in gastric pH an important factor in the overall absorption of drugs? Under what conditions can it be important?
7. Why does food for the most part retard the rates of absorption and not the total bioavailability?
8. Cite some examples of drugs commonly used in psychiatric care and their effects of gastrointestinal blood flow.
9. Why is it sometimes necessary to administer a decarboxylase inhibitor with levodopa in Parkinsonism?
10. Cite examples of drugs whose absorption is significantly reduced in gastrectomy.
11. What is the etiology of steatorrhea?
12. How does the mechanism of digoxin absorption differ from the mechanisms of ampicillin, cephalexin, and phenoxymethylpenicillin absorption?
13. What abnormalities occur in gluten enteropathy and Crohn's disease that affect drug absorption? List the nature of these effects.
14. What is the mechanism of the decreased absorption of pivampicillin in coeliac diseases?
15. Absorption efficiency is greater in children but the blood levels are generally comparable with those of adults (on an adjusted dose basis). Explain.
16. Why would the site of injection affect the rate of absorption?
17. How can Colace increase the absorption of vitamin A and cholesterol?
18. How do cytotoxic agents decrease the absorption of drugs?
19. Comment on the statement, "Divided doses of poorly water-soluble drugs may provide higher bioavailability than a single dose."
20. What is the approximate change in the body fat content between the ages of 18 and 55? Does this differ according to sex?
21. In newborns the total body water represents a higher percentage of body weight than in adults. Why?
22. If the clearance of a drug remains constant between individuals with different volumes of distribution, what factor must change?
23. Why does the half-life of lipid-soluble drugs increase in the elderly?
24. List the disease states which can lead to hypoproteinemic disorders.
25. If the plasma albumin concentration decreases below 2.5 g/100 ml, serious side effects of prednisolone therapy are noted. Why?

26. What kinds of changes occur in protein content and characteristics in renal failure? Why?
27. Which chemical species, weak acid or base, is more affected in terms of protein binding in disease states? Why?
28. Would you expect drug-tissue binding characteristics to change in disease states?
29. In one study a direct correlation between bilirubin concentration and the activity of phenytoin was reported. How do you explain this observation?
30. What is the normal free fatty acid concentration? What effects can be expected at elevated levels?
31. In one study the steady state levels of a drug were found to be identical between patients who had normal and patients with less than normal protein contents (plasma), but the side effects were higher in the patients with hypoalbuminemia. Explain.
32. How does stress affect the distribution of drugs?
33. Can a slight change in the binding (say 1 percent) affect the pharmacologic response significantly?
34. Cite examples of drug interactions leading to increased binding to plasma proteins.
35. Can drug interactions take place in binding to tissues?
36. What is the underlying mechanism in the correlation of Cytochrome P-450 levels and the half-lives of the drugs?
37. What makes antipyrine an ideal drug for the study of drug biotransformation? Which metabolite of antipyrine can be used to correlate liver function?
38. What is the etiology of gray syndrome in children?
39. Why does one obtain polymodal distribution curves for biotransformation in a discontinuous genetic population?
40. How are people classified as slow acetylators or fast acetylators? Cite examples of drugs which are subject to polymorphic acetylation.
41. Cite examples of drugs whose biotransformation is influenced by polygenic factors.
42. Is it true that serious effects of liver disease on drug biotransformation occur only when the liver impairment is severe?
43. Is the enzyme induction property altered in liver disease states?
44. At what level of glomerular filtration rate impairment can a significant effect on drug clearance be expected?
45. What is the effect of age on the glomerular filtration rate?
46. The use of methenamine mandelate and nalidixic acid is not recommended in patients who have creatinine clearance of less than 20 ml/min. What is the main reason for this?
47. What are the various statuses of renal function and how are they correlated with the creatinine clearance?
48. Cite at least five examples of drugs whose dosage regimens should be changed even in mild renal impairment.

49. Which drugs should be avoided in moderate renal impairment and in severe renal impairment?
50. Cite examples of commonly used drugs which are given regardless of kidney function.
51. By plotting the data on creatinine clearance and the half-life of tetracycline, determine the nature of their correlation.
52. Which pharmacokinetic parameters, besides the renal excretion constant, are subject to change when the glomerular filtration rate decreases?
53. Can uremia cause alteration in the gastrointestinal absorption of drugs?
54. What is the potential problem in adjusting dosages according to plasma levels in disease states when the site of action of a drug may lie in deeper tissues?
55. The biotransformation of phenytoin increases in uremia but its activity does not change significantly. Explain.
56. Can the sensitivity of drugs change in the uremic state?
57. What effects of tetracyclines and corticosteroids are magnified in renal impairment?
58. The use of Clomid should be carefully made in patients suffering from biliary duct obstruction. Why?
59. What are possible reasons for a circadian rhythm in the disposition of drugs?
60. How important is patient compliance in determining the variability in human drug response?

CHAPTER 9

The Clinical Pharmacokinetic Basis of Drug Therapy

A successful drug therapy begins with the selection of an optimum starting dose and subsequent maintenance of a desired response through successive doses with a minimum of toxicity or side-effects. The applications of clinical pharmacokinetic principles in determining an appropriate dose are dependent mainly on the concentration-effect relationships of various drugs. The quantity of drug administered determines the levels of the drug reached in the plasma, but these levels may or may not relate to the concentration of drug at the site of action. In many instances, drug action is a result of the indirect effect of a drug on a body system or its constituents. Such cases are difficult to handle in terms of dosage calculation, as is demonstrated with coumarin anticoagulants. However, the principles of clinical pharmacokinetics, when applied correctly, can help solve most of the dosage-related problems.

For a majority of drugs, the pharmacologic response and toxicity are proportional to the levels of drug in the body (not necessarily in the plasma). In many such instances, a linear log dose-response relationship can be established within at least 20 to 80 percent of the response range. The underlying reason for any such correlation between the dose and the response is the distribution-dependent phenomena. A drug must distribute to various body tissues before it can effect a response. Thus the volume of distribution plays a very important role in the effectiveness of a given dose. Fortunately, it is possible to estimate the distribution characteristics of a drug from such parameters as body weight, body surface area, and age. Most of the empirical formulae traditionally based on these properties are in fact related to the volume of distribution, though the term itself and its appreciation by clinicians is very recent.

Note that the application of pharmacokinetic principles to drug therapy begins with the distribution characteristics, whereas the absorption should be the first parameter studied. Unfortunately, it is very difficult to predict the extent of drug absorption in patients for reasons discussed in the previous chapter. This is not, therefore, an important consideration in the initial dose recommendation and is later confronted when the plasma concentrations are monitored (as discussed later). However, in those instances where a clear variability in the absorption profiles is expected—as in gastric or ileojejunal bypass, ostomy, diarrhea, severe gastrointestinal infection, reduced blood volume and flow to the site of absorption, and other specific circumstances—dosage adjustments can be made to account for variability in the absorption properties. These adjustments are in most instances based on empirical findings in specific situations.

The most significant consideration in disease states is the elimination profile of a given drug. In most diseases the elimination function does not change. However, diseases leading to changes in the blood flow to the kidneys, liver, and other sites of drug elimination; to structural modifications in these organs; and to pathologic changes affecting the efficiency of these organs to remove the drug from the body require dosage adjustments based on indirect parameters relating to the efficiency of these systems. For example, creatinine clearance is often used to determine the degree of renal impairment and is related to the disposition half-lives of drugs.

It is a well-appreciated fact that disease states lead to variability in the disposition profiles of drugs, which are often difficult to predict. For example, nephrotic patients can show not only a decreased urinary drug clearance but also altered distribution properties due to changes in the plasma protein binding characteristics. The effect of drug absorption on the blood levels is also difficult to predict. In order to account for all the possible factors affecting the therapeutic response, plasma levels are monitored in individual patients and the dose is corrected according to individual need. This is referred to as individual titration and is probably the most effective method for assuring a desired response. However, the difficulties involved in the analysis of a drug in the plasma make this method less attractive especially in outpatient settings and in small clinical facilities. It should be noted that not all drugs need plasma analysis for dosage adjustment. This is effective only for those drugs with which a reasonable correlation exists between the plasma levels (or tissue levels) and the pharmacologic response. Drugs acting by indirect mechanisms require dosage adjustment based on either their pharmacologic response or other physiologic parameters, such as prothrombin time evaluation in patients maintained on coumarin anticoagulants.

DISTRIBUTION-DEPENDENT DOSAGE ADJUSTMENT

If a patient does not suffer from renal impairment, liver diseases, or other pathophysiologic states which might affect the disposition function, the body surface area can be used to calculate the dosage regimen. It is suggested that the volume of drug distribution is generally proportional to the total body surface area. Since the volume of distribution directly affects the steady state plasma levels of drugs, dosage adjustments based on body surface area generally provide good initial estimates (Table 9.1). The relationships between

Table 9.1. DOSAGE CALCULATION BASED ON BODY SURFACE AREA

KG	WEIGHT (lb)	APPROX. AGE	SURFACE AREA (m^2)	% OF* ADULT DOSE
3	6.6	Newborn	0.2	10
6	13.2	3 months	0.33	17
10	22	1 year	0.45	25
20	44	5.5 years	0.8	42
30	66	9 years	1	56
40	88	12 years	1.3	69
50	110	14 years	1.5	82
60	143	Adult	1.7	93
70	154	Adult	1.76	104

* Percent of adult dose = $4.7\ (\text{kg})^{0.73}$.

body weights, age, and body surface area are based on the general population characteristics which do not take into consideration abnormal patients who may be extremely lean or fat with respect to their age groups. These tables should therefore be used with some allowances for the individual patient.

It should be noted that on a mg/kg basis the dose requirement increases with decreasing age. For example, children between the ages of 7 and 12 receive 125 percent of the adult dose, those between the ages of 1 and 7 receive 150 percent, and those between the ages of 2 weeks and 1 year receive about 200 percent of the adult dose. This increase in the dose is due mainly to two factors, incomplete calcification of the bones and a significantly higher specific body surface area (total body surface/weight).

Although a large volume of data is available on children and adults, very little information is available on newborns, infants, and the elderly. For example, the volume of distribution in neonates is significantly higher than in adults for various drugs.[1] This observation is linked to a decreased binding of drugs to plasma proteins due to qualitative differences in neonate protein characteristics, high concentrations of bilirubin and fatty acids, lowered blood pH, and competition with other endogenous substrates. Total proteins and binding generally reach the adult level at about 10 to 12 months of age.[2]

The newborns also have a higher extracellular volume and often show increased red blood cell: plasma ratios of drugs, as is seen with phenytoin, whereby the ratio in newborns is 0.84 compared to 0.38 in adults.[3]

The expanded volume of distribution in neonates means that at a given plasma concentration higher tissue levels can be expected. An increased activity can therefore be expected for drugs whose site of action lies in the deep tissue compartment(s). Thus the dosage calculations based entirely on body surface area or on body fluid volume may not provide good estimates, requiring information about the site of action which may not be available in many instances.

Pediatric drug therapy is further complicated by the continuous change in the body weight and body area during infancy and childhood. Furthermore, the dearth of reliable data complicates the dosage calculations. For example, the volume of distribution on a liter/kg basis keeps decreasing with age—it reaches a plateau in adults and then changes again significantly in the elderly. The volume of distribution of theophylline in children between the ages of 4 and 15 is about 0.404 liters/kg, which decreases with age.[4] The volume of distribution of digoxin is about 7.5 liters/kg for infants within the age group of 3 to 9 days—this increases to approximately 16.3 liters/kg after one year of age.[5] The latter finding is meant to caution against any blanket statements regarding volume of distribution and age.

Significant changes in the volume of distribution occur in the elderly and since the body composition also changes from lean to fat, appropriate consideration of lean body mass is often recommended in dosage calculations for the elderly. However, since the volume of distribution is dependent on the lipid solubility of the drug itself, it is possible to observe an increased volume of distribution, as with diazepam, warfarin, acetaminophen, etc.

DISPOSITION FUNCTION-DEPENDENT DOSAGE ADJUSTMENT

If a disease state leads to variation in the disposition (biotransformation and excretion) function, an appropriate dosage adjustment is needed. The most common source of disposition function variation is renal impairment, which leads to prolonged half-lives of drugs and their metabolites. Table 8.10 lists the drugs requiring dosage adjustment in renal impairment, and Table 8.11 lists the drugs with no such requirement.

The half-life of a drug is the most important pharmacokinetic parameter in calculating dosage regimens. If, for example, the minimum effective amount of drug in the body is known, a quick method of dosage calculation is to use a loading dose which is twice the maintenance dose, and make the dosing interval equal to the half-life of the drug. This approach has long been used without appreciation of its pharmacokinetic nature. It is, however, mainly applicable in single compartment model systems.

Problem 9.1:

Given a half-life of 8 hours, an effective concentration of 10 μg/ml, and a V_d of 12 liters, calculate a dosage regimen based on the principle above.

The minimum effective amount in the body is:

$$\text{concentration} \times \text{volume of distribution} = 10\ \mu\text{g/ml} \times 12\ \text{liters} \times 1000 = 120\ \text{mg}.$$

Thus, 240 mg of loading dose followed by 120 mg every 8 hours will provide adequate therapy:

Time (hr)	*Amount in the body (mg)*
0	240
8	120 + (120) = 240
16	120 + (120) = 240

Note that at all times the amount of drug in the body remains above 120 mg, the effective level. It reaches 120 mg only at the time of the next dose. These calculations are applicable to the oral administration of drugs provided the absorption half-life is much longer than the elimination half-life.

Notice that in the above computations, the maximum level in the body is twice that of the lowest level (a 200 percent fluctuation is observed). This ratio changes with the dosing interval, as is shown below for intravenous injection:

Dosing interval ($t_{0.5}$)	*Maximum amount:Minimum amount*
0.1	1.072
0.5	1.414
1.0	2.00
1.5	2.828
2.0	3.999
5.0	31.976

Therefore, in order to decrease the maximum amounts of drug in the body during steady state, the dosing interval has to be decreased. However, for drugs with short half-lives it may not be feasible to suggest a dosing interval, except in continuous infusion, where no fluctuation occurs.

Problem 9.2:
If a drug is administered every 4 hours instead of every 8 hours in the preceding problem, calculate the loading and maintenance doses.

Since the maximum amount or the loading dose is 1.414 times the minimum amount, which is 120 mg, the loading dose is 169 mg and the maintenance dose is simply 169 – 120 = 49 mg. You should calculate the loading and maintenance doses for other dosing intervals. Note that you need only Equation 7.4 for all of your calculations.

In the calculations performed above, the minimum effective amount of drug in the body or the minimum effective concentrations were used as a basis for dosage calculation. Another useful parameter is the average steady state level, which is given by the well-known Wagner Equation:

$$\overline{C}_\infty = F X_0/K\tau V = AUC/\tau \qquad \text{(Eq. 9.1)}$$

where AUC can be the area under the curve following a single dose.

Problem 9.3:
Calculate the dose, X_0, which upon multiple dosing (q 8h) will yield an average steady state concentration of 10 μg/ml in a 60 kg patient. The drug has volume of distribution of 0.25 liters/kg. The bioavailability is 45 percent and the half-life is 10 hours.

$$X_0 = \frac{10\ (\mu g/ml) \times (0.693/10) \times 8 \times 0.25\ (liters/kg) \times 60\ (kg) \times 1000\ (ml)}{0.45}$$

$$= 184.8 \text{ mg}$$

Thus a 184.8 mg dose given q 8h will yield a plasma concentration of 10 μg/ml on the average. Note that if the dosing interval is changed, say to q 4h, the dose will decrease by one-half or in direct proportion to the dosing interval. Also notice that in the above equation the dose is proportional to the elimination rate constant ($0.693/t_{0.5}$). Any change in the elimination rate constant will require an adjustment of dosage. However, the changes in the elimination rate constants of drugs are difficult to estimate from the data that is generally available in a patient history. Additional tests that may be required to assess the elimination function include indirect measurements through creatinine, inulin, or bromosulfonphthalein clearance, or through other specific tests, depending on the nature of the elimination routes. The most desirable test, however, is direct estimation of elimination rate constants, which will be discussed later.

In the case of renal impairment, dosage adjustment can be made either by changing the dosing interval or by changing the dose and keeping the interval the same:

$$X_{0r} = \frac{K_r X_0}{K} \quad \text{(Eq. 9.2)}$$

$$\tau_r = \frac{K\,\tau}{K_r} \quad \text{(Eq. 9.3)}$$

where τ_r, X_{0r} and K_r are the parameters in renal malfunction. An estimate of K_r is generally made from either the serum creatinine concentration or by measuring creatinine clearance. Creatinine is an end-product of muscular metabolism and is eliminated primarily through the kidneys, mainly by glomerular filtration. The normal range of the serum creatinine concentration is between 1 and 2 mg%. Since the creatinine levels in the body are dependent on the muscle mass, normal values are generally based on the body surface area and are often used after normalization to a body surface area of 1.73 m^2. Since the renal function decreases with age, when the muscular metabolism is also slowed down, an appropriate correction for age must be made. The effect of gender is also important due to anatomic differences. These considerations are necessary since in many instances the serum creatinine concentration is converted to creatinine clearance by a standard nomogram such as the one shown in Figure 9.1.[6]

The use of the creatinine clearance to determine the kidney function is based more on convenience than on anything else. Not only is creatinine

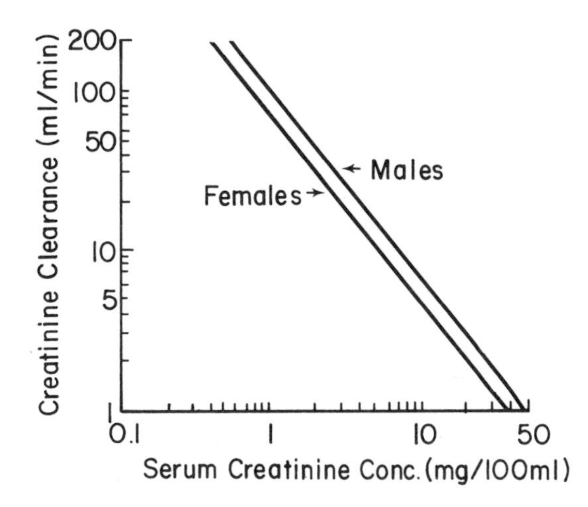

Figure 9.1. Relationship between creatinine clearance and concentration. For females: Q_c = antilog (1.8883 − 1.20 log C_c). For males: Q_c = antilog (2.0080 − 1.19 log C_c). (From Wagner: Fundamentals of Clinical Pharmacokinetics, 1975. Courtesy of Drug Intelligence Publications)

filtered but it also undergoes active secretion in the tubules, and therefore its clearance relates only indirectly to the glomerular function. Inulin is a compound of choice for glomerular filtration studies, but its tedious analytic process along with the need to administer this compound to patients makes it a less desirable choice. Another problem with creatinine clearance estimations is the variability in analysis, depending on the method used. Most methods analyze apparent creatinine, which includes other chromogens which appear in the urine in very small amounts. The result is that the plasma concentration of creatinine is overestimated, resulting in lower clearance values since the clearance is the ratio of the excretion rate and the plasma concentration. True creatinine measurements can also be made by adsorbing onto Lloyd's reagents or by using high pressure liquid chromatography techniques.[7] Values obtained with true creatinine always give a higher clearance (compared to inulin) than when the total chromogens are analyzed.

Although there are good correlations between the plasma creatinine level and its renal clearance, direct measurement of creatinine clearance is still a desirable procedure for several reasons. First, there is significant diurnal variation in the plasma concentration of creatinine. Second, the relationships obtained between the clearance and the plasma concentration are often nonlinear due to population variations. Creatinine clearance can be measured by collecting the urine for a period of time, preferably for 24 hours to eliminate diurnal variation, and calculating the excretion rate. The plasma concentration is determined simultaneously to calculate the clearance.

The relationship between creatinine clearance and the disposition half-lives of drugs is derived on the basis that the excretion rate constant, k_e, is a function of creatinine clearance or the glomerular filtration rate:

$$k_e \propto Q_c \quad \text{(Eq. 9.4)}$$

$$k_e = aQ_c \quad \text{(Eq. 9.5)}$$

where Q_c is the creatinine clearance and "a" is a proportionality constant which can be obtained from a large number of data relating the excretion constant to Q_c.

The half-life of a drug is represented in terms of excretion and other rate constants:

$$K = k_e + k' \qquad \text{(Eq. 9.6)}$$

where k' is the sum of all nonrenal excretion constants, including k_b for biotransformation, which in most cases accounts for a significant fraction of k'.

Substituting Equation 9.5:

$$K = aQ_c + k' \qquad \text{(Eq. 9.7)}$$

Wagner uses the term K%, which is simply 100K, to express Equation 9.7 in order to convert the numbers into manageable dimensions.[6] A large number of drugs have been studied to fit Equation 9.7, and Table 9.2 lists some of these. The dose calculation from this table is made in accordance with the following relationship:

$$\text{Patient dose} = \text{Normal dose} \times \text{Patient K\%/Normal K\%} \qquad \text{(Eq. 9.8)}$$

Problem 9.4:

Given the creatinine clearance of 50 ml/min, calculate the patient dose in terms of percent of normal dose for digoxin, ampicillin, and gentamicin.

From Table 9.2:

	Normal K%:	*Patient K%:*	*Patient dose as % of Normal Dose:*
Digoxin	1.7/hr	0.8 + 0.009 × 50 = 1.25/hr	(1.25/1.7)100 = 73.5%
Ampicillin	70/hr	11 + 0.59 × 50 = 40.5/hr	(40.5/70)100 = 57.9%
Gentamicin	30/hr	2 + 0.28 × 50 = 16.0/hr	(16/30)100 = 53.3%

A large number of studies have reported relationships between the disposition constants and the creatinine serum concentration or its clearance in various modified forms of the preceding equations:

$$t_{0.5} = \frac{0.693}{aQ_c + k'} \qquad \text{(Eq. 9.9)}$$

$$\frac{1}{t_{0.5}} = \frac{ak_o}{0.693\, C_c} + \frac{k'}{0.693}$$

where k_o is the rate of creatinine production:

$$k_o = C_c Q_c \qquad \text{(Eq. 9.10)}$$

Table 9.2. DOSE ADJUSTMENT IN PATIENTS WITH RENAL IMPAIRMENT

DRUG	PATIENT K% ($100K = 100aQ_c + 100k'$)*		NORMAL K%
	($100k'$)	($100a$)	
α-Acetyldigoxin	1	0.02	3
Ampicillin	11	0.59	70
Carbenicillin	6	0.54	60
Cephalexin	3	0.67	70
Cephaloridine	3	0.37	40
Cephalothin	6	1.34	140
Chloramphenicol	20	0.10	30
Chlortetracycline	8	0.04	12
Ciba 36278	3	0.67	70
Colistin	8	0.23	31
Digitoxin	0.3	0.001	0.4
Digoxin	0.8	0.009	1.7
Doxycycline	3	0	3
Erythromycin	13	0.37	50
5-Flucytoxine	0.7	0.243	25
Gentamicin	2	0.28	30
Isoniazid (fast)	34	0.19	53
Isoniazid (slow)	12	0.11	23
Kanamycin	1	0.24	25
Lincomycin	6	0.09	15
Methicillin	17	1.23	140
Methyldigoxin	0.7	0.009	1.6
Oxacillin	35	1.05	140
Penicillin G	3	1.37	140
Polymyxin B	2	0.14	16
Rolitetracycline	2	0.04	6
Streptomycin	1	0.26	27
Strophanthin G	1.2	0.038	5
Sulfadiazine	3	0.05	8
Sulfamethoxazole	7	0	7
Sulfasomidine (children)	1	0.14	15
Tetracycline	0.8	0.072	8
Thiamphenicol	2	0.24	26
Trimethoprim	2	0.04	6
Vancomycin	0.3	0.117	12

* Q_c in ml/min. See text for details.
Adapted from Wagner: Fundamentals of Clinical Pharmacokinetics, 1975. Courtesy of Drug Intelligence Publications.

The equation above assumes that a steady state has been reached in the production of creatinine, which may not be always true, as a result, for instance, of excercise and diurnal variation.

The use of creatinine measurements in dosage adjustment should be made with caution for drugs which exhibit multicompartment pharmacokinetics.

The term k_{10} is comparable to k_e in single compartment systems if the elimination takes place from the central compartment. Since the use of creatinine measure accounts only for k_{10}, it may not reflect real change in the biologic half-life, which is given by β in two compartment models. The reason for this lack of proportionality is the fact that the half-life in multicompartment models is a result of both elimination and distribution, whereas this is only a function of elimination in single compartment models.

The use of creatinine measurements is also complicated by the changes in the volumes of distribution of drugs in disease states (as discussed in earlier chapters). An increase in the volumes of distribution can increase the half-life of a drug, which may not be apparent from creatinine measurements. Further caution is needed with drugs that may be extensively removed by renal processes other than glomerular filtration, such as tubular secretion, whereby correlations may be quite difficult to establish due to the nonlinear nature of these processes.

Creatinine measurements, despite their drawbacks, offer an excellent method of dosage adjustment where their utility has been clearly demonstrated. In many cases specific nomograms have been drawn to take into account a multitude of factors pertinent to a specific drug. Figure 9.2 shows a nomogram for kanamycin which provides good estimates of dosage.

These specific nomograms are necessary for accurate conversion of serum creatinine levels to clearance values. Several studies have attempted to relate these parameters by an empirical and sometimes extensive mathematical approach. For example, age and body weight are considered:[9]

$$Q_c\,(\text{ml/min}) = \frac{(140 - \text{age})\,(\text{wt})}{72C_c} \qquad \text{(Eq. 9.11)}$$

where weight is expressed in kg, and C_c in mg%. The values for females are 0.85 times those for males. Another approach utilizes the creatinine production rates, the volume of distribution, and the lean body weight:[10]

$$Q_c\,(\text{ml/min}) = \frac{[P_c - V(C_2 - C_1)]\,100}{1440\,(C_1 - C_2)/2} \qquad \text{(Eq. 9.12)}$$

where P_c is the daily corrected creatinine production (mg/day):

$$P_c = P\,[1.037 - 0.0338\,(C_1 + C_2)/2] \qquad \text{(Eq. 9.13)}$$

$$P = LBW\,(29.3 - 0.203\ \text{age})\ \text{for males} \qquad \text{(Eq. 9.14)}$$

$$P = LBW\,(25.3 - 0.175\ \text{age})\ \text{for females} \qquad \text{(Eq. 9.15)}$$

where P is the uncorrected creatinine production (mg/day), LBW is the lean body weight in kg, and age is given in years. The volume of distribution of creatinine is generally taken to be 0.4 liters/kg. For calculation purposes it is expressed in multiples of 100 ml.

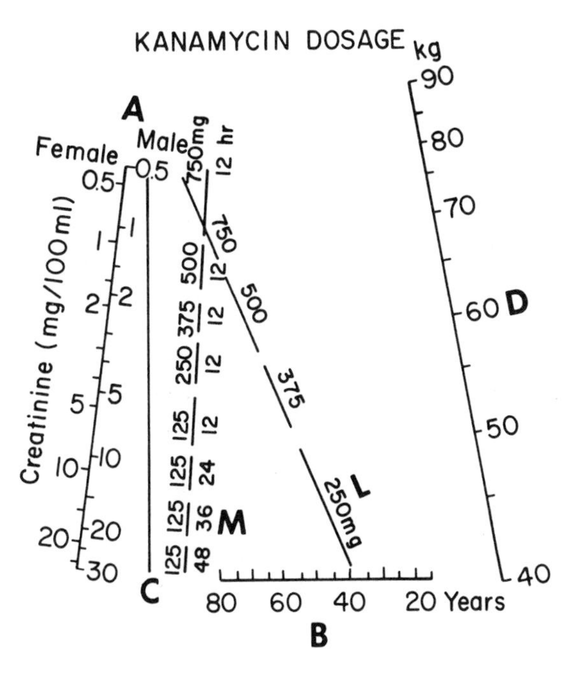

Figure 9.2. Nomogram to calculate kanamycin dosage schedule.

A. Patient not receiving dialysis treatment.
1. Join with a straight line the serum creatinine concentration appropriate to the sex on scale A and the age on scale B. Mark the point at which the straight line cuts line C.
2. Join with a straight line the mark on line C and the body weight on scale D. Mark the points at which this line cuts the dosage lines L and M.
3. The loading dose (mg) is written against the marked part of line L. The maintenance dose (mg) and the appropriate interval (hours) between doses are written against the marked part of line M.
4. The nomogram is designed to give serum concentrations of kanamycin within the range 10–30 μg/ml 2 hours after each dose. In patients with renal insufficiency it is still desirable to perform check assays and to make appropriate dose adjustment.

EXAMPLE: Male, serum creatinine 5.0 μg/100 ml, 45 years, 55 kg; loading dose 375 mg, maintenance dose 125 mg, interval between doses 12 hours.

B. Patient receiving dialysis treatment.
5. When the patient is severely oliguric or anuric do not use the serum creatinine and age scales. To determine the dose schedule join with a straight line the bottom end of line C and the body weight on scale D. Then proceed as in 2 and 3 above.
6. Peritoneal dialysis. In addition to the dose schedule add kanamycin to the dialysis fluid. A concentration of 20 μg/ml is suitable.
7. Hemodialysis. In addition to the dose schedule give a booster dose after dialysis. Half the loading dose is suitable after a 10 hour Kiil dialysis.

(From Mawer et al.: Lancet 2:45, 1972)

The lean body weight is:[11]

$$LBW = (1 - \text{fat fraction}) \text{ body weight} \quad \text{(Eq. 9.16)}$$

$$\text{fat fraction} = [90 - 2(\text{height} - \text{girth})]/100 \quad \text{(Eq. 9.17)}$$

where height and girth are given in inches and girth is the measurement at the umbilical level at expiration. These formulae are given here to demonstrate the degree of complexity in converting serum creatinine levels to clearance values. However, in those instances where urine output is decreased or where it is inconvenient to collect sufficient urine, these approaches become very valuable.

PLASMA LEVELS AS DETERMINANTS OF DOSAGE LEVEL

The ultimate goal of all the dosage adjustment methods discussed previously is to achieve a desired plasma concentration. Whereas indirect estimates of the plasma concentration are often made from disposition and distribution functions, along with the measurement of therapeutic response, the only direct method is to analyze the plasma concentrations of drugs in patients undergoing therapy.

The measurement of plasma concentrations becomes a useful guide only when a therapeutically effective range of concentration can be defined. Currently, few drugs can be listed with a definite concentration range for treating an ailment (Table 9.3). Generally, when a concentration range is established

Table 9.3. GENERAL BLOOD CONCENTRATION (mg/liter) GUIDE TO THERAPEUTICS

DRUG	*THERAPEUTIC CONCENTRATION*	*TOXIC CONCENTRATION*	*LETHAL CONCENTRATION*
Acetaminophen	10–20	400	1500
Acetazolamide	10–15	—	—
Acetohexamide	21–56	—	—
Aminophylline	10–20	—	—
Amitriptyline	0.05–0.20	0.4	10–20
Amphetamine	0.02–0.03	—	2
Barbiturates			
Short-acting	1	7	10
Intermediate-acting	1–5	10–30	30
Long-acting	~10	40–60	80–150
Bromide	50	500–1500	2000
Carbamazepine	2	8–10	—
Chloral hydrate	10	100	250
Chlordiazepoxide	1–3	5.5	20
Chlorpromazine	0.5	1–2	3–12
Chlorpropamide	30–140	—	—
Chlorprothixine	0.04–0.3	—	—
Codeine	0.025	—	—
Desipramine	0.59–1.4	—	10–20
Dextropropoxyphene	0.05–0.2	5–10	57
Diazepam	0.5–2.5	5–20	>50
Digitoxin	0.02–0.035	—	0.32
Digoxin	0.0006–0.0013	0.002–0.009	—

Table 9.3. *(CONTINUED)*

DRUG	*THERAPEUTIC CONCENTRATION*	*TOXIC CONCENTRATION*	*LETHAL CONCENTRATION*
Diphenhydramine	5	10	
Ethinamate	5–10	—	—
Ethchlorvynol	5	20	150
Ethosuximide	25–75	—	—
Glutethimide	0.2	10–80	30–100
Imipramine	0.05–0.16	0.7	2
Lidocaine	2	6	—
Lithium	4.2–8.3	13.9	13.9–34.7
Meperidine	0.6–0.65	5	30
Meprobamate	10	100	200
Methadone	0.48–0.86	2	>4
Methapyrilene	0.002	30–50	>50
Methaqualone	5	10–30	>30
Methyprylon	10	30–60	100
Morphine	0.1	—	0.05–4
Nitrofurantoin	1.8	—	—
Nortriptyline	0.0012–0.0016	5	13
Papaverine	1	—	—
Paraldehyde	50	200–400	500
Pentazocine	0.14–0.16	2–5	10–20
Phensuximide	10–19	—	—
Phenylbutazone	100	—	—
Phenytoin	5–22	50	100
Primodone	10	50–80	100
Probenecid	100–200	—	—
Procainamide	6	10	—
Propoxyphene	0.05–0.2	5–20	57
Propranolol	0.025–0.2	—	8–12
Quinidine	3–6	10	30–50
Salicylate	20–100	150–300	500
Sulfonamides	30–150	—	—
Theophylline	20–100	—	—
Thioridazine	1–1.5	10	20–80
Trimethobenzamide	1–2	—	—
Warfarin	1–10	—	—
Zoxazolamine	3–13	—	—

After Winek: Clin Chem 22:832, 1976.

it is a narrow range for most potent drugs, with increasing toxic effects expected above this range. However, interindividual differences also apply here, whereby a toxic concentration in one individual may be an ineffective concentration in another. The therapeutic index, which is the ratio between the effective plasma concentration and the toxic concentration, generally varies from as little as one, where the effective concentration is also producing toxic symptoms, to several thousand, where the two are far apart. For example, the index for nortriptyline is about 3500, whereas for procainamide it is only

1.5. Whereas the drugs with low indices (Table 9.4) require greater care in their dosing, the main determinant is the nature of the toxic response that is obtained. For example, the therapeutic index of digoxin is generally higher than that of aspirin, but the toxic symptoms of the former require much more care in dosing than is needed with aspirin. The complicating factor in the utility of blood or plasma levels of drugs remains the uncertainty associated with the concentration–effect relationship among individuals. Whereas the pharmacokinetic basis of this variability can be deciphered from blood level studies, inherent differences in the response to a given concentration remain in the domain of the clinician, where the dose can be determined from an apparent response.

The measurement of plasma drug concentrations as a guide to dosage adjustment is especially beneficial when a drug fails to produce a response or results in an unanticipated toxic symptom, especially in patients with renal, hepatic, cardiovascular, or gastrointestinal disease. A full exploitation of the plasma levels can be made if such additional information is known as the binding of the drug to red blood cells, plasma proteins, and tissue proteins and the possible interactions in these bindings. The binding of the drug to blood constituents is also important in the calculation of the free drug concentration, which equilibrates with the body tissues. If the free drug concentration is a more or less constant fraction of the total blood concentration then the total drug levels will be indicative of the pharmacologic effect. However, significant interpatient variability in the binding of drugs has been reported,

Table 9.4. THERAPEUTIC INDICES OF VARIOUS DRUGS

INDEX GREATER THAN 10:	*INDEX BETWEEN 5 AND 10:*
Acetaminophen	Barbiturates
Bromide	Diazepam
Chloral hydrate	Digoxin
Dextropropoxyphene	Imipramine
Glutethimide	Meperidine
Meprobamate	Paraldehyde
Methapyrilene	Primodone
Nortriptyline	Thioridazine
Pentazocine	
Propoxyphene	
INDEX LESS THAN 5:	
Amitriptyline	Methaqualone
Carbamazepine	Methyprylon
Chlordiazepoxide	Phenytoin
Diphenhydramine	Procainamide
Ethchlorvynol	Quinidine
Lidocaine	Salicylates
Methadone	

some reasons for which were discussed in an earlier chapter. A good example of interindividual differences in the binding is that of nortriptyline, with which an almost 100 percent variation is quite common, regardless of the level achieved.[13] Phenytoin on the other hand shows a decrease in binding in such disease states as azotemia and uremia, where the seizures can be controlled at a much lower dosage level than is otherwise necessary.[14] There is, therefore, a need to utilize the free plasma concentration instead of the total concentration in dosage adjustments. In recent studies by the Levy group this aspect has been emphasized repeatedly.[15]

The drugs for which the action is not directly proportional to the plasma concentration can also be monitored by using the plasma concentration as a therapeutic guide, provided the underlying mechanism of action is known. For example, warfarin does not show a direct correlation between its log of dose and the observed effect (Fig. 9.3). Warfarin acts by inhibiting the synthesis of vitamin K-dependent clotting factors II, VII, IX, and X, which in turn decreases the prothrombin complex activity as reflected in the blood clotting time. Therefore, a certain lag time is involved in the peaking of drug activity to allow for the depletion of the remaining clotting factors. A direct relationship beween the warfarin concentration and its activity cannot therefore be expected. However, since the inhibition of clotting factor synthesis is directly affected by warfarin, a relationship between the synthesis rate of prothrombin complex activity and the plasma warfarin concentration has been demonstrated[16] (Fig. 9.4). Unfortunately, a clear understanding of such mechanisms of drug action is rare, making the use of plasma concentrations as therapeutic guides less useful for many potent compounds.

Another category of drugs involves the mediation of action through an active metabolite. In such cases a direct correlation between the pharmacol-

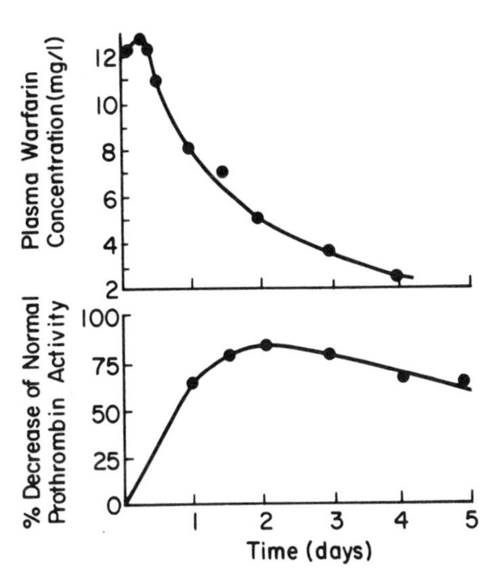

Figure 9.3. Plasma-warfarin concentration and depression of prothrombin complex activity as a function of time after oral administration of 1.5 mg warfarin sodium per kg body weight. Average of five normal subjects. (From Nagashima et al.: Clin Pharmacol Ther 10:22, 1969)

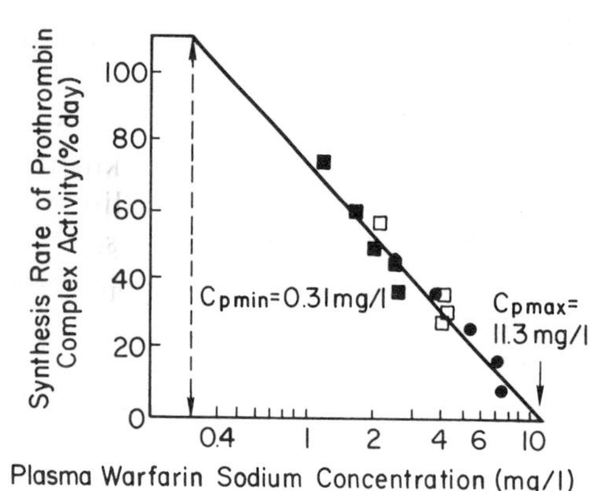

Figure 9.4. Relationship between the synthesis rate of prothrombin complex activity and the plasma warfarin concentration in six normal subjects after a single oral dose of 1.5 mg/kg (closed circles); 10 mg daily for 5 days (closed squares); and 15 mg daily for 4 days (open squares). (From Nagashima et al.: Clin Pharmacol Ther 10:22, 1969)

ogic response and the plasma concentration of the parent drug cannot be expected if the disposition property of the active metabolite is not proportional to that of the parent drug. For example, the active metabolite may be released more slowly than the parent drug, as with methyldopa, or the active metabolite may be formed during the first pass through the liver. The best example of the latter occurrence is that of propranolol, which shows a linear relationship between the log of concentration and the effect (exercise tachycardia).[17] The same effect is, however, much more pronounced when the drug is given orally, due to the transformation of propranolol to its active 4-hydroxy metabolite, which has a significant beta adrenergic blocking effect. The 4-hydroxy metabolite disappears rapidly from the plasma and thus within a few hours after oral administration the activity of propranolol appears to be equivalent to that obtained from intravenous administration.[18] It should be noted that since the first pass effect involving the transformation of propranolol to its hydroxy metabolite is a saturable process, significant interindividual and dose-dependent effects can be anticipated.

The dosage regimen of some drugs is currently determined with the plasma level as an index. This approach has been strongly suggested for a variety of other drugs as a sound alternative to conventional approaches. The following discussion of specific drugs alludes to the specific nature of the properties of and the approaches for some of these drugs.

CARDIOTONIC DRUGS

Digoxin

Digoxin is a cardiac glycoside which is used for congestive heart failure, atrial fibrillation, paroxysmal tachycardia, atrial flutter, and other diseases related to the heart. Digoxin is about 75 to 85 percent absorbed upon oral administration (a recent formulation using soft gelatin capsule reported almost 100 percent absorption); it is little metabolized in the body and is excreted in the kidneys primarily unchanged.

The therapeutic and toxic levels of digoxin lie within a narrow range and, since a large degree of variation in the absorption and disposition kinetics of digoxin is possible among individuals, a great degree of variability in the pharmacologic response to digoxin has been noted. It is a well-established fact that the serum concentration of digoxin can be directly related to its therapeutic and toxic effects. Most patients exhibit a satisfactory inotropic effect at the steady-state levels of digoxin between 0.7 and 1.5 ng/ml.[19] Development of cardiac arrhythmias is most likely to occur at concentrations above 3 ng/ml. (Fig. 9.5).

The plasma concentration monitoring of digoxin is useful in many situations, as discussed below. These conditions can be extended to other drugs as well:[20]

1. When the history of cardiac glycoside ingestion is inadequate or unavailable: A clinician can easily determine the loading and maintenance dose based on the existing plasma levels of the drug, if any.
2. When noncompliance is suspected: The problem of noncompliance is one of the most serious considerations in therapeutic management with potent drugs. It is also very difficult to detect. However, with routine plasma concentration monitoring this problem is easily solved.
3. When gastrointestinal problems may affect the absorption of orally administered digoxin: It has often been demonstrated that a variety of disease states can affect the absorption of digoxin. It is very difficult to predict these effects in individual patients and plasma concentration monitoring provides the only method of accounting for possible differences in the degree of absorption.

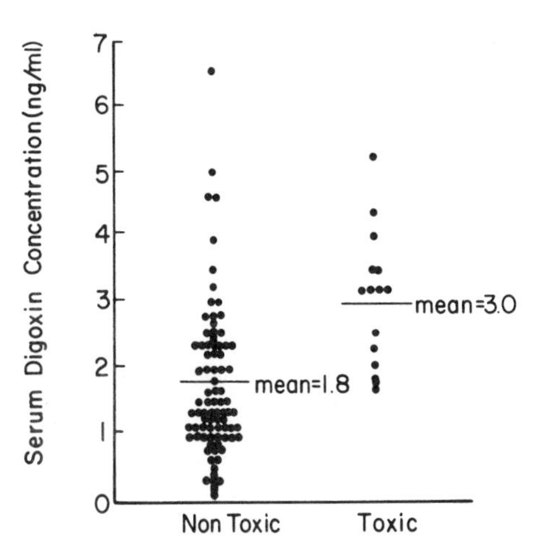

Figure 9.5. Results of 100 serum digoxin radioimmunoassay measurements. Sixteen patients were believed to be clinically toxic; their mean serum was 3.0 ng/ml. Seven patients thought to be nontoxic also had serum levels of more than 3.0 ng/ml. Overlap of normal and toxic values does occur; therefore judgment must be used when evaluating results. (From Doherty: Ann Intern Med 79:229, 1973)

4. When concomitant administration of other drugs will interfere with the absorption: Several drugs, such as antacids, cholestyramine, and neomycin have been seen to interfere with digoxin absorption.[21–23]
5. When the bioavailability of a particular digoxin product may be a source of variation in the therapeutic response: The dosage adjustment principles discussed above included primarily the disposition function considerations, since the absorption aspects are difficult to predict and quantitate. The measurement of the plasma concentration provides a direct estimate of the bioavailability as well as of the disposition function parameters.
6. When prophylactic doses of digoxin are given before surgery: Since there is no true endpoint for such therapy, the achievement of a specific plasma concentration may be useful as a guideline for such prophylactic therapy.
7. When the clinical response to digoxin is questionable: If adequate plasma levels are achieved without the control of such cardiac symptoms as atrial fibrillation, the etiology of symptoms may be questioned. On the other hand, if the concentration is low, the first aspect that should be questioned is the absorption and/or disposition characteristics of the drug.

The drawbacks in the use of the serum concentration of digoxin to monitor drug therapy include the lack of a simple but specific assay method. The radioimmunoassay is neither very specific nor reproducible. Another drawback is the wide variation in the volumes of distribution of digoxin among individuals, especially in the disease states, such as renal malfunction. The therapeutic range discussed previously generally provides an adequate safety margin, but in some patients concentrations above 3.0 ng/ml may not prove toxic and in neonates and infants such a range may not apply at all. Few studies have addressed this aspect. For example, comparative studies show that adults with concentrations above 3.0 ng/ml show toxic symptoms, but the mean value of 3.45 ng/ml in neonates is not associated with any toxic symptoms at all.[24]

A variety of equations has been proposed to calculate the dosage regimen of digoxin. These equations are based on the predicted or actual creatinine clearance or on other related parameters, such as the blood urea nitrogen (BUN). For example, rapid and slow digitalization can be achieved on the basis of the blood urea nitrogen:[25]

For rapid digitalization:

$$\text{Total body dose} = \text{lean body weight (lb)} \times 0.0075\ \text{mg/lb} \qquad \text{(Eq. 9.18)}$$

The total body dose is given in three divided doses at six-hour intervals. Rapid digitalization should be avoided if previous digitalis intake is unknown and in patients treated with digitalis in the preceding week. The total body dose is approximate and may be reduced in the presence of hypokalemia, hyponatremia, hypercalcemia, or hypothyroidism.

For slow digitalization or maintenance dose the total body dose is calculated by Equation 9.18 and the daily maintenance dose is calculated according to the following measure of renal function:

BUN	*Daily Maintenance Dose (% of total body dose)*
<20	33.3
30	30.0
40	27.0
50	24.0
60	21.0
70	18.0
>80	15.0

More elaborate approaches include consideration of various parameters, including sex, age, body weight, lean body weight, serum creatinine on two successive days, the elimination rate constant, and the volume of distribution.[26] The serum creatinine is converted to creatinine clearance by using Equation 9.12, for which the lean body weight can be calculated as:

$$LBW = 50 + 2.3\,[\text{height (in)} - 60] \text{ for males} \qquad \text{(Eq. 9.19)}$$

$$LBW = 45.5 + 2.3\,[\text{height (in)} - 60] \text{ for females} \qquad \text{(Eq. 9.20)}$$

If the *LBW* is greater than the body weight then the actual body weight can be used in place of the *LBW*. If, however, the body girth is available, the *LBW* can be calculated using Equation 9.16. The elimination rate constant is dependent on the Q_c:

$$K = (\text{normal } K)\left[\left(\frac{Q_c}{Q_c \text{ normal}} - 1\right)F_{el} + 1\right] \qquad \text{(Eq. 9.21)}$$

where F_{el} is the fraction excreted, usually taken to be 0.76; normal $Q_c = 120$ ml/min; and normal $K = 0.0169\ \text{hr}^{-1}$. The volume of distribution is calculated as:[27]

$$V_d(L) = V_A + \frac{V_N Q_c}{K_D + Q_c} \qquad \text{(Eq. 9.22)}$$

where V_A = minimum digoxin volume of distribution in severe renal dysfunction = 226 liters /1.73 m² = 130.63 liters/m²; V_N and K_D are Michaelis-Menton constants; V_N = 298 liters/1.73 m² = 172.25 liters/m²; and K_D = (29 ml/min)/1.73 m² = (16.76 ml/min)/m².

The dose is finally calculated as:

$$\text{Dose} = \frac{\bar{C}\, V_d K \tau}{F} \qquad \text{(Eq. 9.23)}$$

where F is the fraction absorbed, usually equal to 0.8.

Problem 9.5:
Calculate the dose of digoxin in a 69-year-old male weighing 59 kg, with a height of 70 inches and with a creatinine plasma concentration of 6.0 and 6.2 mg% on successive days. The desired dosing interval is 24 hours and the desired concentration is 2.0 ng/ml.

$$LBW = 50 + 2.3\,(70 - 60)$$
$$= 73 \text{ kg; thus use } LBW = 59 \text{ kg} \qquad \text{(see Eq. 9.19)}$$

$$P = 59(29.3 - 0.203 \times 69)$$
$$= 902.29 \text{ mg/day} \qquad \text{(see Eq. 9.14)}$$

$$P_c = 902.29[1.037 - 0.0338(6.0 + 6.2)/2]$$
$$= 749.64 \text{ mg/day} \qquad \text{(see Eq. 9.13)}$$

$$Q_c\,(\text{ml/min}) = \frac{[749.64 - (4 \times 59)(6.2 - 6.0)]100}{1440 \times (6.0 + 6.2)/2}$$
$$= 9.07 \text{ ml/min} \qquad \text{(see Eq. 9.12)}$$

$$Q_c\,(\text{ml/min}) = \frac{(140 - 69)59}{72 \times 6.1}$$
$$= 9.54 \text{ ml/min} \qquad \text{(see Eq. 9.11)}$$

Note that the use of the simpler form of Equation 9.11 gives a fairly good estimate of renal function when compared with that obtained by using creatinine production and other body parameters.

The body surface area is determined by:[28]

$$S = W^{0.425} \times H^{0.725} \times 71.84 \qquad \text{(Eq. 9.24)}$$

where S = body surface area in cm²; W = weight in kg; and H = height in cm.

$$S = 59^{0.425} + 178^{0.725} \times 71.84$$
$$= 17399.34 \text{ cm}^2 = 1.74 \text{ m}^2$$

$$V_A = 130.63 \times 1.74 = 227.3 \text{ liters}$$

$$V_N = 172.25 \times 1.74 = 299.7 \text{ liters}$$

$$K_D = 0.01676 \times 1.74 = 0.02916 \text{ liters/min}$$

$$V_d\,(\text{liters}) = 227.3 + \frac{299.7 \times 0.12}{0.02916 + 0.12}$$
$$= 468.41 \text{ liters} \qquad \text{(see Eq. 9.22)}$$

$$K = 0.0169\left[\left(\frac{9.07}{120} - 1\right)0.76 + 1\right]$$
$$= 0.0050268 \text{ hr}^{-1} \qquad \text{(see Eq. 9.21)}$$

$$\text{Dose (mg)} = \frac{2.0 \text{ ng/ml} \times 468.41 \text{ liters} \times 0.0050268 \text{ hr}^{-1} \times 24 \text{ hr}}{0.8}$$
$$= 0.14 \text{ mg} \qquad \text{(see Eq. 9.23)}$$

A large number of equations have been proposed to calculate the dosage regimen of digoxin based on the renal function and other parameters given. It is not within the scope of this book to discuss all of these approaches. However, the calculation given here is the most complicated one and gives a fairly good idea of the various factors that might affect the choice of loading and maintenance doses. In most instances the loading dose will be equal to the product of the average concentration desired and the volume of distribution of the drug.

Digitoxin

Digitoxin is not used as frequently as digoxin. It is almost 100 percent absorbed upon oral administration.[29] Digitoxin is more strongly and extensively bound than digoxin (95 percent binding compared with about 23 percent for digoxin). It is metabolized primarily in the liver (about 60 percent), converting to a major inactive metabolite and to digoxin (about 8 percent). It also undergoes significant enterohepatic recycling (about 26 percent compared to 7 percent for digoxin). About 30 percent of the drug is excreted unchanged and the disposition half-life is about 5.7 days, compared to 1.6 days for digoxin. The volume of distribution of digitoxin is approximately 0.6 liters/kg and varies significantly between individuals.

The therapeutic range of levels is between 8 and 38 ng/ml and toxic symptoms appear at levels above 35 to 40 ng/ml.[30] Most of the discussion for digoxin regarding the rationale for plasma concentration monitoring also applies to digitoxin, except that there is much less likelihood of intersubject variation in the absorption of digitoxin.

Digitalization in adults involves a recommended dose of 0.6 mg followed by 0.4 mg and then 0.2 mg at 4- to 6-hour intervals. The maintenance dose is generally 10 percent of the loading dose, since the K for digitoxin is approximately 0.1 day^{-1}. In children, however, digitalization must be controlled by individual titration of the effect and/or plasma level monitoring.

ANTIARRHYTHMIC DRUGS

Lidocaine, propranolol, procainamide, and quinidine are the most commonly used antiarrhythmic drugs. These drugs are used to prevent or terminate paroxysmal ventricular tachycardia, except where it complicates AV block. They are also indicated, in place of digoxin, for the prevention or termination of paroxysmal atrial tachycardia. Quinidine is often used to terminate atrial fibrillation after mitral valvulectomy and is less effective in terminating chronic atrial fibrillation in the presence of mitral insufficiency or where the underlying cardiac disease cannot be removed. Lidocaine is often used as an alternate to quinidine in paroxysmal ventricular or atrial tachycardia.

Most antiarrhythmic drugs are potent in nature and have narrow therapeutic indices. Coupled with large variations in their absorption and subsequent disposition, they require monitoring of plasma concentrations for their safe and effective use. An additional problem in the use of lidocaine, propranolol, and procainamide is due to the activity of their metabolites, requiring monitoring of both the parent drug and the metabolites to predict a therapeutic response or toxic reaction.

Lidocaine

Plasma concentrations of lidocaine decline with a biexponential decay. An early decay with a mean half-life of about 8.3 minutes correlates well with the rapid decline of its pharmacologic activity after an intravenous bolus dose. However, the terminal phase of disposition has a half-life of about 107 minutes which determines the elimination of lidocaine from the body.[31] Its volume of distribution is approximately 1.6 liters/kg and its total body clearance is about 10 ml/kg/min.

Plasma levels of 1.2 to 6.0 μg/ml effectively suppress ventricular arrhythmias.[32] Concentrations below 1.2 μg/ml are ineffective and above 6 μg/ml adverse effects quickly appear, especially at the central nervous system (CNS) level.[32] There is some indication that the metabolites of lidocaine, such as its monodeethylated product, may be responsible for its CNS toxicity. In patients showing excessive toxicity an altered pattern of lidocaine metabolism may exist.[33,34] Patients with heart failure exhibit higher plasma concentrations of lidocaine than normal subjects and patients with liver disease show lower clearance and higher volumes of distribution.[33]

The liver is the major site of lidocaine elimination and therefore a dose adjustment is needed in patients with hepatic diseases. In anephric or oliguric patients there is no need to adjust lidocaine doses. Even though the metabolites of lidocaine may accumulate in renal failure, this seems to have little effect on the therapeutic management.

The short pharmacologic half-life of lidocaine requires frequent administration. Generally, a rapid bolus dose of 0.5 mg/kg provides a plasma concentration of approximately 1 μg/ml. The therapy is begun with a 1 mg/kg dose to provide a 2 μg/ml concentration. If arrhythmias are not suppressed at this dose level an additional dose may be needed. If these initial doses provide a concentration of 5 μg/ml and arrhythmias are still not controlled, then the patient might be refractory to the drug. Once a control of arrhythmias is achieved, constant intravenous infusion is begun to maintain the therapeutic level. The infusion rate is determined by:[34]

$$\text{Infusion rate} = \text{Plateau concentration} \times \text{Clearance} \qquad \text{(Eq. 9.25)}$$

Since the clearance of lidocaine is approximately 10 ml/kg/min, the infusion rate for a 3 μg/ml plateau concentration is 30 μg/kg/min.

If the status of the patient changes during therapy, necessitating higher levels, a bolus dose may be given during infusion and the rate of infusion increased correspondingly. It should be noted that it will take a long time to reach steady state with infusion alone. Once patients start to improve, the infusion rates are tapered off in 10 μg/kg/min steps. This is necessary to identifying a therapeutic level if arrhythmias appear during the tapering-off. If arrhythmias appear during tapering, a 0.5 mg/kg dose is administered immediately, followed by appropriate changes in the infusion rate. This approach also determines whether the plateau maintained was much above the therapeutic range or just near it.

Problem 9.6:

A 59 kg male required two 0.5 mg/kg doses following a single 1 mg/kg dose of lidocaine to control arrhythmias. The level was then maintained by continuous intravenous infusion. After the patient improved, tapering off by 10 μg/kg/min resulted in arrhythmias. What is the initial infusion rate and how would you bring the patient back to therapeutic level?

The patient needed a total of 2 mg/kg × 59 = 118 mg to control arrhythmias. An infusion rate of 1.18 mg/min would control arrhythmias by maintaining the plateau level of approximately 2 μg/ml. Tapering off the infusion in 10 μg/kg/min steps results in a plasma concentration of 1 μg/ml, with signs of arrhythmias. A bolus dose of 0.5 mg/kg will bring the concentration back to 2 μg/ml and it will require restoring the infusion rate to its initial rate. The level reached initially was the desired effective level.

Plasma concentration monitoring of lidocaine in cardiac arrhythmias is not very useful because of the rapidity of response generated by the drug. Whereas it is highly beneficial to correlate the response to a given plasma concentration, the lag time involved in analyzing a plasma sample (at least one hour) is much too long to be relied upon as a measure of therapeutic response. In those instances where the rule of thumb that 0.5 mg/kg dose results in a plasma concentration of 1 μg/ml is not applicable, as in the case of liver diseases (where the volume of distribution increases by about 75 percent and clearance decreases by 40 percent), it is beneficial to obtain steady state plasma levels of the drug in order to roughly estimate these parameters. Another concern is the role of lidocaine metabolites in the pharmacologic response, which remains uncertain. However, patients showing abnormal metabolic patterns need to be identified in order to reduce such toxicity risks as may arise due to toxic metabolites. Plasma concentration monitoring, however, is a distinct advantage when long term therapy is desired and especially when decreased clearance is suspected.

Propranolol

Propranolol has been in use for the last 15 years as a potent β-adrenoceptor antagonist in the treatment of angina pectoris, arterial hypertension, and cardiac arrhythmias. The plasma concentration monitoring of propranolol is necessitated by the large variation in the plasma concentrations obtained from the same dose in different individuals, and within an individual. Differences are observed of as much as 700 percent in the peak plasma levels.[35] These differences are attributed to bioavailability and the first pass effect, as discussed earlier. The liver extraction of propranolol is effected by two systems, one with high affinity but low capacity and the other with lower affinity but much greater capacity. A large degree of difference can exist between individuals regarding the relative function of these two systems, resulting in highly variable first pass transformations.[36]

An accepted therapeutic range of propranolol is between 50 and 100 ng/ml.[37] It is, however, necessary to estimate also the metabolite concentration (4-hydroxy) since it is also pharmacologically active as an adrenoceptor antagonist.[38] It should also be noted that propranolol is normally given as a racemate and the β-adrenoceptor antagonism is due to the levo isomer. In fluorescent analysis both isomers provide equal fluorescence, but the terminal half-life of the levo isomer is longer than that of the dextro isomer.[39] Total plasma concentration monitoring does not reveal the difference in the isomer or any changes in their relative proportion with time.

Procainamide

Procainamide controls ventricular arrhythmias in the plasma concentration range of 4 to 8 μg/ml. Major toxicity appears above 8 μg/ml and concentrations above 16 μg/ml have catastrophic effects on cardiovascular function.[40] In most deaths attributed to procainamide, the plasma concentration varied between 17 and 32 μg/ml. Generally, a critical concentration is needed for a specific arrhythmia.[41]

In subjects with normal renal function, 40 to 67 percent of procainamide is excreted unchanged in the urine.[42] A major mechanism of the biotransformation of procainamide is N-acetylation. The rate of this transformation is dependent on the genetically determined activity of N-acetyltransferase enzyme in the body. Fast acetylators show higher nonrenal clearance of procainamide.[40] The acetyl metabolite of procainamide also accumulates in patients with impaired renal function.[43] The acetyl metabolite (NAPA) is also pharmacologically active and reaches plasma concentrations that are generally comparable to and often greater than those of the parent drug.[44]

It is therefore imperative that the plasma concentration monitoring of procainamide be coupled with the levels of NAPA. It is also necessary to determine whether the patient is a slow or fast acetylator in order to adjust the dose to the plasma concentrations.

Quinidine

Therapeutic plasma levels of quinidine lie between 1 and 2 μg/ml. Above 8 μg/ml serious toxic symptoms such as cardiovascular disturbances, proxysmal ventricular tachycardia, and heart block appear. The main problem in the use of the plasma concentration as a therapeutic guide for quinidine is a lack of a specific and sensitive analytic method.

BRONCHODILATORS

Theophylline or aminophylline is a potent drug for the treatment of pulmonary diseases involving reversible bronchoconstriction. A large degree of variation exists in the disposition half-lives of theophylline in adults and children.[45,46] For example, in adults with an average age of 56, the half-life of theophylline ranged from 3 to 9.5 hours.[45] In children it was reported to range between 1.2 and 10 hours.[46] The main reason for such variation is the metabolic clearance of the drug. The renal clearance, which is approximately 6 ml/min, remains constant whereas the overall plasma clearance of 72 ± 21 ml/min remains variable.[45] An additional problem in theophylline therapy arises from the controversy on the anhydrate versus the hydrate form of the drug. In the opinion of the author, these differences are not important in therapeutic management and it is recommended that regardless of the product used, a change in the brand is not suggested during long-term therapy.

The therapeutic range of the theophylline level is between 8 and 20 μg/ml. Good correlation has been reported between improvement in pulmonary function (first-second forced expiratory volume) and plasma theophylline concentration.[47,48] The distribution phase of theophylline takes about 30 minutes, followed by a log-linear fall in the plasma concentration.[47] Since the most common dosing interval for theophylline is 6 hours, sufficient blood levels should be collected within this time to characterize the pharmacokinetics of theophylline or to predict toxic response. Saliva levels have also been used to monitor theophylline in the body. An almost linear relationship exists between salivary and plasma levels up to 14 μg/ml in the plasma, which corresponds to about 7.5 μg/ml in the saliva.[49,50] In disease states, however, salivary output may be seriously affected, hampering the utility of this method.

ANTIPSYCHOTIC DRUGS

Plasma level monitoring of antipsychotic drugs, such as chlorpromazine, offers a distinct advantage in the assessment of a clinical response since the patients are often unable to communicate effectively. It is especially important to determine whether a patient is refractory to a given drug or whether the lack of response is due to abnormal disposition properties. Analytic methods have not yet been perfected which can be routinely used and therein lies a serious problem in the monitoring of plasma levels.

A wide range of plasma concentrations are observed in response to a given dose of antipsychotic drugs, due mainly to differences in the biotransformation variability. This is true of chlorpromazine, with which the metabolite level may be determined genetically.[51] A therapeutic range of 35 to 350 ng/ml has been suggested for chlorpromazine.[52] About 11 metabolites of chlorpromazine have been identified and some metabolites seem to have activity similar to or higher than chlorpromazine itself.[53] For example, a higher ratio of 7-hydroxy chlorpromazine to chlorpromazine is related to increased activity, whereas increased levels of sulfoxide of chlorpromazine yield lowered activity.

ANTICOAGULANTS

Warfarin is one of the most commonly used coumarin-type anticoagulants. It acts by inhibiting the synthesis of blood clotting factors. The relationship between its plasma concentration and its activity was discussed earlier (Fig. 9.3). Although no definite concentration of warfarin can be suggested for a therapeutic response, plasma concentration monitoring of warfarin offers many advantages in detecting drug interactions, abnormal disposition properties, and plasma protein binding characteristics.[54]

The significant variation in the required dose of warfarin is typified by an example in which 145 mg of warfarin was needed to achieve an anticoagulant response; this high requirement was not associated with increased biotransformation since the blood levels were 25 to 30 times higher than normal (about 2 μg/ml).[55]

Plasma level monitoring will reveal any changes in warfarin metabolism, such as those brought about by increased microsomal enzyme levels due to induction with such drugs as barbiturates (Fig. 6.7). Inherent resistance to warfarin can also be determined by the plasma levels—the resistance may occur due to excessive levels of clotting factors, whose rate of synthesis is affected by warfarin.

It should also be noted that warfarin exists as a mixture of two enantiomers which have different disposition half-lives and different potencies as well.

It has also been suggested that the free warfarin concentration correlates better with the pharmacologic response, and since in many disease states the plasma protein binding characteristics change drastically, it is fruitful to study the binding of warfarin to the components of a patient's plasma.

ANTIBIOTICS

Plasma concentration monitoring of antibiotics is important since in many instances a minimum concentration in the plasma must be reached to inhibit microorganisms. Since all microorganisms require a certain minimum inhibitory concentration, which is easily calculated using in vitro incubation experiments, a rationale for a therapeutically effective plasma concentration can be established. However, it should be noted that if the site of infection is not within the highly perfused tissues, where the concentration will be comparable to that in plasma, significant differences may be expected between the observed plasma concentration and the concentration at the site of action. The duration of action will also be difficult to assess from the plasma concentration sampling alone unless tissue level calculations are performed. If the site of action lies in a region where the transport is mainly due to active processes, such as the urinary excretion of penicillin, no direct comparison between the plasma concentration and therapeutic effectiveness can be made.[56]

It is important that the plasma concentration be maintained just above the minimum effective level, since elevating the level will not increase the effectiveness but will increase the toxic symptoms. It is also important to note that if at any time the plasma concentration is allowed to fall below the effective levels, such as during the dosing interval, the microorganisms will be exposed to a subminimal concentration, making the success of therapy questionable.

An important classification of antibiotics is aminoglycosides, which are amino sugars linked to other moieties by a glycoside bond—examples are streptomycin, neomycin, paromonycin, kanamycin, gentamicin, tobramycin, amikacin, sisomicin, and netilmicin. These compounds accumulate in the sensitive cells by a complex series of active transport steps in which they bind irreversibly to bacterial ribosomes and thus destroy the cells through interference in the genetic code.[57–59] Most aminoglycosides have comparable volumes of distribution, are primarily eliminated through the kidneys, are minimally bound to plasma proteins, and have blood levels mostly proportional to the dose and dosage regimen, which can be modified to provide desired plasma levels.[60]

A dosage adjustment is generally necessary in impaired renal function for most of aminoglycoside antibiotics. It is also necessary to provide sufficient peaks and troughs of concentrations of these antibiotics to partially relieve the problems of their toxicity. One of the most serious symptoms due to these drugs is ototoxicity as a result of drug penetration into the inner ear during periods of high serum concentration. This would dissipate slowly if the plasma concentration decreases.[61,62] If sufficiently low trough levels are not provided then a high degree of accumulation occurs in the inner ear, resulting in excessive toxicity.[63–66] Generally, levels below 2 μg/ml serve this purpose.

Aminoglycosides also damage the proximal tubular cells of the kidneys, sparing the glomeruli. The mechanism of this toxicity is not fully understood—however, these antibiotics do accumulate in renal cortical tissue in concentrations which are higher than that achieved in the serum.[67–69] Streptomycin has a much lower affinity for kidney tissue and as a result shows very little renal toxicity.[70] The importance of peaks and troughs is not fully understood in determining the toxicity of these drugs to the kidneys.[71]

Desirable serum levels of gentamicin, tobramycin, sisomicin, and netilmicin are 5 to 8 μg/ml at peak and 1 to 2 μg/ml at trough. Toxicity symptoms appear at concentrations above 12 μg/ml. The usual dose in normal renal function is 1.7 mg/kg q 8h, which is significantly reduced in impaired renal function. For kanamycin and amikacin, the desired levels are 20 to 25 μg/ml at peak and 5 to 10 μg/ml at trough; toxicity appears above 35 μg/ml. The usual dose is 5 mg/kg q 8h and adjustments are necessary in renal malfunction. For streptomycin the peak levels are 5 to 20 μg/ml and trough levels are less than 5 μg/ml; toxicity appears at above 50 μg/ml. The usual dose is about 500 mg q 12h, which is adjusted according to renal function.[60]

ANTIDEPRESSANTS

Lithium carbonate is highly active in the treatment of manic depression. The plasma level monitoring of lithium is important since the therapeutic levels lie close to toxic levels.[72] A desirable range of therapeutic concentration is between 0.7 and 1.3 mEq/liter; serious toxicity is common above 2 mEq/liter.[73] A significant advantage in lithium monitoring is that the salivary levels of lithium are generally higher than the plasma levels and a good correlation exists between the two, affording a convenient method for continuously monitoring the plasma concentration.[74]

Nortriptyline provides an interesting example of antidepressant drugs, with which an optimum range of plasma concentrations is required to elicit a desired response. Two factors should be considered. First, there is significant variation in the plasma levels of nortriptyline due to genetic differences in the metabolism of enzymes.[75] Second, the activity of nortriptyline decreases when the concentration in the plasma is too high (above 140 ng/ml). The bell-shaped level-response phenomenon is attributed to possible blockade of adrenergic receptors at higher concentrations, similar to that observed for phenothiazines.[75]

ANTICANCER COMPOUNDS

Methotrexate is one of the most important anticancer drugs which require plasma concentration monitoring. Methotrexate acts cytotoxic by binding to the enzyme dihydrofolate reductase. This action decreases tetrahydrofolate levels, and thus the biosynthesis of deoxyribonucleic acid, ribonucleic acid, and other cellular proteins.[76] Since the action of methotrexate is phase-specific, effective concentration in the plasma should be maintained at all times (approximately 10^{-8} M). The high dose methotrexate therapy involves giving toxic doses and then rescuing the patient by administering citrovorum factor, which reverses the action of methotrexate. However, a certain lag time is involved in this reversal process due to the active nature of drug transport across cell membranes and the fact that a minimum concentration of citrovorum factor must be established in the cell to be effective.

ANTIRHEUMATICS

Salicylate concentrations between 150 and 300 μg/ml provide good antirheumatic properties. Salicylism, the event of toxic symptoms, generally appears within the 250 to 350 μg/ml range; children and "sensitive" individuals show these symptoms at even lower concentrations.[77]

A good correlation exists between toxicity and plasma levels (Fig. 9.6) if plasma concentrations are extrapolated to the time of ingestion. Poor correlations can be expected at later times since salicylates undergo saturable biotransformation and show wide variation in their degree of distribution and renal clearance among individuals.

The maximum tolerated salicylate levels have frequently been determined by increasing the dose until tinnitus appears and then decreasing the dose to a no-symptom level. The problem with this method is that in patients with existing hearing loss this method may prove fatal. Another approach suggests measurement of plasma levels 12 hours after an oral dose of 1.2 g of aspirin. If the levels are below 5 μg/ml, therapeutic levels will probably not be reached upon long-term therapy with 4.8 g doses every day. Plasma concentration levels above 10 μg/ml will provide indication of salicylism in similar long-term therapy.[78]

Salivary monitoring of salicylates has also been suggested; however, the saliva:plasma ratio for salicylates is very low, about 0.03, decreasing the

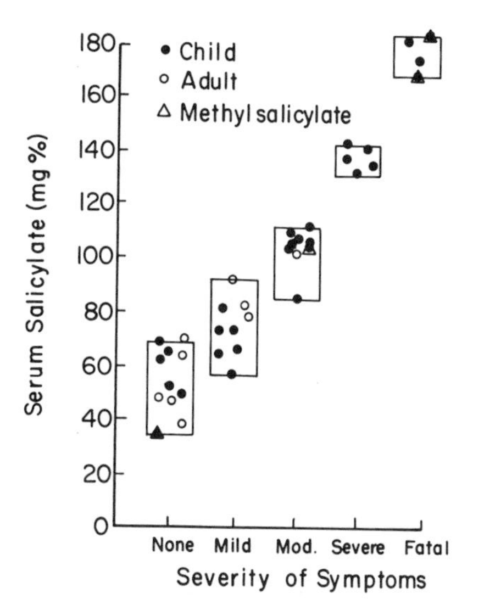

Figure 9.6. The relationship between serum salicylate concentrations extrapolated back to the time of ingestion and the clinical severity of intoxication. (From Prescott et al.: in Davies and Prichard, eds., Biological Effects of Drugs in Relation to their Plasma Concentrations, 1972. Courtesy of University Park)

utility of this method if plasma concentrations are below 50 μg/ml. Another complexity arises at higher plasma levels of salicylates. At concentrations above 150 μg/ml the binding of salicylates to plasma proteins becomes saturated, increasing the free salicylates and thus the saliva levels. It is therefore suggested that the plasma:saliva ratio of salicylates may be a function of concentration, making it a less desirable alternative to direct plasma concentration monitoring.[80]

ANTICONVULSANTS

Phenytoin

Phenytoin is one of the most commonly used anticonvulsants. It is also a most potent and toxic compound. The concentration of phenytoin in plasma is generally reflective of its concentration in the brain and if plasma concentration is maintained between 10 and 20 μg/ml, seizures can be adequately controlled. Unfortunately, a given dose results in varying plasma concentrations between subjects and even within a subject (Fig. 9.7).[81] Not only is this wide variation in the plasma concentration responsible for inadequate therapeutic response, it results in toxic symptoms which are directly proportional to the plasma concentration (Fig. 9.8).[82,83] The main reason for the observed variability in the plasma concentration is the nonlinear biotransformation of phenytoin in humans. The conversion of phenytoin to its p-hydroxy derivative is a saturable process in which the plasma concentration is described by

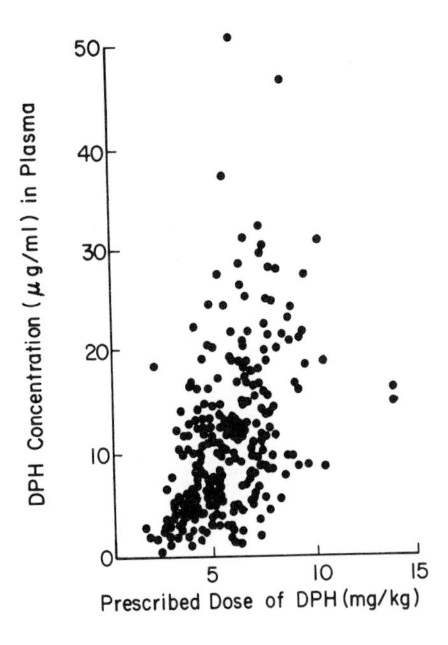

Figure 9.7. The relationship between the prescribed daily dose of phenytoin (DPH) in mg/kg of body weight and the concentration in μg/ml in 294 patients with epilepsy. (From Lund: in Davies and Prichard, eds., Biological Effects of Drugs in Relation to their Plasma Concentrations, 1972. Courtesy of University Park)

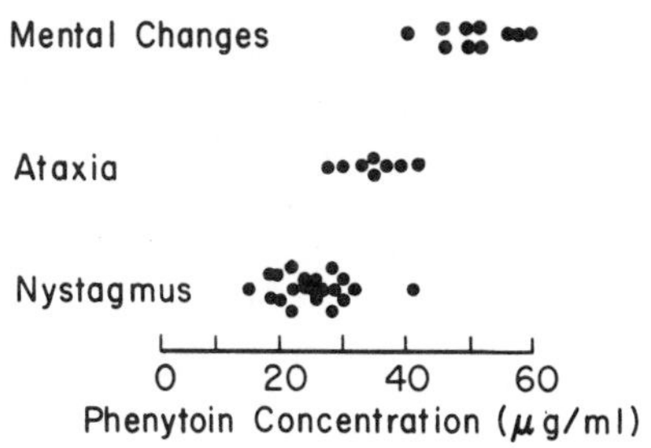

Figure 9.8. Phenytoin blood levels at the time individuals experienced certain adverse effects. (After Kutt et al.;[82] and Gibaldi[83])

Equation 7.100. However, if the drug is continuously administered:[84,85]

$$dC/dt = k_0/V - \frac{V_m C}{K_m + C} \qquad \text{(Eq. 9.26)}$$

where k_0 is the zero order infusion rate (such as 100 mg qid). At the steady state, $dC/dt = 0$ and thus:

$$k_0/V = \frac{V_m C_{ss}}{K_m + C_{ss}} \qquad \text{(Eq. 9.27)}$$

where C_{ss} is the steady plasma level.

Thus a linear relationship between the dose and the steady state plasma concentration cannot be expected; the plasma concentration is likely to increase more than expected from a straight line relationship. Phenytoin is also highly bound to plasma proteins—the extent of binding varies in the disease states, leading to further variability in the plasma concentration profiles. Various drugs interact with phenytoin with regard to its binding to proteins as well as to the extent of its metabolism, inducing great variability in the observed plasma concentration in response to a given dose.

The dosage calculations for phenytoin are therefore based on the plasma concentrations, from which the pharmacokinetic parameters V_m and K_m can be determined. Approaches based on population parameters have not been very successful, due to unexpected variability in these parameters between individuals and even within an individual as the disease state changes, or upon concomitant administration of other drugs. However, if K_m and V_m are determined in each patient, accurate dosage calculation can be easily made:[85]

$$k_0 = \frac{V_m C_{ss}}{K_m + C_{ss}} \qquad \text{(Eq. 9.28)}$$

Notice that V_m has the units of amount/time, such as mg/day, which are the same as the units of k_0. A rearrangement of the above equation gives:

$$k_0 = V_m - \frac{k_0}{C_{ss}} K_m \qquad \text{(Eq. 9.29)}$$

Thus a plot of k_0 against k_0/C_{ss} yields a straight line with the slope of K_m and

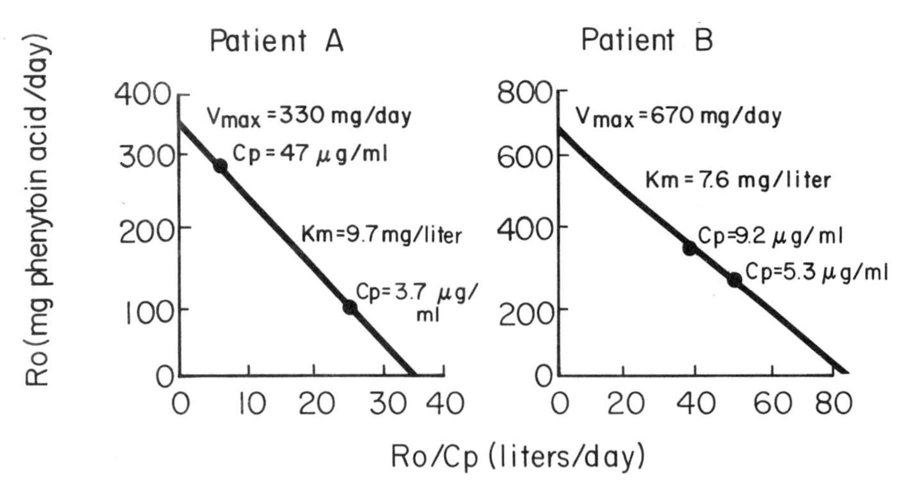

Figure 9.9. Graphical estimation of individual V_m and K_m. (From Ludden et al.: Clin Pharmacol Ther 21:287, 1977)

intercept V_m (Fig. 9.9).[85] This requires at least two data points of the steady state plasma level against two rates of administration to calculate these parameters. Good correlations have been reported between calculated and observed plasma levels using the method described above (Fig. 9.10).

A recently proposed method employs a linear transformation of the Michaelis-Menton relationship:[86,87]

$$k_0 = V_m - \frac{k_0}{C_{ss}} K_m \qquad \text{(Eq. 9.30)}$$

where A is sodium phenytoin per day, B has units of concentration analogous

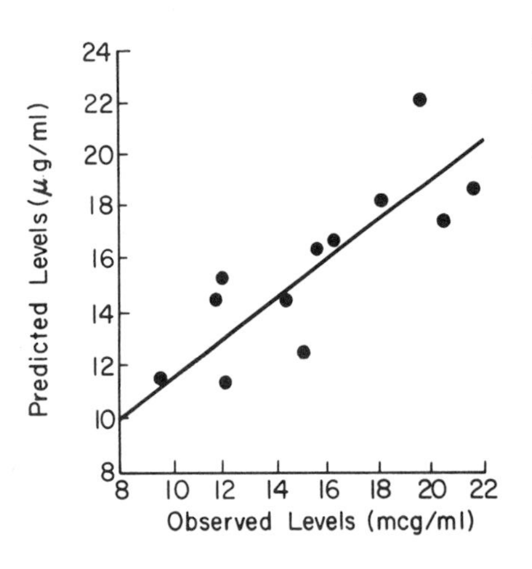

Figure 9.10. Correlation of phenytoin serum levels predicted from individual K_m and V_m values with observed levels. (From Ludden et al.: Clin Pharmacol Ther 21:287, 1977)

to V_m and K_m for Michaelis-Menton constants, and C_{ss}^{min} is the concentration just before the next dose. The calculations are made by first establishing an individual nomogram for each patient:

> On a sheet of ordinary graph paper a set of coordinate (x,y) axes are drawn to provide the first two quadrants. The x and y axes are then labeled Concentration and Dose, respectively. Each known steady-state concentration is arbitrarily assigned a negative value and marked on the concentration axis, while the corresponding-dose values are denoted on the y-axis. The paired concentration-dose values are then joined with a straight line which is extended into the first quadrant (Fig. 9.11). Theoretically, assuming a true Michaelis-Menton relationship between steady-state concentrations and dose, all lines extended into the first quadrant should intersect at a common point. Lines drawn from this intersection point parallel to the two axes will intercept the concentration and dose axes at B and A, respectively.[86]
>
> The dose required to produce a desired steady-state concentration can be determined simply by drawing a straight line from the intersection point in the first quadrant to the desired absolute value on the negative part of the concentration axis. The point where this line cuts the dose axis gives the required dose.[87]

The approach presented above provides very good estimates of plasma levels. Initially, instead of three steady state levels, two levels can be used to predict the dose required to achieve a desired (third) concentration. This data can then be added to the graph to obtain a more reliable estimate of the dose. This practice can be continued to "fine tune" the dosage regimen. This record will also help clinicians to determine any changes in the disposition characteristics, as well as where drug and disease interactions are suspected.

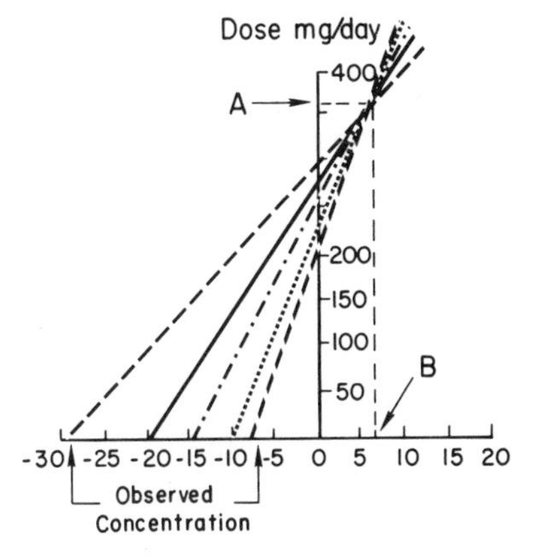

Figure 9.11. The direct linear plot as used to relate steady-state serum phenytoin concentrations and the dose of sodium phenytoin. See text for details. (From Mullen: Clin Pharmacol Ther 23:228, 1978)

Carbamazepine

Carbamazepine seems to be about as effective as phenytoin in the treatment of grand mal and psychomotor epilepsy. It is the drug of choice in trigeminal neuralgia.[88] The correlation between dose and plasma levels achieved is very poor for carbamazepine, due to variation in its biotransformation between individuals, self-induction of biotransformation by carbamazepine, and the effect of such other drugs as phenytoin or phenobarbital (frequently given concomitantly with carbamazepine) which lower its level. These factors necessitate monitoring of the plasma concentration of carbamazepine since therapeutic efficacy is achieved at levels between 5 and 10 μg/ml; higher levels invariably lead to toxic effects.[88]

SALIVARY MONITORING OF DRUG CONCENTRATION

Most of the discussion presented above is applied to the monitoring of the plasma concentration as a means of adjusting the dosage regimen. However, this is an invasive method which is often difficult to establish in outpatient situations. Salivary drug concentration monitoring is noninvasive and less painful to the patient, especially when several samples have to be obtained to characterize a pharmacokinetic profile.[80] A significant advantage of the use of the saliva concentration is that it represents, in most instances, the equilibrium concentration with the free or unbound concentration in the plasma. Any changes in the protein binding of drugs would be immediately reflected in the saliva concentration and, since in many instances the pharmacologic response is proportional to the unbound fraction, salivary concentrations are more relevant pharmacologically.

There is evidence that many organic compounds enter the saliva by a passive diffusion process, but these may also be secreted actively.[74] An essential prerequisite in the use of salivary concentration to monitor drug therapy is that the saliva:plasma (s:p) ratio remain constant throughout the course of treatment. This ratio is often determined by using the following equation.[89] For weakly acidic compounds:

$$s:p = \frac{1 + 10^{(pH_s - pK_a)}}{1 + 10^{(pH_p - pK_a)}} \times \frac{f_p}{f_s} \qquad \text{(Eq. 9.31)}$$

where pH_s and pH_p are salivary and plasma pH, and f_p and f_s are the unbound fractions of the total drug concentration in the plasma and in saliva. For weakly basic drugs the exponents are negative.

Several examples of salivary drug concentration monitoring have been discussed previously. An interesting example is that of phenytoin, whereby salivary concentrations one-tenth the plasma concentration can be easily detected. Intersubject variability is also smaller in the s:p ratio,[90] which is close to unity when calculated with respect to the unbound plasma fraction.[91]

All of these properties make salivary drug monitoring an ideal system for phenytoin and for all drugs which exhibit these properties.

A recent report provides excellent recommendations on the use of salivary concentrations in patient monitoring.[92] The predictability of s:p ratios (see Equation 9.31) is generally very good for drugs which are un-ionized in plasma and saliva, such as theophylline (base, pK_a 0.7, f_p 0.85) or antipyrine (base, pK_a 1.4, f_p 0.92). However, drugs whose pK_a values are closer to the pH of saliva (a range of 6 to 8) give highly variable s:p ratios. Thus good correlations are possible for such drugs as antipyrine, phenobarbital, phenytoin, theophylline, and amylobarbital, all of which show very little variability in their theoretically calculated s:p ratios as a function of normal saliva pH. Based on this argument, compounds such as chlorpropamide, propranolol, tolbutamide, and procainamide are not suitable for salivary concentration monitoring. The crucial parameter is therefore the pH of saliva, which is not only variable between and within individuals, but is also dependent on salivary flow rates. Increased salivary flow as a result of artificial or natural stimulation results in an increase in the pH of saliva.

References

1. Morselli PL: Clinical pharmacokinetics in neonates. Clin Pharmacokinet 1:81, 1976
2. Windorfer A, Kuenzer W, Urbanek R: The influence of age on the activity of acetyl salicylic acid-esterase and protein salicylate binding. Eur J Clin Pharmacol 7:227, 1974
3. Borondy P, Dill WA, Chang T, et al: Effect of protein binding on the distribution of 5,5-diphenylhydantoin between plasma and red blood cells. Ann NY Acad Sci 220:82, 1973
4. Maselli R, Casal GL, Ellis EF: Pharmacologic effects of intravenously administered aminophylline in asthmatic children. J Pediatr 76:777, 1970
5. Morselli PL, Assael BM, Gomeni R, et al: Digoxin pharmacokinetics during human development. In Morselli PL, Garattini —, Serini F (eds): Basic and Therapeutic Aspects of Perinatal Pharmacology. New York, Raven, 1975, p. 377
6. Wagner JG: Fundamental of Clinical Pharmacokinetics. Hamilton, Ill, Drug Intelligence Publications, 1975, p 161
7. Rowe JW, Andres R, Tobin JD, et al: The effect of age on creatinine clearance in men: A cross-sectional and longitudinal study. J Gerontol 31:155, 1976
8. Mawer GE, Lucas SB, McGough JG: Kanamycin dosage schedule. Lancet 2:45, 1972
9. Cockcroft DW, Gault MH: Prediction of creatinine clearance from serum creatinine. Nephron 16:31, 1976
10. Jelliffe RW, Jelliffe SM: A computer program for estimation of creatinine clearance from unstable serum creatinine levels, age, sex, and weight. Math Bio Sci 14:17, 1972
11. Weisberg HF: Water, Electrolytes and Acid-Base Balance, 2nd Ed. Baltimore, Williams and Wilkins, 1962
12. Winek CL: Tabulation of therapeutic, toxic, and lethal concentrations of drugs and chemicals in blood. Clin Chem 22:832, 1976
13. Alexanderson B, Borga O: Interindividual differences in plasma protein binding of nortriptyline in man—a twin study. Eur J Clin Pharmacol 4:196, 1972

14. Odar-Cederlof I, Borga O: Kinetics of diphenylhydantoin in uremic patients: Consequences of decreased plasma protein binding. Eur J Clin Pharmacol 7:31, 1974
15. Yacobi A, Udall JA, Levy G: Serum protein binding as a determinant of warfarin body clearance and anticoagulant effect in patients. Clin Pharmacol Ther 19:552, 1976
16. Nagashima R, O'Reilly RA, Levy G: Kinetics of pharmacologic effects in man: the anticoagulant action of warfarin. Clin Pharmacol Ther 10:22, 1969
17. Coltart DJ, Shand DG: Plasma propanolol levels in the quantitative assessment of β–adrenergic blockade in man. Br Med J 3:731, 1970
18. Cleveland CR, Shand DG: The effect and duration of drug administration on the relationship between β–adrenergic blockade and plasma propanolol concentration. Clin Pharmacol Ther 13:181, 1972
19. Doherty JE: Digitalis glycosides: Pharmacokinetics and their implications. Ann Intern Med 79:229, 1973
20. Weintraub M: Interpretation of the serum digoxin concentration. Clin Pharmacokinet 2:205, 1977
21. Brown DD, Juhl RP: Decreased bioavailability of digoxin due to antacids and Kaolin-pectin. N Engl J Med 295:1034, 1976
22. Hall WH, Shappell SD, Doherty JE: Effect of cholestyramine on digoxin absorption and excretion. Am J Cardiol 39:213, 1977
23. Lindenbaum J, Maulitz RM, Bottler VP: Inhibition of digoxin absorption by neomycin. Gastroenterology 71:399, 1976
24. McCredie RM, Chia BL, Knight PW: Infant versus adult plasma digoxin levels. Aust NZ J Med 4:223, 1974
25. Ogilvie RI, Ruedy J: An educational program in digitalis therapy. JAMA 222:50, 1972
26. Conklin D: Digoxin Dose Calculation. Hewlett Packard Library Program, No 00791D (For HP-97)
27. Jusko WJ, Szeflzer SJ, Goldfarb AL: Pharmacokinetic design of digoxin dosage regimens in relation to renal function. J Clin Pharmacol 14:525, 1974
28. DuBois D, DuBois EFL: A formula to estimate the approximate surface area if height and weight be known. Arch Intern Med 17:863, 1916
29. Perrier D, Mayersohn M, Marcus FI: Clinical pharmacokinetics of digitoxin. Clin Pharmacokinet 2:292, 1977
30. Lukas DS, Peterson RE: Double isotope dilution derivative assay of digitoxin in plasma, urine, and stool of patient maintained on the drug. J Clin Invest 45:782, 1966
31. Melmon KL, Rowland M, Sheiner L, Trager W: Clinical implications of the disposition of lidocaine in man. In Davies DS, Prichard BNC (eds): Biological Effects of Drugs in Relation to their Plasma Concentrations. British Pharmacological Society Symposium. London, University Park, 1972
32. Gianelly R, Von der Groeben JO, Spivack AP: Effect of lidocaine on ventricular arrhythmias in patients with coronary heart disease. N Engl J Med 277:1215, 1967
33. Thompson PD, Rowland M, Melmon KL: The influence of heart failure, liver disease, and renal failure on the disposition of lidocaine in man. Am Heart J 82:417, 1971
34. Rowland M: Clinical pharmacokinetics of lidocaine: General approach to drug administration and individualized therapy. In Levy G (ed): Clinical Pharmacokinetics, a Symposium. Washington DC, American Pharmaceutical Assoc, 1974
35. Shand DG, Nuckolls EM, Oates JA: Plasma propanolol levels in adults. Clin Pharmacol Ther 7:112, 1970
36. Evans GH, Shand DG: Disposition of propanolol. VI: Independent variation in steady state circulating drug concentrations and half-life as a result of plasma binding in man. Clin Pharmacol Ther 13:114, 1973
37. Nies AS, Shand DG: Clinical pharmacology of propanolol. Circulation 52:6, 1975

38. Paterson JW, Connolly ME, Dollery CT, et al: The pharmacodynamics and metabolism of propanolol in man. Eur J Clin Pharmacol 2:127, 1970
39. George CF, Fenyvesi T, Connolly ME, Dollery CT: Pharmacokinetics of dextro-, laevo-, and racemic propanolol in man. Eur J Clin Pharmacol 4:74, 1972
40. Koch-Weser J: Clinical applications of pharmacokinetics of procainamide. Cardiovas Clin 6:2, 1974
41. Koch-Weser J: Correlation of serum concentration and pharmacologic effects of antiarrhythmic drugs. Proceedings 5th International Congress of Pharmacology. Vol 3. Basel, Karger, 1973 p 69
42. Koch-Weser J: Pharmacokinetics of procainamide in man. Ann NY Acad Sci 179:370, 1971
43. Drayer DE, Lowenthal DT, Woosley RL, et al: Cumulation of N-acetylprocainamide, an active metabolite of procainamide, in patients with impaired renal function. Clin Pharmacol Ther 22:63, 1977
44. Elson J, Strong JM, Lee WK, Atkinson AJ: Antiarrhythmic potency of NAPA. Clin Pharmacol Ther 17:134, 1975
45. Jenne JW, Wyze E, Rood FS, MacDonald FM: Pharmacokinetics of theophylline. Applications to adjustment of the clinical dose of aminophylline. Clin Pharmacol Ther 13:357, 1972
46. Ellis EH, Yaffe SS, Levy G: Pharmacokinetics of theophylline in asthmatic children. American Academy of Allergy 30th Annual Meeting Abstracts, 1974, p 9
47. Levy G: Pharmacokinetic control of theophylline therapy. In Levy G (ed): Clinical Pharmacokinetics, a Symposium. Washington DC, American Pharmaceutical Assoc, 1974, p 103
48. Maselli R, Casal GL, Ellis EF: Pharmacologic effects of intravenously administered aminophylline in asthmatic children. J Pediatr 76:777, 1970
49. Koup JR, Schentag JJ, Vance JW, et al: System for clinical pharmacokinetic monitoring of theophylline therapy. Am J Hosp Pharm 33:949, 1976
50. Levy G, Ellis EF, Koysooko R: Indirect plasma-theophylline monitoring in asthmatic children by determination of theophylline concentration in saliva. Pediatrics 53:873, 1974
51. Green DE, Forrest IS, Forrest FM, Serra MT: Interpatient variation in chlorpromazine metabolism. Exp Med Surg 23:278, 1965
52. Curry SH: Gas chromatographic methods for the study of chlorpromazine and some of its metabolites in human plasma. Psychopharmacol Comm 2:1, 1976
53. Sakalis G, Chan T, Gershon S, Park S: The possible role of metabolites in therapeutic response to chlorpromazine treatment. Psychopharmacologia 32:297, 1973
54. Breckenridge A, Leorme M: Measurement of plasma warfarin concentration in clinical practice. In Davies DS, Prichard BNC (eds): Biological Effects of Drugs in Relation to their Plasma Concentrations. British Pharmaceutical Society Symposium. London, University Park, 1972, p 145
55. O'Reilly RA, Aggeler PM, Hoag MS, Leong L: Hereditary transmission of exceptional resistance to coumarin anticoagulant drugs. N Engl J Med 271:809, 1964
56. Rolinson GN: Plasma concentrations of penicillin in relation to the antibacterial effect. In Davies DS, Prichard BNC: Biological Effects of Drugs in Relation to their Plasma Concentrations. British Pharmacological Society Symposium. London, University Park, 1972, p 183
57. Bryan LE, Vander Elzen HM: Streptomycin accumulation in susceptible and resistant strains of E. coli and Pseudomonas aeruginosa. Antimicrob Agents Chemother 9:928, 1976
58. Davis BD, Dulbecoo R, Eisen HM: Protein synthesis. In Microbiology, 2nd ed. Hagerstown Md, Harper and Row, 1973, p 300
59. Davies J: Bacterial resistence to aminoglycoside antibiotics. J Infect Dis 124(S):7, 1971
60. Barza M, Scheife RT: Antimicrobial spec-

trum, pharmacology, and therapeutic use of antibiotics. IV: Aminoglycosides. Am J Hosp Pharm 34:273, 1977

61. Tjernstrom O, Banck G, Belfrage S: The ototoxicity of gentamicin. Acta Pathol Microbiol Scand (B)81 241:73, 1973
62. Logan TB, Prazma J, Thomas WG: Tobramycin ototoxicity. Arch Otolaryngol 99:190, 1974
63. Banck G, Belfrage S, Juhlin I: Retrospective study of the ototoxicity of gentamicin. Acta Pathol Microbiol Scand (B)81 241:54, 1973
64. Nordstrom L, Banck JP, Belfrage S: Prospective study of the ototoxicity of gentamicin. Acta Pathol Microbiol Scand (B)81 241:58, 1973
65. Jackson GG, Arcieri G: Ototoxicity of gentamicin in man. A survey and controlled analysis of clinical experience in the United States. J Infect Dis (Suppl) 124:130, 1971
66. Anonymous: Serum gentamicin. Lancet 2:1185, 1974
67. Alfthan O, Renkonen OV, Sivonen A: Concentration of gentamicin in serum, urine, and urogenital tissue in man. Acta Pathol Microbiol (B)81 241:92, 1973
68. Edwards CQ, Smith CR, Baughman KL: Concentrations of gentamicin and amikacin in human kidneys. Antimicrob Agents Chemother 9:925, 1976
69. Luft FC, Patel V, Yum MN: Nephrotoxicity of cephalosporin-gentamicin combinations in rats. Antimicrob Agents Chemother 9:831, 1976
70. Luft FC, Kleit SA: Renal parenchymal accumulation of aminoglycoside antibiotics in rats. J Infect Dis 127:299, 1973
71. Dahlgren JG, Anderson ET, Hewitt WL: Gentamicin blood levels: A guide to nephrotoxicity. Antimicrob Agents Chemother 8:58, 1975
72. Prien RF, Cuffy EM Jr, Klett CJ: Relationship between lithium serum level and clinical response in acute mania treated with lithium. Br J Psychiatry 120:409, 1972
73. Schou M, Amidsen A, Berastrup PC: The practical management of lithium treatment. Br J Hosp Med 6:53, 1971
74. Groth U, Prelluitz W, Jahnchen E: Estimation of pharmacokinetic parameters of lithium from saliva and urine. Clin Pharmacol Ther 16:490, 1974
75. Balzar A, Marie A, Dick T: Relationship between steady-state plasma concentration of nortriptyline and some of its pharmacologic effects. In Davies DS, Prichard BNC (eds): Biological Effects of Drugs in Relation to their Plasma Concentrations. British Pharmacological Society Symposium. London, University Park, 1972, p 191
76. Shen DD, Azarnoff DL: Clinical pharmacokinetics of methotrexate. Clin Pharmacokinet 3:1, 1978
77. Prescott LF, Roscoe P, Forrest JAH: Plasma concentration and drug toxicity in man. In Davies DS, Prichard BNC (eds): Biological Effects of Drugs in Relation to their Plasma Concentrations. British Pharmacological Society Symposium. London, University Park, 1972, p 72
78. Graham GC, Champion GD, Day RO, Pull PD: Patterns of plasma concentrations and urinary excretion of salicylates in rheumatoid arthritis. Clin Pharmacol Ther 22:410, 1977
79. Graham G, Rowland M: Applications of salivary salicylate data to biopharmaceutical studies of salicylates. J Pharm Sci 61:1219, 1972
80. Danhof M, Breimer DD: Therapeutic drug monitoring in saliva. Clin Pharmacokinet 3:39, 1978
81. Lund L: Effect of phenytoin in patients with epilepsy in relation to its plasma concentration. In Davies DS, Prichard BNC (eds): Biological Effects of Drugs in Relation to their Plasma Concentrations. British Pharmacological Society Symposium. London, University Park, 1972, p 227
82. Kutt H, Winter W, Kokenge R: Diphenylhydantoin metabolism, blood levels and toxicity. Arch Neurol 11:642, 1964
83. Gibaldi M: Biopharmaceutics and Clini-

cal Pharmacokinetics. Philadelphia, Lea and Febiger, 1977, p 159
84. Arnold K, Gerber N: The rate of decline of diphenylhydantoin in human plasma. Clin Pharmacol Ther 11:121, 1970
85. Ludden TM, Allen JP, Valutsky WA, et al: Individualization of phenytoin dosage regimens. Clin Pharmacol Ther 21:287, 1977
86. Cornish-Bowden A, Eisenthal R: Statistical considerations in the estimation of enzyme kinetic parameters by the direct linear plot and other methods. Biochem J 139:721, 1974
87. Mullen PW: Optimal phenytoin therapy: a new technique for individualizing dosage. Clin Pharmacol Ther 23:228, 1978
88. Bertilsson L: Clinical pharmacokinetics of carbamazepine. Clin Pharmacokinet 3:128, 1978
89. Matin SB, Wan SH, Karam JH: Pharmacokinetics of tolbutamide: Prediction by concentration in saliva. Clin Pharmacol Ther 16:1052, 1974
90. Paxton JW, Whiting B, Stephen KW: Phenytoin concentrations in mixed, parotid and submandibular saliva and serum measured by radioimmunoassay. Br J Clin Pharmacol 4:185, 1977
91. Bochner F, Hooper WD, Sutherland JM, et al: Diphenylhydantoin concentration in saliva. Arch Neurol 31:57, 1974
92. Mucklow JC, Bending MR, Kahn GC, Dullery CT: Drug concentration in saliva. Clin Pharmacol Ther 24:563, 1978

Questions:

1. What is the basis for classifying drugs into such groups as directly or indirectly acting?
2. The relationship between pharmacologic response and the log of the dose can be linear within which range and why?
3. What is the role of drug distribution in determining the dose-response relationships?
4. Which parameter—body weight, body surface area, or age—can most closely describe the volume of distribution of a drug?
5. What is the meaning of "individual patient titration"?
6. On a mg/kg basis infants and children receive significantly higher doses than adults. Why?
7. The volume of distribution of some drugs is much different in neonates. Discuss the reasons for this which are based on blood composition.
8. At what age do total protein and binding characteristics in neonates reach the adult level?
9. Is the volume of distribution in infants always higher than in adults?
10. Cite examples of drugs whose volume of distribution may be higher in the elderly.
11. If the dosing interval is equal to 1.5 times the half-life of a drug, calculate the ratio of the maximum and minimum amount of drug in the body at steady state.
12. What is the well-known Wagner Equation?
13. What is the dose needed to achieve a steady state concentration of 10 μg/ml in a 60 kg male for a drug which distributes throughout the body water and is administered q 8h?

14. The AUC for a 100 mg dose was 500 μg·hr/ml. If a 125 mg dose is given q 4h what will be the steady state concentration in the blood? What is the TBC of this drug?
15. What is the normal value of serum creatinine? What physiologic factors affect this?
16. What are the major advantages and disadvantages in using serum creatinine or creatinine clearance to estimate the efficiency of renal function?
17. What is the most appropriate method of calculating the creatinine clearance from direct urine collection?
18. What will be the slope and intercept of a plot (linear) of $1/t_{1/2}$ against Q_c?
19. Rework Problem 9.4 with a Q_c of 35 ml/min.
20. Give specific reasons why the creatinine clearance may not provide a good estimate of the disposition half-life of a drug which undergoes multicompartment disposition.
21. What is the effect of change in a drug's volume of distribution on its renal clearance? Is this reflected in the creatinine clearance?
22. Calculate Q_c using Equations 9.11 and 9.12 for a 50-year-old male weighing 300 pounds. C_c = 3.0 mg% on day 1 and 3.3 mg% on day 2; girth = 50 inches. Comment.
23. Calculate the percentage of your body that is fat.
24. What is the main prerequisite for using the plasma concentration as a therapeutic guide?
25. What is the therapeutic index? Does a low index correspond with serious side-effects?
26. Under what circumstances is plasma concentration monitoring the most advantageous? What additional parameters are needed to fully exploit these data?
27. What is the rationale for choosing total blood concentration or free drug concentration in the plasma for correlation with the pharmacologic response?
28. How can the anticoagulant effect of warfarin be related to its plasma concentration?
29. When is the monitoring of drug metabolites warranted?
30. The increased activity of propranolol after oral administration levels off after a few hours and becomes comparable to the activity after intravenous administration. Why?
31. What specific situations necessitate monitoring the plasma concentration of digoxin?
32. Discuss possible reasons for the lower toxicity of digoxin in infants when compared to adults.
33. If the LBW is 50 kg, what is the total body dose of digoxin and how should it be given? If the BUN is 40, what should the maintenance dose be?
34. Rework Problem 9.5 for an 81-year-old female weighing 48 kg with a height of 60 inches.

35. Compare the protein binding and disposition characteristics of digoxin and digitoxin.
36. Suggest reasons for the lack of bioavailability problems in the use of digitoxin, whereas digoxin has serious problems in its absorption.
37. What is the approximate daily and weekly loss of digitoxin from the body at steady state?
38. What is the role of metabolites in determining the antiarrhythmic activity of lidocaine, propranolol, and procainamide?
39. What metabolite of lidocaine is supposedly related to its CNS toxicity?
40. Does a C_c of 3.5 mg% indicate a need of a dosage adjustment for lidocaine?
41. Why must lidocaine be administered by continuous intravenous infusion?
42. Why is plasma concentration monitoring of lidocaine not always recommended?
43. What is the main reason for large variations in the observed plasma concentrations of propranolol following oral administration of identical doses?
44. Which isomer of propranolol has the longer biologic half-life?
45. How are the disposition kinetics of procainamide controlled by genetic factors?
46. What is the main reason for differences between adults and children in the disposition half-life of theophylline?
47. Which metabolite of chlorpromazine is related to its increased activity?
48. Even though no direct correlation between plasma concentration and prothrombin time has been achieved for warfarin, monitoring its plasma concentration is still recommended. Why?
49. Can the plasma concentration be of any use for an antibiotic used as a urinary antiseptic if it is secreted actively in the kidneys?
50. Why is it necessary to allow plasma concentrations to decrease to a certain minimum level (trough) when treating a patient with aminoglycoside antibiotics?
51. How close are the toxic and effective levels of lithium?
52. Provide an explanation for the bell-shaped dose-response curve of nortriptyline.
53. What is the role of the citrovorum factor in high-dose methotrexate therapy?
54. What is the conventional approach to determining the maximum tolerable levels of salicylates in the treatment of rheumatism? What is the 12-hour sample method?
55. What is the disadvantage in the use of saliva for monitoring the salicylate levels in the body?
56. What is the source of the nonlinear disposition kinetics of phenytoin?
57. Why are the population disposition parameters for phenytoin not always suitable for dosage calculation?
58. Compare and contrast the conventional Michaelis-Menton method and

the linear transformed Michaelis-Menton method in the calculation of a dosage regiment for phenytoin.

59. How is the dosage adjustment of phenytoin "fine tuned"?
60. Following administration of 100, 150, and 200 mg/day, the steady state plasma levels obtained were 5, 10, and 25 μg/ml. Using the direct linear plot method calculate the dose of phenytoin needed to maintain a 20 μg/ml concentration in the patient.
61. The following urinary excretion data were collected following intravenous administration of a 500 mg dose of vancomycin to a 50 kg 30-year-old male whose renal function has been found to be within normal limits (V = 40 percent of body weight). Vancomycin has been shown to result in deafness when the plasma level exceeds the 80 to 100 μg/ml range for any significant period of time. Does this patient risk deafness from a 500 mg q 6h IV dose? If this patient's renal function decreased to 20 percent of normal would the risk change?

Time interval (hr)	*Urine volume (ml)*	*Conc. (mg/ml)*
0–1	100	0.520
1–2	80	0.578
2–3	75	0.548
3–4	110	0.333
4–5	90	0.363
5–7	175	0.157

62. The following plasma levels were obtained from a single dose:

Time (hr)	*C (μg/ml)*
0	0
0.25	1.2
0.50	2.35
0.75	3.1
1	3.76
1.5	7.5
2	7.18
4	6
6	5.1
12	3.0

If the same dose is given q 8h and the toxic concentration is 14.5 μg/ml, will this regimen provide a toxic response? How would you suggest correcting the dosage regimen so that the plasma concentration never exceeds 12 μg/ml?

63. What are two distinct advantages in the use of the salivary concentration to monitor body levels of drugs?
64. If the s:p ratio for a drug is unity; $f_s = 1$; $pH_s = 8.4$; $pH_p = 7.4$; and $pK_a = 6.4$, what is the percent binding in the blood? Calculate for both acid and base.

Appendices

Appendix A: Definitions

ABSOLUTE BIOAVAILABILITY: Total amount of drug absorbed into general circulation from a given route of administration.

ABSORPTION: Penetration of free drug molecules through a biologic membrane.

ACCUMULATION: Increase in the plasma or tissue concentration of a drug—i.e., the total amount of drug in the body following repetitive drug administration when intake exceeds elimination.

ACTIVE TRANSPORT: Transfer of drug molecules through a membrane against the concentration gradient by making use of energy processes of the membrane.

ACTIVITY: Measure of the intensity of a given drug action.

AFFINITY CONSTANT (K, l/mol): Reciprocal of the dissociation constant of a chemical or enzymatic reaction described by the law of mass action or the Michaelis-Menton equation. Also termed *Association Constant.*

AFTER-EFFECT: A biologic drug effect which lasts longer than the drug concentration at the site of action: usually the result of any type of irreversible or slowly reversible reaction.

AMORPHISM: Lack of any organized structure. See *Polymorphism.*

AREA UNDER THE CURVE (AUC): Product of concentration and time on a plasma or blood concentration profile, obtained by kinetic integration, trapezoidal rule, or such physical means as weighing the paper.

ASSOCIATION CONSTANT: See *Affinity Constant.*

BILIARY EXCRETION: Removal of intact drug molecules or their metabolites in the bile.

BINDING STRENGTH: The strength of association between the drug molecules and proteins (such as plasma proteins), as evidenced by the ratio of association and dissociation rate constants.

BIOAVAILABILITY: The rate and extent of drug absorption into the general circulation.

BIOEQUIVALENT DRUG PRODUCTS: Pharmaceutical equivalents or alternatives whose rates and extents of absorption do not show a significant difference when administered at the same molar dose.

BIOPHARMACEUTICS: Study of the relationship between the nature and intensity of the biologic effects of a dosage form and the formulation and manufacturing factors involved in its preparation.

BIOTRANSFORMATION: Chemical alteration of a drug through its sojourn in a biologic system. Biotransformation, which involves the chemical effects of the body on a drug, must be distinguished from metabolism, detoxification, and pharmacodynamics.

BLOOD UREA NITROGEN: Nitrogen content of the blood due to urea content (BUN)—about 272 mg/liter for men and 241 mg/liter for women. The urea content increases when protein breakdown is increased, as well as in disturbances of kidney excretion and obstruction of the urinary tract.

BUCCAL: Pertaining to the cheek and sides of the mouth.

BUFFER: Combinations of weak acids/bases and their salts which resist change in the pH.

CENTRAL COMPARTMENT: Components of the body (organs or parts thereof) which become instantaneously equilibrated with the drug molecules in the plasma.

CHEMICAL EQUIVALENTS: Those multiple-source drug products which contain essentially identical dosage forms, and which meet existing physicochemical standards of official compendia.

CHEMICAL POTENTIAL: Tendency of drug molecules to escape from one phase to another phase.

CHRONOPHARMACOLOGY: Study of drug action as affected by the time of day or a specific cycle of time-dependent physiology.

CIRCADIAN RHYTHM: Twenty-four hour cycling of events.

CLEARANCE: Volume of fluid which is totally cleared of the drug in a unit time. *Renal* clearance refers to the removal of drug by the kidneys and *intercompartmental* clearance refers to intercompartmental transfer of drug molecules.

CLINICAL PHARMACOKINETICS: Application of pharmacokinetics to the safe and effective therapeutic management of the individual patient.

CLINICAL RESPONSE: Change in the symptoms and course of a disease, as ascertained by laboratory findings of anatomical change.

COMPARATIVE BIOAVAILABILITY: Amount and rate of drug absorption obtained with one dosage form as compared either to other dosage forms or to units of the same dosage form without reference to the actual rate and extent of absorption.

COMPARTMENT: Anatomic spaces in the body which have kinetic homogeneity (not necessarily concentration homogeneity).

COMPENDIA: A book of specifications, such as *USP* or *NF.*

COMPETITIVE BINDING: Binding of drug molecules to various elements of the body, such as plasma proteins, where two molecules compete for the same site of binding.

COMPLEXATION: Binding, mainly physical, between the drug molecules and formulation components on the one hand and such physiologic components as mucin on the other.

CONCENTRATION GRADIENT: Difference in the concentration between two phases.

CONTENT UNIFORMITY TEST: A test for the evaluation of variability in the chemical content in each unit of a given batch.

CREATININE CLEARANCE: *Apparent* creatinine clearance is the ratio of the urinary excretion rate of creatinine to the plasma concentration of creatinine and similar chromogens. *True* creatinine clearance takes only creatinine into account.

DE-EQUILIBRATION: A decrease in the drug concentration in one phase as a result of decrease in the concentrations in other phases which are in mutual equilibrium.

DELAYED DISTRIBUTION EQUILIBRIUM: Noninstantaneous equilibration between various body tissues and the plasma, resulting in initial sharp decline in the plasma concentration as a result of drug distribution.

DIFFUSION LAYER: Liquid layer around drug particles or biologic membranes through which the drug molecules pass as a result of the concentration gradient, as affected by their volume, fluid viscosity, temperature, and other factors affecting the kinetics of transport.

DILUENTS: Inert components used to increase the bulk of a dosage form.

DISINTEGRANT: Formulation components that help in the disintegration of large particles into smaller particles.

DISINTEGRATION: Breakdown of a large particle into smaller particles.

DISPOSITION: Loss of drug from the blood compartments, due for instance to distribution and/or elimination.

DISSOLUTION: Kinetic process leading to merging of two phases, such as solid and liquid, to form a thermodynamically stable solution.

DIURNAL VARIATION: See *Circadian Rhythm.*

DOSAGE FORM: The pharmaceutic preparation of a drug for therapeutic use.

DOSAGE REGIMEN: The schedule of drug doses and dosage intervals for specific therapeutic use.

DOSE DEPENDENCE: Dependence of the kinetic processes of absorption, distribution, biotransformation, and excretion on the amount of drug in the body. Generally refers to Michaelis-Menton kinetics.

DOSE-EFFECT CURVE: The graph which relates dose to the intensity of effect with any drug-object system.

DRIVING FORCE: The physical or chemical cause of change of a compartment's drug concentration.

DRUG THERAPY: The treatment of diseases with drugs.

ELECTROSTATIC CHARGE: Stagnant electrical charges on the surface of particles, generally the result of friction.

ELIMINATION: The combined effect of biotransformation and excretory processes.

ENDOGENOUS: Belonging to the body naturally—amino acids and sugars, for example.

ENTERAL ROUTES: Routes of administration directly involving the gastrointestinal tract.

ENTEROHEPATIC CYCLING: Excretion of drugs in the bile and subsequent reabsorption.

ENZYME INDUCTION: Increase in the enzyme content as a result of direct drug effect.

EQUILIBRIUM: State of zero net transfer between compartments.

EXCRETION: Removal of intact drug or metabolites from the body.

EXCRETION RATIO: Ratio of organ clearance of the total drug to the blood flow rate.

EXOGENOUS: Alien to the body.

EXTRANEOUS MATERIAL: Undesirable or unsuspected contaminants or byproducts in a formulation.

FACILITATED TRANSPORT: Carrier-mediated transport in the direction of the concentration gradient.

FAST DISPOSITION: Biologic half-life of 1 to 4 hours.

FEATHERING: Graphical method for the separation of exponents. Also called *Residual Method.*

FIRST ORDER REACTION: A reaction in which the rate of change in drug amount is proportional to the concentration.

FIRST PASS EFFECT: Clearance of drug before reaching the general circulation. This generally refers to the biotransformation of drugs in the liver upon oral administration before the drug can reach the circulation.

FLIP-FLOP MODEL: A case in which the half-life of the input function is longer than the half-life of the disposition function.

FREE DRUG CONCENTRATION: See *Plasma Water Concentration.*

GLOMERULAR FILTRATION: Filtration of drugs in the glomeruli with the plasma water.

HALF-LIFE: In a first order process, this is the time needed to decrease the drug concentration to one-half of its initial value at any given time.

HEMODYNAMIC: Relating to the physical aspects of the blood circulation.

HOMEOSTASIS: The tendency of a system to maintain internal stability through the coordinated response of its parts to any situation or stimulus tending to disturb its normal condition or function.

HYBRID RATE CONSTANT: A combination rate constant.

HYDROPHILIC: Water loving or water miscible.

HYDROPHOBIC: Water hating or water immiscible.

IMPACTION: Deposition of particles as a result of their momentum in the respiratory tract.

INITIAL DOSE: First dose of a multiple-dose regimen of treatment. Also called *Priming Dose* or *Loading Dose.*

INSTANTANEOUS DISTRIBUTION: Attainment of an equilibrium immediately after the introduction of drug into the system, as in single compartment models.

INTERSTITIAL SPACE: Space between the cellular elements of a structure or part.

LEAN BODY MASS: Fat-free body weight.

LIPOPHILIZING MOIETIES: Chemical groups that impart lipid-soluble characteristics to the molecule.

LOADING DOSE: See *Initial Dose.*

LUBRICANT: Formulation component used to facilitate granule flow and tablet ejection from die. Primarily hydrophobic in nature.

MAINTENANCE DOSE: One of the doses following the initial dose in a multiple-dose treatment regimen, designed to maintain a certain minimal or average drug concentration in the blood plasma for the duration of therapy.

METABOLISM: Biotransformation of materials essential to an adequate nutritional state of the body. According to this definition, the term *metabolism* does not apply to the biotransformation of most drugs.

METASTABLE: Of uncertain stability. In a condition to pass into another phase when slightly disturbed, as with metastable polymorphs converting to stable polymorphic form.

MICHAELIS-MENTON EQUATION: A kinetic equation used to describe such phenomena as binding of drugs to proteins, adsorption, and nonlinear or dose-dependent disposition kinetics for drugs.

MICRONIZATION: Reduction of particles to micron range diameter sizes.

MICROSOMAL ENZYMES: Enzymes obtained from one of the small vesicles derived from the endoplastic reticulum after disruption of cells by centrifugation. These enzymes bear the primary responsibility for the biotransformation of drugs in the body.

MODERATE DISPOSITION: Biologic half-life of 4 to 8 hours.

NEPHROSIS: Degeneration of renal tubular epithelium.

NONENTERAL ROUTES: Routes of administration which do not directly involve the gastrointestinal tract.

NOMOGRAM: A graphic technique for the calculation of the dosage from two or more given parameters.

NONLINEAR PHARMACOKINETICS: Kinetics of drug absorption, distribution, biotransformation, and excretion which are dependent on the dose administered.

NORMAL DISTRIBUTION: A continuous-distribution, bell-shaped curve which describes population characteristics.

PARAMETERS: Variable terms of an equation which determine the overall position within the axes and configuration of the line or figure described by the equation. Parameters should be distinguished from independent and dependent variables and from constants. In pharmacokinetics, the parameters are the volume of distribution, rate constants, clearance, and time constants.

PARTITION COEFFICIENT: Ratio of concentrations at equilibrium between two phases.

PASSIVE DIFFUSION: Movement of drug molecules as a result of the concentration gradient.

PERFUSION MODEL: A pharmacokinetic model based on the blood flow to various organs and the rate of drug equilibration between various organs.

PERIPHERAL COMPARTMENT: Tissue of the delayed distribution compartment.

PHARMACEUTICAL ALTERNATIVE: A drug product containing an identical therapeutic moiety, or its precursor, but not necessarily in the same amount or dosage form or as the same salt or ester as the original drug.

PHARMACEUTICAL EQUIVALENT: A drug product that contains an identical amount of the same active ingredient in an identical dosage form, but not necessarily containing the same inactive ingredients as the original drug.

PHARMACOKINETIC PARAMETERS: See *Parameters*.

PHARMACOKINETICS: Study of the kinetics of absorption, distribution, biotransformation, and excretion of drugs, poisons, and exogenous chemicals, and of some endogenous substances as well.

PHARMACOLOGIC ENDPOINT: A reference point used in recording a pharmacologic response.

PHARMACOLOGIC RESPONSE: Action of drugs on various body organs or systems.

PHENOTYPING: Classification according to genetic make-up.

PINOCYTOSIS: Absorption process characterized by engulfment of a substance by membrane movement and subsequent release.

PLACEBO: An inert compound, identical in appearance to material being tested in experimental research, which may be indistinguishable from the test material. The patient and/or researcher may not know the identity. *Placebo effect* refers to the physical/psychological effects of a placebo.

PLASMA CONCENTRATION: The concentration of the drug in the blood plasma—i.e., the sum of the plasma protein bound and the free drug concentration.

PLASMA WATER CONCENTRATION: The concentration of a drug or metabolite in the plasma ultrafiltrate, also called the *Free Drug Concentration.*

PLATEAU CONCENTRATION: The accumulated concentration of a drug in the plasma or tissue when, after chronic medication, a steady state is reached between intake and elimination of the drug.

POLYMORPHISM: Existence of various internal structures of a crystalline state. *Amorphism* refers to lack of any organized structure.

PRIMING DOSE: See *Initial Dose.*

PRODUCT INHIBITION: The case in which the product of a reaction inhibits the rate of its own formation.

PSEUDODISTRIBUTION EQUILIBRIUM: A state of drug distribution indicating kinetic homogeneity, in which the plasma concentration can be described by a monoexponential equation.

RATE CONSTANT: A parameter governing the rate of drug translocation (per unit of time)—e.g., the rate constant for drug elimination, for distribution, etc.

RATE-LIMITING STEP: The process with slowest turnover rate in a series of kinetic processes.

RELAPSE: The return of a disease after it has already run its course.

REMISSION: A temporary or permanent decrease or subsidence of manifestations of a disease.

RESIDUAL METHOD: See *Feathering.*

SCALING UP: Extrapolation of a pharmacokinetic or pharmacologic model from animals to humans based on their respective physiologic parameters.

SECONDARY PLASMA PEAKS: Peaks in the plasma concentration profiles due to enterohepatic circulation.

SINK CONDITION: The state whereby the drug concentration is equal to or less than 10 percent of the maximum possible concentration in the system—e.g., the saturation solubility.

SLOW DISPOSITION: Biologic half-life of 8 to 24 hours.

SOLID DISPERSION: Mixture of a drug and a carrier in the solid state. The carrier effects the drug release from the mixture.

SOLID SOLUTION: Molecular dispersion of a drug in a carrier compound in the solid state.

SOLUBILITY: Maximum achievable drug concentration in a given solvent at a given temperature and pressure.

SOLUBILIZATION: Increase in the solubility as a result of entrapment of drug molecules in the micelles of surfactants.

SOLVATES: Molecular structures containing molecules of such solvents as water.

SOLVENT DRAG: Increase in the diffusion rate of a substance across a membrane as a result of water flux in the same direction.

SPECIFIC SURFACE AREA: Total surface area divided by the weight.

STEADY STATE: A state in which the rate of change of a property is equal to zero. A dynamic equilibrium state.

SUBLINGUAL: Underneath the tongue.

SUSTAINED DELIVERY: Prolonged rate of drug input.

SYSTEMIC ACTION/RESPONSE: An action or drug response effected on the whole body system as a result of access to the general circulation (as opposed to localized action).

THERAPEUTIC INDEX: The ratio of the toxic dose to the minimally effective dose of a drug.

THERAPEUTIC SYSTEMS: Active drug in a delivery module consisting of a drug reservoir, a rate controller, and an energy source to bring about the release of drug molecules.

THERMODYNAMIC STABILITY: A state characterized by a low level of free energy, as with a solution which is physically stable.

TISSUE COMPARTMENT: Peripheral or delayed distribution compartment. This often includes all tissues which have low rates of blood perfusion.

TISSUE LOCALIZATION: Selective uptake of drugs in the tissues—e.g., localization of thiopental in the body fat and muscles.

TITRATION: Determination of the exact dose needed by a patient by continuously monitoring the pharmacologic response and/or the plasma concentration.

TOTAL BODY CLEARANCE: Fraction of the total volume of distribution for a drug which is cleared off in a given unit of time.

TOTAL PHARMACOLOGIC ACTIVITY: Area under the pharmacologic response versus time curve up to infinite time.

TRAPEZOIDAL RULE: Calculation of area under the curve by breaking up a curve into a large number of trapezoids.

ULTRA-FAST DISPOSITION: Biologic half-life less than 1 hour.

VERY SLOW DISPOSITION: Biologic half-life greater than 24 hours.

VOLUME OF DISTRIBUTION: Ratio of the amount of drug in the body to the plasma concentration. In some instances this term also denotes the physiologic area of drug distribution.

Appendix B: Glossary of Mathematical Terms

α — Apparent first order hybrid rate constant for the distribution of drug in the body.

β — Apparent first order hybrid rate constant describing the elimination of drugs in a two compartment model.

τ — Dosing interval.

τ_r — Dosing interval in renal failure.

A — Area of absorption surface.

AUC — Total area under the plasma concentration versus time curve.

C — Drug concentration in the plasma at time t.

C_{gut} — Drug concentration in the gastrointestinal tract.

dC/dt — Rate of change of drug concentration in the plasma.

$\int_0^\infty C \cdot dt$ — Total area under the plasma drug concentration versus time curve.

$\int_0^t C \cdot dt$ — Area under the plasma drug concentration versus time curve from time zero to t.

$\overline{C}_\infty$ — "Average" drug concentration in the plasma at steady state.

C_{max} — Maximum drug concentration in the plasma following a single dose.

C_{min} — Minimum effective drug concentration in the plasma.

C_n — Drug concentration in the plasma at any time t during the nth dosing interval.

$\overline{C}_n$ — "Average" drug concentration in the plasma during the nth dosing interval.

C_{ss} — Drug concentration in the plasma at steady state following zero order infusion.

C_0 — Drug concentration in the plasma immediately following intravenous injection.

$(C_1)_{max}$ — Maximum drug concentration in the plasma after the first of a series of repetitive doses.

$(C_1)_{min}$ — Minimum drug concentration in the plasma after the first of a series of repetitive doses.

$\overline{C}_1$ — "Average" concentration of drug in the plasma after the first of a series of repetitive doses.

C_∞ — Drug concentration in the plasma at any time t during a dosing interval at steady state with multiple dosing.

$\int_0^\tau C_\infty \cdot dt$ — Area under the drug concentration versus time curve during a complete dosing interval at steady state.

$(C_\infty)_{max}$ — Maximum concentration of drug in the plasma during a dosing interval at steady state.

$(C_{\infty})_{min}$	Minimum concentration of drug in the plasma during a dosing interval at steady state.
$[D]$	Free drug concentration.
E	Intensity of a pharmacologic effect.
$\int_0^{\infty} E \cdot dt$	Total area under the pharmacologic effect intensity versus time curve. Also termed TPA, or total pharmacologic activity.
F	A general term used to define the extent of absorption—i.e., the fraction of the dose available.
F_b	Fraction of the total available dose biotransformed in the body.
f_p	Fraction of unbound drug in the plasma.
f_s	Fraction of unbound drug in the saliva.
f_{ss}	Fraction of the steady-state plasma concentration after any given dose during multiple dosing.
K	Apparent first-order elimination rate constant in a one compartment model system.
K_a	Association constant.
K_d	Dissociation constant.
k_a	Apparent first-order absorption rate constant.
k_e	Apparent first-order renal excretion rate constant in a single compartment model system.
k_b	Apparent first-order rate constant for metabolite formation in a single compartment model system.
k_{lu}	Apparent first-order rate constant for drug clearance in the lungs in a single compartment model system.
k_{ij}	Apparent first-order intercompartmental transfer rate constants, where $i = 1, 2 \ldots ; j = 1, 2 \ldots ; i = k$.
K_m	Michaelis constant.
K_r	Disposition rate constant in renal failure.
k_0	Zero-order input or infusion rate constant.
k_{10}	Apparent first-order elimination rate constant from the central compartment.
$k_{12}, k_{21} \ldots$	See k_{ij}.
LBW	Lean body weight.
M	Amount of metabolite in the body at time t.
dM/dt	Rate of change in the amount of metabolite in the body.
M_u	Amount of metabolite excreted in the urine to time t.
dM_u/dt	Rate of appearance of metabolite in the urine.
M_u^{∞}	Total amount of metabolite ultimately excreted in the urine.
N	Number of binding sites on protein molecules.
n	Number of doses administered in a multiple-dosing regimen.
$[P]$	Free protein concentration.
$[PD]$	Concentration of drug-protein complex.
$[P_t]$	Total protein concentration.
P_c	Corrected creatinine production rate.
P	Uncorrected creatinine production rate.
pK_a	Logarithm of $1/K_a$.
Q	Total body clearance.
Q_e	Renal clearance.
Q_c	Creatinine clearance.
S	Total body surface area.

S_t	Total solubility.
S_u	Solubility of the un-ionized form.
S_i	Solubility of the ionized form.
t	Time.
t_d	Duration of action.
t_p	Time at which a maximum concentration of drug in the plasma occurs after a single dose.
t_p'	Time at which a maximum concentration of drug in the plasma occurs during a dosing interval at steady state with multiple dosing.
$t_{\frac{1}{2}}$ or $t_{0.5}$	Half-life.
V	Apparent volume of distribution of a drug in a single compartment model system.
V_d	Apparent volume of distribution of a drug in a multicompartment model system. A proportionality constant between the amount of drug in the body and the plasma concentration in the postdistributive phases.
V_c	Apparent volume of the central compartment.
V_m	Theoretical maximal rate of process describable by Michaelis-Menton kinetics.
V_T	Apparent volume of distribution in a tissue compartment.
X	Amount of drug in the body at time t.
dX/dt	Rate of change of drug level in the body.
$\overline{X}$	"Average" amount of drug in the body at steady state.
X_A	Amount of drug absorbed into the systemic circulation at time t.
X_A^∞	Amount of drug absorbed into the systemic circulation up to time=infinity.
X_a	Amount of drug at the absorption site.
dX_a/dt	Absorption rate.
X_c	Amount of drug in the central compartment at time t.
X_E	Cumulative amount of drug eliminated by all routes to time t.
dX_E/dt	Rate of drug elimination from the body.
X_{min}^*	Minimum amount of drug in the body needed to elicit a measurable pharmacologic effect.
X_n	Amount of drug in the body at any time t during the nth dosing interval of a series of repetitive doses.
X_t	Amount of drug in the peripheral compartment.
dX_t/dt	Rate of change of drug level in the peripheral compartment.
$(X_t)_\infty$	Amount of drug in the peripheral compartment at any time during a dosing interval at steady state.
X_{ss}	Amount of drug in the body at steady state upon continuous intravenous infusion.
X_u	Cumulative amount of unchanged drug excreted in the urine.
dX_u/dt	Rate of renal excretion of unchanged drug.
$\Delta X_u/\Delta t$	Average rate of renal excretion of unchanged drug over a period of finite time.
X_u^∞	Cumulative amount of unchanged drug ultimately excreted in the urine.
X_o^*	Loading or priming dose.
X_∞	Amount of drug in the body at any time during a dosing interval at steady state.

Appendix C: Mathematical and Computational Aids

Throughout this book, calculations are performed with differential equations in several modified forms. Graphic techniques are often employed to calculate pharmacokinetic parameters, along with numerical manipulations. The most important differential equation is that of a first order process:

$$dX/dt = -KX \qquad \text{(Eq. C.1)}$$

which can be rearranged to:

$$\frac{dX}{X} = -Kdt \qquad \text{(Eq. C.2)}$$

An integration between limits of time zero and time t:

$$\int_{x_0}^{x_t} \frac{dX}{X} = \int_0^t -K \cdot dt \qquad \text{(Eq. C.3)}$$

$$ln\, X - ln\, X_0 = -Kt \qquad \text{(Eq. C.4)}$$

The term ln represents the natural logarithm of X and it appears because the integral of $1/X$ is $ln\, X$.

The meaning of logarithms must be clearly understood. The logarithm of a number (abbreviated as log or ln) is the exponent or the power to which a given base (10 or e, respectively) must be raised to equal a given number. For example, the log of 10 is 1 since:

$$10^1 = 10$$
$$10^2 = 100$$
$$10^3 = 1000 \text{ and so on .....}$$

Instead of using a base 10, base e can be used, which is equal to 2.718281828 . . .

$$e^{2.302585093} = 10$$
$$e^{4.605170186} = 100$$
$$e^{6.907755279} = 1000$$

Notice that in each successive logarithm conversion the exponents are added, such as 1 + 1 = 2 or 2.302 . . . + 2.302 . . . = 4.605 . . . In other words 10 × 10 is simply given in logarithmic form by adding the exponents of 10, which forms the first property of logs—i.e., exponents are additive upon mutiplication. Similarly, 1000/10 can be calculated in terms of logs by subtracting the respective exponents.

The natural base of a log can be converted to a log to the base 10:

$$\ln X = 2.3025 \ldots \log X \qquad \text{(Eq. C.5)}$$

where log refers to log to the base 10. Notice that e raised to the power of 2.3025 . . . (ordinarily taken as 2.303) is equal to 10.

The use of ln provides an exponential function which only at one point ($\ln e$) becomes equal to one. This exponential function is a familiar function used in the decay of plasma concentration, population growth, radioactive decay, and many other properties which depend on an absolute value of their own, such as concentration, population, radioactivity, etc. In all of these situations, log or ln of the dependent variable plotted against an independent variable results in a straight line, and this is termed a log–linear function.

The differential equations frequently encountered in pharmacokinetic studies can also be solved by convenient Laplace transformations, which give a mathematical opportunity to solve these equations without going through the integration process.[1] The time domain is expressed by a complex function operator termed s, followed by algebraic problem-solving and conversion of the Laplace back to linear form. For example, a Laplace solution to first order absorption can be obtained as follows:

$$dX/dt = K_a X_a - KX \qquad \text{(Eq. C.6)}$$

$$dX_a/dt = -k_a X_a \qquad \text{(Eq. C.7)}$$

Taking the Laplace transform of both equations:

$$s\overline{X} = k_a \overline{X}_a - K\overline{X} \qquad \text{(Eq. C.8)}$$

$$s\overline{X}_a - FX_0 = -k_a \overline{X}_a \qquad \text{(Eq. C.9)}$$

where all terms with a bar represent Laplace transforms of the respective terms, s is the Laplace operator, and FX_0 is the initial amount, which will all be absorbed. Notice that the transformation of a differential operator is $s\overline{X}$ minus initial value. The above equations can be solved by substitution:

$$\overline{X}_a = \frac{FX_0}{(s + k_a)} \qquad \text{(Eq. C.10)}$$

$$\overline{X} = \frac{k_a F X_0}{(s + k_a)(s + K)} \qquad \text{(Eq. C.11)}$$

Table C.1. LAPLACE TRANSFORMS OF SOME USEFUL FUNCTIONS

FUNCTIONS	LAPLACE TRANSFORMS
F(t)	f(s)
1	$1/s$
A	A/s
t	$1/s^2$
t^m	$m!/s^{m+1}$
Ae^{-at}	$A/(s+a)$
Ate^{-at}	$A/(s+a)^2$
$\frac{A}{a}(1-e^{-at})$	$A/s(s+a)$
$\frac{A}{a}e^{-(b/a)t}$	$A/(as+b)$
$\frac{(B-Aa)e^{-at}-(B-Ab)e^{-bt}}{b-a}\ (b \neq a)$	$(As+B)/(s+a)(s+b)$
$\frac{A}{b-a}(e^{-at}-e^{-bt})$	$A/(s+a)(s+b)$
$e^{-at}[A+(B-Aa)t]$	$(As+B)/(s+a)^2$
$-\frac{1}{PQR}[P(Aa^2-Ba+C)e^{-at}+Q(Ab^2-Bb+C)e^{-bt}+R(Ac^2-Bc+C)e^{-ct}]$ $(P=b-c,\ Q=c-a,\ R=a-b)$	$(As^2+Bs+C)/(s+a)(s+b)(s+c)$
$A\left[\frac{1}{ab}+\frac{1}{a(a-b)}e^{-at}-\frac{1}{b(a-b)}e^{-bt}\right]$	$A/s(s+a)(s+b)$
$\frac{A}{a}t-\frac{A}{a^2}(1-e^{-at})$	$A/s^2(s+a)$
$\frac{B}{ab}-\frac{Aa-B}{a(a-b)}e^{-at}+\frac{Ab-B}{b(a-b)}e^{-bt}$	$(As+B)/s(s+a)(s+b)$
$\frac{B}{ab}-\frac{a^2-Aa+B}{a(b-a)}e^{-at}+\frac{b^2-Ab+B}{b(b-a)}e^{-bt}$	$(s^2+As+B)/s(s+a)(s+b)$
t^re^{-at}	$r!/(s+a)^{r+1}$
$e^{-at}\cos bt$	$(s+a)/[(s+a)^2+b^2]$
$e^{-at}\sin bt$	$b/[(s+a)^2+b^2]$
$t^re^{-at}\cos bt$	$(-1)^r d^r/ds^r \{(s+a)/[(s+a)^2+b^2]\}$
$t^re^{-at}\sin bt$	$(-1)^r d^r/ds^r \{b/[(s+a)^2+b^2]\}$

Adapted from Resigno A, Segri G: Drug and Tracer Kinetics. Waltham, Mass., Blaisdell, 1966

The anti-Laplace (Table C. 1) of the above equation yields:

$$X = \frac{k_a F X_0}{(k_a - K)}(e^{-Kt} - e^{-k_a t}) \qquad \text{(Eq. C.12)}$$

In the above equation, a direct substitution is possible. However, more complex equations require the use of a matrix to solve for X—the reader is here referred to another text with a more complex approach to the topic.[2]

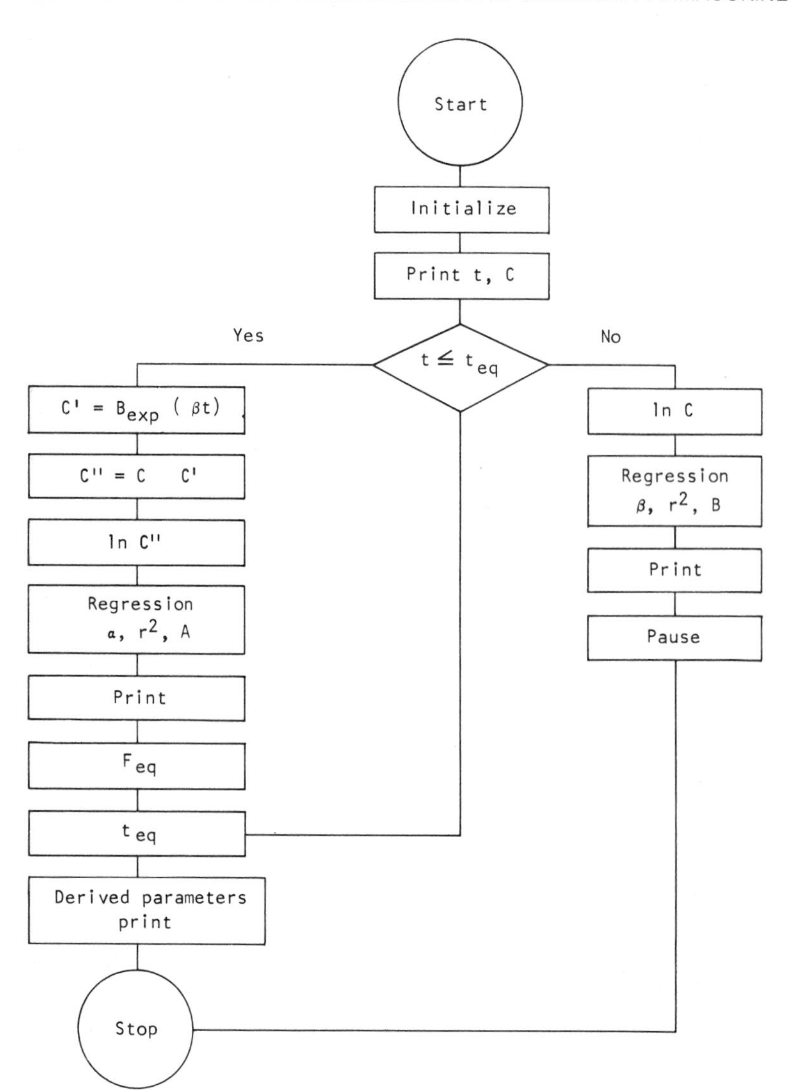

Figure C.1. Program flow in modeling and parameter calculations.

COMPUTATIONAL AIDS

The old slide rule has now been entirely replaced by the electronic calculator, which machine can work wonders. The new breed of programmable calculators, more aptly termed minicomputers, finds a great many applications in pharmacokinetic calculations. The following is a brief review of various programs and machines available to perform these calculations.

A variety of digital and analog computer programs are available for pharmacokinetic analyses. These programs fall into two categories, *simulating* and *integrating*. Simulating programs, such as CSMP and MIMIC, simulate the data directly from the differential equations without integration.[3,4] The computer integrations are performed via such methods as the

Table C.2. COMPUTING STEPS

001	*LBLA	045	÷	089	*LBLC	133	+	177	RCLA
002	X↔Y	046	PRTX	090	SF1	134	RCLA	178	ROL0
003	PRTX	047	RCL6	091	P↔S	135	−	179	−
004	X↔Y	048	RCL4	092	0	136	RCLB	180	÷
005	PRTX	049	RCLi	093	STO4	137	−	181	RCLD
006	SPC	050	CHS	094	0	138	PRTX	182	×
007	F1?	051	ISZI	095	STO5	139	SPC	183	ENT↑
008	GSa	052	×	096	0	140	STOC	184	1
009	LN	053	−	097	STO6	141	P↔S	185	+
010	X↔Y	054	RCL9	098	0	142	RCL0	186	RCLE
011	Σ+	055	÷	099	STO7	143	P↔S	187	÷
012	RTN	056	e-	100	0	144	RCL3	188	LN
013	*LBLB	057	STOi	101	STO8	145	RCL1	189	RCL2
014	SPC	058	ISZI	102	0	146	+	190	RCL0
015	P↔S	059	PRTX	103	STO9	147	÷	191	−
016	RCL8	060	P↔S	104	P↔S	148	PRTX	192	÷
017	RCL4	061	SPC	105	RTN	149	RCLB	193	P↔S
018	RCL6	062	R/S	106	*LBLb	150	×	194	PRTX
019	×	063	GSBb	107	DSP5	151	PRTX	195	RTN
020	RCL9	064	RTN	108	P↔S	152	RCL0	196	*LBLc
021	÷	065	*LBLE	109	RCL1	153	÷	197	P↔S
022	−	066	CLRG	110	RCL2	154	PRTX	198	STO4
023	ENT↑	067	P↔S	111	×	155	RCL3	199	X↔Y
024	ENT↑	068	CLRG	112	RCL3	156	RCL2	200	STO5
025	RCL4	069	P↔S	113	RCL0	157	÷	201	RCL3
026	X^2	070	CF1	114	×	158	RCL1	202	RCL2
027	RCL9	071	STO0	115	+	159	RCL0	203	÷
028	÷	072	RTN	116	RCL1	160	÷	204	RCL1
029	RCL5	073	*LBLa	117	RCL3	161	+	205	RCL0
030	X↔Y	074	P↔S	118	+	162	PRTX	206	÷
031	−	075	STO3	119	÷	163	P↔S	207	+
032	÷	076	X↔Y	120	PRTX	164	RTN	208	P↔S
033	CHS	077	STO2	121	SPC	165	*LBLD	209	ENT↑
034	STOi	078	RCL0	122	STOA	166	P↔S	210	RCL0
035	PRTX	079	×	123	RCL0	167	STOD	211	÷
036	CHS	080	CHS	124	RCL2	168	CHS	212	P↔S
037	×	081	e-	125	×	169	ENT↑	213	RCL4
038	RCL6	082	RCL1	126	RCLA	170	1	214	×
039	X^2	083	×	127	÷	171	+	215	RCL5
040	RCL9	084	ST−3	128	PRTX	172	STOE	216	÷
041	÷	085	RCL2	129	SPC	173	P↔S	217	P↔S
042	CHS	086	RCL3	130	STOB	174	RCL2	218	PRTX
043	RCL7	087	P↔S	131	RCL0	175	RCLA	219	RTN
044	+	088	RTN	132	RCL2	176	−	220	R/S

Runge-Kutta, the fifth-order Milne predictor-corrector, Simpson's, the trapezoidal, and other methods. Nonlinear least square regression is also performed by several programs, such as NLIN,[5] SAAM,[6] and NONLIN.[7] NONLIN has been the program most extensively used and it can handle most pharmacokinetic modeling siutations, such as zero order processes and nonlinear pharmacokinetic modeling. Another recent program made available is AUTOAN,[8] which performs the least square fit of raw blood data to determine whether elimination is best described by Michaelis-Menton kinetics or first order kinetics. When coupled with NONLIN, AUTOAN provides an excellent tool for pharmacokinetic analysis.

Most of the programs described above require elaborate computer set-ups and are often quite expensive to run, especially for a small scale facility. In order to reduce the cost, stripping-type computer programs (such as CSTRIP or HP–97) and the SR–52 model programs reported by the author can be used.[9–11] CSTRIP requires access to a digital computer, whereas the use of such inexpensive calculators as HP–97 and SR–52 provides model fitting with an accuracy comparable to that of CSTRIP. To illustrate, the HP–97 program for the fitting of a two compartment open model with intravenous administration is reproduced as follows.

HP–97 PROGRAM FOR TWO COMPARTMENT ANALYSIS AND SIMULATIONS

The flow diagram of the program is given in Figure C.1. Turn the calculator on and go into program mode. After clearing the existing program memory, enter the steps listed in Table C.2. Using the data in Table C.3, calculate the pharmacokinetic parameters. The first step involves calculation of approximate parameters in order to obtain an estimate of the time at which the distribution process will be fairly well completed:[11]

$$t_{eq} = \frac{ln\left[\left(F_{eq}\frac{\alpha - k_{21}}{k_{21} - \beta} + 1\right) \Big/ (1 - F_{eq}) \right]}{(\alpha - \beta)} \qquad \text{(Eq.C.13)}$$

Table C.3. PLASMA CONCENTRATION OF WARFARIN FOLLOWING ADMINISTRATION OF A 200 MG INTRAVENOUS DOSE

t (hr)	C (μg/ml)	*t* (hr)	C (μg/ml)
0.25	41.3	37.00	15.6
0.50	33.8	48.00	13.0
0.75	30.2	72.00	9.0
1.00	28.4	90.00	7.0
3.00	26.2	117.00	4.5
6.00	24.0	145.00	2.9
8.50	25.0	168.00	2.0
12.50	23.0	192.00	1.4
24.00	19.0		

where F_{eq} is the fraction of distribution equilibrium, generally taken as 0.95 for these calculations, and t_{eq} is the time needed to reach a given F_{eq}. The raw parameters are obtained by entering the first three and the last three data points. The final analysis is performed by entering all blood level data while calling subroutine for feathering at time equal to or less than t_{eq}.

Table C.4. TYPICAL RUN OF DATA IN TABLE C.3

DESCRIPTION	*INPUT*	*KEY*	*PRINT OUT*
Load program			
Initialize enter dose	200	E	
Enter last 3 data points	192	↑	
	1.4	A	192
			1.4
	168	↑	
	2	A	168
			2
	145	↑	
	2.9	A	145
			2.9
Calculate raw β, r^2 and B		B	.01549 β
			0.99942 r^2
			27.2640 B
Initiate feathering		C	
Enter first 3 data points	0.25	↑	
	41.3	A	0.25
			41.3
	0.5	↑	
	33.8	A	0.5
			33.8
	0.75	↑	
	30.2	A	0.75
			30.2
Calculate raw α, r^2 and A		B	2.9403 α
			0.9999 r^2
			29.4448 A
Calculate derived parameter		R/S	
			1.42166 k_{21}
			0.03204 k_{10}
			1.50213 k_{12}
			3.52678 V_c
			0.11299 TBC
			7.29417 V_d
			1770.13650 $\int_0^\infty C \cdot dt$
Calculate t_{eq}	0.95	D	1.266
Go back to step 2, initiating feathering at $t < 1.266$			
Steady state C as function of dose and dosing interval:			
Dosing interval (τ)		↑	
Dose		f C	$\overline{C}$
Example	4	↑	445
	200	f C	

Thus, in using this program one not only obtains all of the pharmacokinetic parameters, but one can also simulate the blood levels, given a dosing interval and the dose. Programs of this nature have great applicability in clinical situations, where the individualization of dosage regimen may be based on the patient's pharmacokinetic parameters. Following are some programs available from the Hewlett-Packard Company:[12]

Program Number	*Program Title*
00791D	Digoxin Dose Calculation
00792D	Gentamicin Dose Calculation
01786D	Nitroprusside Dosage
01785D	Gentamicin Dosages, Peak, Trough, and Average Serum Levels
01784D	Amikacin Dosages, Peak, Troughs, and Average Serum Levels
01567D	Aminophylline, IV, Loading and Maintenance Dosage
01788D	Dopamine Dosage
01787D	Digoxin Loading and Maintenance and Serum Levels—Critical Care Program
01962D	Multicompartment Pharmacokinetic Analysis and Simulations

This list is continuously growing and the reader must contact Hewlett-Packard for more recent information.

References

1. Strum RD, Ward JR: Laplace Transformations Solutions of Differential Equations: A Programmed Text. Englewood Cliffs, N.J., Prentice-Hall, 1968
2. Gibaldi M, Perrier D: "Pharmacokinetics," Drugs and the Pharmaceutical Sciences. New York, Marcel Dekker, 1975
3. System/360 Continuous System Modeling Program, Program No. 360A-CX-16X, 5th edition. White Plains, N.Y., IBM Corporation Technical Publications, 1972
4. MIMIC, Publication No. 44610400. St. Paul, Minn., Control Data Corporation, Special Systems Publication, 1968
5. Marquardt DW: DPE–NLIN. Share General Library Program No. 7-1354
6. Berman M, Weiss MF: SAAM. Bethesda, Md., National Institute of Health
7. Metzler CM: NONLIN. Kalamazoo, Mich., Upjohn
8. Sedman AJ: AUTOAN. Ann Arbor, Mich., Publications Distribution Service

9. Sedman AJ, Wagner JG: CSTRIP, a Fortran IV computer program for obtaining initial polyexponential parameter estimates. J Pharm Sci 65:1006, 1976
10. Niazi S: Hewlett-Packard Library Program No. 01962D. Hewlett-Packard, Corvallis, Ore.
11. Niazi S: Application of a programmable calculator in data fitting according to one and two compartment open models in clinical pharmacokinetics. Computer Programs Biomed 7(1):41, 1977
12. Hewlett-Packard Library of Programs for HP-97. Corvallis, Ore. In Europe, Hewlett-Packard S.A. User's Library, Geneva, Switzerland

Index

Index